Vulnerability, Risks, and Complexity
Impacts of Global Change on Human Habitats

Dedication

On April 21, 2011, Gabriel Moser, the great *spiritus rector* of people-environment studies, died. Having a background in environmental psychology, he was always open-minded to the contributions of other disciplines as well as interdisciplinary and transdisciplinary approaches. During his time as President of IAPS (2004–2008) he founded the series ***Advances in People-Environment Studies*** together with David Uzzell. We dedicate this 3rd volume of the series to him.

Advances in People-Environment Studies

David Uzzell; PhD, FBPsS, FRSA, Prof., Surrey, UK, Past-President of the International Association for People-Environment Studies (IAPS).
(Series Editor)

The new book series ***Advances in People-Environment Studies***, published in collaboration with the International Association for People-Environment Studies (IAPS; www.iaps-association.org), is a timely initiative to provide researchers with up-to-date reviews and commentaries on the diverse areas of people-environment studies that are of current concern. The series focuses on significant and currently debated themes. The books are interdisciplinary, drawing on expert authors from the social, environmental, and design disciplines, especially those who are working at the interface between the design (e.g., architects, landscape planners, urban designers, urban planners) and the social sciences (e.g., environmental psychologists, sociologists, geographers). Each volume reports on the latest research and applications of research in the field. The series is meant to provide a bridge, not only between disciplines but also between cultures. The authors and contributors come from many different countries and are undertaking research and practicing in culturally diverse environments. Books in the series are therefore a precious source for those who want to know what is going on in a specific field elsewhere and to find ideas and inspiration for their own work.

Advances in People-Environment Studies Vol. 3

Vulnerability, Risks, and Complexity
Impacts of Global Change on Human Habitats

Sigrun Kabisch
Anna Kunath
Petra Schweizer-Ries
Annett Steinführer
(Editors)

HOGREFE

Library of Congress Cataloging-in-Publication Data

is available via the Library of Congress Marc Database under the
LC Control Number 2012931375

Library and Archives Canada Cataloguing in Publication

Vulnerability, risks, and complexity: impacts of global change on human habitats / Sigrun Kabisch ... [et al.], editors.

(Advances in people-environment studies ; v. 3)
Includes bibliographical references and index.
ISBN 978-0-88937-435-5

1. Human ecology. 2. Human beings–Effect of environment on. 3. Human beings–Effect of climate on. 4. Global environmental change. I. Kabisch, Sigrun II. Series: Advances in people- environment studies ; v. 3

GF41.V84 2012 304.2 C2012-900650-5

Cover illustration: Wilfrid Moser (1914–1997). Sortie. 1964, oil on canvas, 65 × 80 cm. Private collection Switzerland. © Wilfrid Moser Foundation

PUBLISHING OFFICES
USA: Hogrefe Publishing, 875 Massachusetts Avenue, 7th Floor, Cambridge, MA 02139
Phone (866) 823-4726, Fax (617) 354-6875; E-mail customerservice@hogrefe-publishing.com
EUROPE: Hogrefe Publishing, Merkelstr. 3, 37085 Göttingen, Germany
Phone +49 551 99950-0, Fax +49 551 99950-425, E-mail publishing@hogrefe.com

SALES & DISTRIBUTION
USA: Hogrefe Publishing, Customer Services Department,
30 Amberwood Parkway, Ashland, OH 44805
Phone (800) 228-3749, Fax (419) 281-6883, E-mail customerservice@hogrefe.com
EUROPE: Hogrefe Publishing, Merkelstr. 3, 37085 Göttingen, Germany
Phone +49 551 99950-0, Fax +49 551 99950-425, E-mail publishing@hogrefe.com

OTHER OFFICES
CANADA: Hogrefe Publishing, 660 Eglinton Ave. East, Suite 119 - 514, Toronto, Ontario M4G 2K2
SWITZERLAND: Hogrefe Publishing, Länggass-Strasse 76, CH-3000 Bern 9

Hogrefe Publishing
Incorporated and registered in the Commonwealth of Massachusetts, USA, and in Göttingen, Lower Saxony, Germany

Printed and bound in Germany
ISBN: 978-0-88937-435-5

Table of Contents

Preface

The content of this volume is based on selected and peer-reviewed contributions from participants of the 21st Conference of the International Association for People-Environment Studies – IAPS on "Vulnerability, Risks, and Complexity: Impacts of Global Change on Human Habitats," held at the Helmholtz Centre for Environmental Research – UFZ in Leipzig, Germany, in 2010. We are grateful to IAPS for giving us the opportunity to host the conference as it has allowed us to also highlight our internationally recognised scientific institute for people-environment studies with its special focus on nature and the environment. It is an integral part of the work of the UFZ to promote close collaboration between natural and social sciences in order to generate interdisciplinary and transdisciplinary synergies. It provides an appropriate arena for the intensive exchange of scientific knowledge, research applications, and good practices at the science-policy interface.

The theme of the 21st IAPS conference and the topics of the individual conference sessions attracted considerable scientific attention: 350 papers were presented. In addition to the ongoing scientific debates, political representatives spoke to the IAPS community. They emphasised the necessity to act in an integrative and holistic manner. The Executive Secretary of the UN Economic Commission for Latin America and the Caribbean (ECLAC), Ms. Alicia Bárcena (Chile), pointed out the interlinkages between economic, social, and environmental progress. Emphasising the Latin American context, she highlighted the close relationship between poverty and vulnerability and extreme events like floods and earthquakes. The Prime Minister of the Free State of Saxony, Mr. Stanislaw Tillich (Germany), underlined the need to develop appropriate strategies for decision-making, behavioural change, and technical improvements as preparation for extreme environmental events. He called for scientifically based coping strategies and tools to deal with the long-term environmental, social, and demographic changes that Germany and other countries are facing. He further stressed the responsibility of political decision-makers to find comprehensive solutions that take the complexity of people-environment interdependencies into consideration.

The theoretical reflections, empirical descriptions, and political statements presented during the conference were confronted with an unexpected piece of reality: The consequences of the March 2011 earthquake in Japan and the ensuing tsunami. The destruction of the nuclear power stations in Fukushima showed the extent to which the existence of humans and other species is interlinked and revealed the interdependencies of human, technological, and ecological systems. In the wake of this event, the world is seen and interpreted differently. It is important to stress that not only extreme events, due to natural hazards, but also human-made risks, in terms of technical uncertainties, have a broad impact on human habitats. The Japanese catastrophe warns us about the irresponsible use of resources and calls for a respectful, responsible, and humane attitude towards our environment. With its scientific and practical input IAPS provides an appropriate framework to deal with these issues, as does this book in your hands.

Sigrun Kabisch
UFZ chief coordinator of the 21st IAPS conference

Introduction

Vulnerability, Risks, and Complexity

Impacts of Global Change on Human Habitats

Sigrun Kabisch,[1] Anna Kunath,[1] Petra Schweizer-Ries,[2] and Annett Steinführer[3]

[1]Department of Urban and Environmental Sociology, Helmholtz Centre for Environmental Research – UFZ, Leipzig, Germany
[2]Saarland University with branch at the Otto von Guericke University, Saarbrücken/Magdeburg, Germany
[3]Institute of Rural Studies, Johann Heinrich von Thünen Institute (vTI), Braunschweig, Germany

Scope of the Volume

Following volumes 1 and 2 of the IAPS series "Advances in People-Environment Studies" (Abdel-Hadi, Tolba, & Soliman, 2010; Bonaiuto, Bonnes, Nenci, & Carrus, 2011), this third volume provides a spectrum of papers that are in direct accordance with the mission of the International Association for People-Environment Studies (IAPS). IAPS was established in 1981 to facilitate interdisciplinary debate and international collaboration in environment-behaviour research and people-environment studies.

This volume offers up-to-date research on the interrelations between social, built, and natural environments. It includes papers investigating human habitats in a variety of settlement structures within various natural settings that are negatively impacted by extreme weather events and natural hazards such as floods or earthquakes, infectious diseases, contaminations, or crime. Further, it provides empirical case studies describing ways to strengthen environmentally sound behaviour and how people behave in very specific living conditions, for example, in post-disaster settlements or in memorial places.

Like the other volumes of this series, the focus of the book is on one main topic: Here we stress the necessity of studying vulnerability, risks, and complexity as expressions of the major impacts of global change on human habitats. These include not only climate and hazard impacts but also new socio-demographic trends and regional economic developments influencing people's quality of life. The topic was intensively discussed at the 21st IAPS Conference in Leipzig, Germany, which took place from 27 June to 2 July, 2010. About 400 scientists from 40 countries participated and 350 papers were presented. The large number of papers and the variety of subjects proved that this focus is highly topical. Many communities and settlements worldwide are facing vulnerability to environmental or human-made risks, which pose a threat to their daily life, livelihood, or even survival (Blaikie, Cannon, Davis, & Wisner, 1994; Brauch et al., 2011; García-Mira, Real Deus, Uzzell, San Juan, & Pol, 2006; Heinrichs, Hansjürgens, Krellenberg, & Martinez, 2011; Lahsen et al., 2010; Oliver-Smith, 2001; Romero Lankao & Qin, 2011; Satterthwaite, Huq, Reid, Pelling, & Romero Lankao, 2007; UN-Habitat, 2007). In a complex and not always linear fashion, these risks overlap with existing patterns of socio-economic exclusion, demographic change, urbanisation processes, and impacts of a globalised economy. These are challenges that encourage us to look for new scientific approaches for comprehensive and integrative problem solutions. They should help to strengthen the resilience of human habitats (Comfort et al., 1999; Couch & Kroll-Smith, 1991; Newman, Beatley, & Boyer, 2009; Vale & Campanella, 2005). Hence, more sophisticated theories, innovative methods, and appropriate empirical findings are needed. Ideally, research should integrate the social, natural, and engineering/planning sciences, and it should be transdisciplinary, that is, conducted in close collaboration with representatives of the civil society, the economy, the private sector, NGOs, and political decision-makers (Kötter & Balsiger, 1999; Lawrence, 2010; Lawrence & Després, 2004; Scholz, Lang, Wiek, Walter, & Stauffacher, 2006; Thompson Klein, 2004). This comprehensive approach enables us to understand cities, which are characterised by high population density, a concentration of economic values, and cultural assets, as habitats particularly vulnerable to various types of hazards and risks (Gross, 2001; Pelling, 2003; Ultramari & Rezende, 2007).

These include:

- Extreme weather events (storms, floods, heat waves, etc.);
- Permanent urban risks (emergence and re-emergence of diseases, climate variability, water scarcity, etc.);
- Natural hazards (earthquakes, tsunamis, landslides, etc.);
- Technological risks (leading to disasters such as the oil spill in the Gulf of Mexico in 2010) and, last but not least;
- Interdependent social-ecological risks (e.g., droughts, fires, or coupled natural-technological risks like those that emerged in wake of the Fukushima disaster in 2011).

These risks and their interactions with socio-economic and demographic changes lead to an increasing vulnerability of social systems, governance structures, and ecosystems in urban agglomerations in a very complex way. Social vulnerability, which also refers to survival strategies in poor and dangerous living environments affected, for example, by heavy traffic and contaminated land, is also on the increase.

Climate change, with its uncertain implications, is regarded as an amplifier of these risks and hazards – be it through increasing numbers or stronger impacts of harmful events or because of changing exposure patterns, for example, due to sea level rise (McEvoy, 2007; Prasad et al., 2009). New strategies of both adaptation and mitigation are required in order to enhance the resilience of social systems to climate change. To achieve application-oriented research results, it is necessary to investigate social patterns of vulnerability and, in particular, the preparedness and coping strategies of many different societal groups. Furthermore, we need new and appropriate policies, regulations, and options for adaptation on the global, regional, and local levels and for various types of human habitats.

In order to reach these aims, system solutions for combined societal-environmental problems need to be developed. This includes the description of the impacts of global change on human habitats, as well as the specification of the anthropogenic contribution to global change. Methodological designs must consider comprehensive approaches and specific local and social contexts. Models and scenarios will promote a creative dialogue between the disciplines and support the application-oriented creation of knowledge. Further, we have to be aware of the limits of our knowledge and learn how to deal with both uncertainties and surprises. Unknown but obviously important factors emerging from global change processes, but also local issues, such as contaminants in the soil and water of former industrial areas, have to be taken into account in decision-making. Furthermore, extended knowledge about the environmental perceptions, environmental attitudes, and behavioural preferences of societal groups will help in the search for strategies and instruments aimed at counterbalancing current societal trends as well as for designing appropriate architectural and city planning concepts. To understand this complexity, new approaches in urban planning and also urban design are being developed to interlink research and education, to encourage young researchers to deal with emerging topics and innovative methods, and to combine empirical case studies with new theoretical insights.

In order to meet the scope of this volume and to focus on relevant and up-to-date topics, the contributions attempt to answer the following essential questions:

- What does global change imply locally for people-environment relations?
- How do climate, environmental, demographic, and economic changes interact?
- Do vulnerable places coincide with vulnerable people and how can social capacities for coping with the risks of and adapting to future hazards be described and developed?
- Are cities specific "risk habitats" in the process of global change, and are they especially affected?
- How can the liveability and functionality of human habitats be ensured under conditions of global change?

The Nature of the Papers

The 22 papers presented in this volume include theoretical-conceptual texts as well as contributions that use theoretical concepts to introduce empirical case studies. Additionally, practice-oriented papers with specific proposals for decision-makers and innovative training methods are presented. The contributions consider a variety of conceptual and geographical scales ranging from global to regional and local down to neighbourhood or even individual building scale. Macro-, meso-, or micro-level perspectives are used to explain the context of each specific research approach. Some contributions reflect on the emergence of risky environments using data and information related to the global scale. Others investigate selected subjects on a local level from a micro-perspective. Many papers focus on a specific case that has been investigated in detail. Nevertheless, the results in terms of models, tested methodological approaches, or concrete insights can be transferred to and used in comparable cases. The issues raised in these case studies, taken from many parts of the world, provide a broad panorama of affected human habitats.

With regard to their disciplinary origin and scientific methodology, the papers come from the fields of environmental psychology, urban planning and design, environmental sociology, urban sociology, human ecology, and urban geography. In spite of the multi-disciplinary character of these contributions (and the entire IAPS community), the volume acknowledges the origins of IAPS in environmental psychology and urban planning. Therefore, a few pilot-studies and original disciplinary papers by PhD students in environmental psychology are also included in this volume. They describe theoretical approaches, research questions, and/or methodological issues related to scientific work in progress. The papers are detailed, methodologically well grounded, and often connected with case studies.

The contributions in this volume cover six broad research fields:

- Reflections on vulnerability and risks in complex environments;
- Coping with climate change and natural hazards;
- Social dimensions of vulnerability and risks;
- Participation and strategies of risk reduction;
- Strengthening pro-environmental behaviour;
- Place making and urban design.

Taking the whole range of contributions into consideration and reflecting upon them in the light of the IAPS mission, we found that most papers are rooted in disciplinary approaches. We therefore see the need for increased interdisciplinary collaboration because of the complexity of the questions and issues involved. Successful cooperation between disciplines from the very beginning of a project to its completion, including the development of a common language and a shared methodological understanding, is still rare, although often stressed. To concentrate, for example, on very narrow methodological questions, using discipline-specific abbreviations and highly specialised statistical calculations and indices, runs the risk of not being understood or misunderstood by a reader who is interested in the subject but not an expert in the specific discipline. Thus, it is necessary to find a good balance between the specialised and the interdisciplinary debate, which is a true challenge. IAPS and its ongoing activities offer an excellent setting for exactly this kind of innovative scientific exchange.

Structure and Contents of the Volume

This volume consists of six sections.

Section 1: Reflections on Vulnerability and Risks in Complex Environments

The first section includes three papers dealing with theoretical reflections and research results about the vulnerability of humans in various contexts. *S. Harris Ali* describes the spread of infectious diseases by worldwide traffic connections as a new health risk in a globalised world. *Christian Kuhlicke et al.* provide a conceptual framework for the investigation of urban vulnerability, and discuss similarities and differences between growing and shrinking cities with respect to the specific factors that contribute to their urban vulnerability. In order to develop an appropriate methodological approach for analysing complex and risky living conditions to reduce vulnerability, *Roderick J. Lawrence* calls for inter- and transdisciplinary concepts and methods.

Section 2: Coping With Climate Change and Natural Hazards

In the second section, empirical results related to coping with the negative impacts of natural hazards as well as climate change are presented. *Amalia Fernandez-Bilbao* describes the efforts of coastal communities to adapt to climate change in terms of being prepared for repeated and intensified flood events and future sea level rise. Her special focus is on ageing populations and fragile economic conditions. *Rachma M. Syam and Ryuzo Ohno* discuss the mixed blessings of donated post-disaster housing, showing the discrepancies between pre-fabricated design and actual usage. *Elisabeth Michel-Guillou* deals with water managers and their perceptions of climate change in terms of appropriate decision-making on the local level. In the final paper of this chapter, *Heiko Lieske* stresses public participation and integrative planning cultures to find acceptable adaptation strategies for cultural heritage buildings threatened by flood hazards.

Section 3: Social Dimensions of Vulnerability and Risks

The third section contains five contributions dedicated to the social dimensions of vulnerability and risks. *Ricardo García-Mira et al.* explain the interdependencies of settlement development and the social-ecological risks related to forest fires, based on human behaviour and decision-making. *Rittirong Chutapruttikorn and Peter Kellet* use the example of informal railway settlements to illustrate difficult survival strategies and to explain changing vulnerabilities in the course of upgrading the living conditions of the poor. *Shaharudin Idrus et al.* present research results of shifting vulnerabilities of residents at risk of housing expansion, inadequate water consumption as well as dengue fever and crime. The last two papers in the section, by *Antônio Tarcísio Reis and Celina Dittmar* and by *Zulmira Áurea Cruz Bomfim et al.*, deal with fear and crime in a residential environment and whether young people are being prepared to resist either one. The latter paper is based on the results of a pilot-study.

Section 4: Participation and Strategies of Risk Reduction

In the fourth section, the concepts of participation as well as risk-reduction strategies in various post-industrial areas are considered. It starts with the paper by *Irina Rau et al.*, which discusses the relationship between participation and acceptance, using the example of renewable energy technologies. *Alena Bleicher and Matthias Gross* propose strategies for risk reduction with a special focus on unknown and surprising events in the remediation of contaminated sites. *Jörn Harfst et al.* describe participative concepts for transforming former mining areas into recreational and environmentally sound areas.

Section 5: Strengthening Pro-Environmental Behaviour

The fifth section includes four psychologically grounded contributions that investigate the strengthening of pro-environmental behaviour. *Christophe Demarque et al.* place their research focus on individual differences in environmental consciousness and stress-binding communication as an essential behavioural mechanism. *Dóra Medvés and Judit Kovács* investigate individual differences in promoting pro-environmental behaviour. The paper by *Kathrin Röderer and Renate Cervinka* focuses on the exploration of mental representations of nature by using a highly elaborated qualitative approach. *Massimiliano Scopelliti et al.* discuss the restorative potential of various kinds of urban nature.

Section 6: Place Making and Urban Design

The closing chapter encompasses papers presenting different perspective on places, their meaning, décor, and design. It starts with the paper by *Mats Lieberg*, in which he describes the phenomenon of spontaneous memorials, their causes, and their multiple functions in urban environments. *Fahed A. Khasawneh et al.* explain an effective common place design using the example of an environmentally friendly university campus. The section is rounded off with a paper by *Ombretta Romice et al.*, in which an innovative training programme for urban design is presented.

The wide range of papers presented in this volume, showing different disciplinary, methodological, and also geographical perspectives, indicates the great variety and complexity of risks and vulnerabilities in a changing world. The analysis and evaluation of the vulnerability of human habitats aim at the improvement and strengthening of their resilience. Present and future IAPS activities will certainly be devoted to these topics.

References

Abdel-Hadi, A., Tolba, M. K., & Soliman, S. (Eds.). (2010). *Environment, health, and sustainable development* (Advances in People-Environment Studies, Vol. 1). Göttingen, Germany: Hogrefe.

Blaikie, P., Cannon, T., Davis, I., & Wisner, B. (1994). *At risk: Natural hazards, people's vulnerability, and disasters*. London, UK: Routledge.

Bonaiuto, M., Bonnes, M., Nenci, A. M., & Carrus, G. (Eds.). (2011). *Urban diversities – Environmental and social issues* (Advances in People-Environment Studies, Vol. 2). Göttingen, Germany: Hogrefe.

Brauch, H. G., Oswald Spring, Ú., Mesjasz, C., Grin, J., Kameri-Mbote, P., Chourou, B., . . . Birkmann, J. (Eds.). (2011). *Coping with global environmental change, disasters and security. Threats, challenges, vulnerabilities and risks* (Hexagon Series on Human and Environmental Security and Peace, Vol. 5). Berlin, Heidelberg, New York: Springer.

Comfort, L., Wisner, B., Cutter, S., Pulwarty, R., Hewitt, K., Oliver-Smith, A., . . . Krimgold, F. (1999). Reframing disaster policy: The global evolution of vulnerable communities. *Environmental Hazards, 1*, 39–44.

Couch, S. R., & Kroll-Smith, J. S. (Eds.). (1991). *Communities at risk: Collective responses to technological hazards*. New York, NY: Peter Lang.

García-Mira, R., Real Deus, J. E., Uzzell, D. L., San Juan, C., & Pol, E. (2006). Coping with a threat to quality of life: The case of the prestige disaster. *European Review of Applied Psychology, 56*, 53–60.

Gross, A. J. (2001). Megacities and small towns: Different perspectives on hazard vulnerability. *Environmental Hazards, 3*, 63–80.

Heinrichs, D., Hansjürgens, B., Krellenberg, K., & Martinez, J. (Eds.). (2011). *Risk habitat megacity*. Berlin, Heidelberg: Springer.

Kötter, R., & Balsiger, P. (1999). Interdisciplinarity and transdisciplinarity. *Issues in Integrative Studies, 17*, 87–120.

Lahsen, M., Sanchez-Rodriguez, R., Romero Lankao, P., Dube, P., Leemans, R., Gaffney, O., & Stafford Smith, M. (2010). Impacts, adaptation and vulnerability to global environmental change: Challenges and pathways for an action-oriented research agenda for middle-income and low-income countries. *Current Opinion in Environmental Sustainability, 2*, 364–374.

Lawrence, R. (2010). Beyond disciplinary confinement to transdisciplinarity. In V. Brown, J. Harris, & J. Russell (Eds.), *Tackling wicked problems through the transdisciplinary imagination* (pp. 16–30). London, UK: Earthscan.

Lawrence, R., & Després, C. (2004). Futures of transdisciplinarity. *Futures, 36*, 397–405.

McEvoy, D. (2007). Climate change and cities. *Built Environment, 33*, 5–9.

Newman, P., Beatley, T., & Boyer, H. (2009). *Resilient cities: Responding to peak oil and climate change*. Washington, DC: Island Press.

Oliver-Smith, A. (2001). Theorizing vulnerability in a globalized world: A political ecological perspective. In G. Bankoff, G. Frerks & D. Hilhorst (Eds.), *Mapping vulnerability: Disasters, development and people* (pp. 10–24). London, UK: Earthscan.

Pelling, M. (2003). *The vulnerability of cities. Natural disasters and social resilience*. London and Sterling, VA: Earthscan.

Prasad, N., Rangieri, F., Shah, F., Trohanis, Z., Kessler, E., & Sinha, R. (2009). *Climate resilient cities. A primer on reducing vulnerabilities to disasters*. Washington, DC: The World Bank.

Romero Lankao, P., & Qin, H. (2011). Conceptualizing urban vulnerability to global climate and environmental change. *Current Opinion in Environmental Sustainability, 1*, 1–8.

Satterthwaite, D., Huq, S., Reid, H., Pelling, M., & Romero Lankao, P. (2007). *Adapting to climate change in urban areas. The possibilities and constraints in low- and middle-income nations* (Human Settlements Working Paper, Series Climate Change and Cities, No. 1). London, UK: IIED.

Scholz, R. W., Lang, D., Wiek, A., Walter, A., & Stauffacher, M. (2006). Transdisciplinary case studies as a means of sustainability learning. *International Journal of Sustainability in Higher Education, 7*, 226–251.

Thompson Klein, J. (2004). Prospects for transdisciplinarity. *Futures, 36*, 515–526.

Ultramari, C., & Rezende, D. (2007). Urban resilience and slow motion disasters. *City & Time, 2*, 47–64. Retrieved from http://www.ct.cecibr.org

UN-Habitat. (2007). *Enhancing urban safety and security. Global report on human settlements.* London, UK: Earthscan.

Vale, L. J., & Campanella, T. J. (2005). *The resilient city. How modern cities recover from disaster.* Oxford, UK: Oxford University Press.

Reflections on Vulnerability and Risks in Complex Environments

Infectious Diseases as New Risks for Human Health

S. Harris Ali

Faculty of Environmental Studies, York University, Toronto, ON, Canada

Abstract

Throughout history, significant changes in the relationships between human beings and the natural and built environment have served the impetus for the emergence and redistribution of new diseases. The last quarter century has been witness to an unprecedented number of such diseases including, for example, HIV, Ebola, West Nile, mad cow disease, hoof-and-mouth disease, antibiotic-resistant strains of bacteria such as those involved in tuberculosis and *Clostridium difficile*, SARS and, more recently, Influenza A/HIN1. The question therefore arises, could recent changes in the human-environment relationship explain this surge in new and emerging diseases? I address this question by discussing how the forces of economic and cultural globalization over the last few decades have resulted in changes conducive to the emergence and spread of new threats to human health.

Key words: emerging infectious disease, globalization, neoliberalism, security, urbanization

Introduction

Throughout history, significant changes in the nature and types of relationships that human beings have had with nature, and particularly animals, have led to changes in the development and spread of infectious diseases. It is from this perspective that McMichael (2001) outlines how, for instance, the change in settlement patterns over 10,000 years, from nomadic hunter-gathering to settled agrarian-village living, enabled countless strains of bacteria and viruses to jump from domesticated herd animals and from rodents to relatively stationary human beings. Indeed, many existing infectious diseases that affect human beings today, including smallpox, measles, tuberculosis, leprosy, influenza, the common cold, malaria, dengue, and the bubonic plague, can be traced to this historic period of transition. Other examples of changes in the human-environment relationship, as it relates to the spreading of diseases, include the increased trade, travel, and military movements of the past 1,000 to 2,500 years, during which time smallpox and measles spread from the Indian Subcontinent to Europe via Roman Empire troops returning from settling unrest in Syria (in 2 BCE). During the same period, trade along the Silk Route facilitated the spread of the same diseases from Europe to China. Furthermore, the introduction of the bubonic plague to 14th century Europe occurred as the caravans and armies of the Mongol Empire entered the European continent through the Black Sea. A third major shift in the human-environment relationship accompanied the practices of the colonialist period that spanned the 17th to 19th century and led to the well-documented trans-oceanic spread of disease via European ships and the decimation of indigenous populations within the Americas. McMichael (2001) concludes by speculating that we may be entering a fourth transitional period in the ever-changing environment-human relationship – a new epoch defined by globalization.

Some supporting evidence exists for the claim that globalization has led to changes in the environment-human relationship which, in turn, has induced changes in infectious disease patterns. Consider the fact that over the last quarter-century, the world has witnessed an unprecedented number of "new and (re)emerging" diseases. These may be thought of as a set of infectious diseases that appear in a population *de novo* or that have been known for some time but are rapidly increasing in incidence or in geographic range (Mayer, 2000; Morse, 1993, 1995). There are numerous examples of these diseases, including HIV/AIDS, Severe Acute Respiratory Syndrome (SARS), *E. coli* 0157:H7, *Clostridium difficile*, West Nile virus, Lyme disease, antibiotic-resistant tuberculosis, the Ebola virus, and avian influenza (Drexler, 2003; Garrett, 1994; Levy & Fischetti, 2003; Nikiforuk, 2006). Not only has the number of the types of such diseases increased dramatically over recent times, but so too has the actual number of disease outbreaks. For instance, outbreaks caused by food and waterborne pathogens have risen substantially from 533 in 1990 to 1,414 in the year 2000 (Nadakavukaren, 2000). This mirrors the dramatic intensification of globalized and vertically integrated large-scale livestock operations – a practice also referred to as factory farming (Ali, 2004; Davis, 2005; Nikiforuk, 2006; Schlosser, 2002). In what follows, I explore various aspects of McMichael's hypothesized relationship between globalization and infectious disease emergence by broadening the focus of discussion. I adopt a more sociological focus, in order to cast some light on the implications that globalization, and its associated developments of urbanization, neoliberalization, and post-9/11 securitization have for understanding new and emerging diseases such as SARS and Influenza A/H1N1.

Globalization

Throughout his writings, Weber (1978, 2003) argued that, from an historical perspective, large-scale societal developments such as the development of capitalism, democracy, and bureaucracy were informed and guided by the process of "rationalization" (see also Merton, 2002, on the relationship of rationalization and the development of modern science). As such, for Weber, rationalization could be viewed as a kind of all-encompassing "master process" that significantly influenced a wide range of values, beliefs, activities, and organizational developments within the European context of the day. Arguably, it could be said that the process of "globalization" exerts a similarly wide-ranging, pervasive, yet underlying, influence on social, organizational, and institutional life in the contemporary era. A useful starting point for the current analysis in this respect is Held, McGrew, Goldblatt, and Perraton's (2002) general conceptualization of globalization. For Held et al. (2002), globalization may be thought of in terms of the transformations involved in the organization of human affairs that occur because of and through the linking together and expansion of human activity across regions and continents. These transformations may be understood to operate through several mechanisms. The first mechanism involves the increased *extensivity* of global networks, that is, how global networks today have a wider and more pervasive reach than in the past. The second mechanism involves increases in the *intensity* of global interconnections. The third mechanism refers to the speeding up in the *velocity* of global flows. And, lastly, globalization is facilitated by an increase in what is referred to as the "*impact propensity*" – that is, the deepening enmeshment of the local and global, such that the impact of distant events is magnified, while local events may have global consequences. Thus, we can see these mechanisms in action when considering the international spread of infectious diseases such as SARS or Influenza A/H1N1 as a globalized phenomenon. First, it can be noted that the networks of social contacts involved in the spatial diffusion of these diseases were quite extensive, involving long chains of transmission connecting Southern China to Toronto for SARS, or Mexico City to Northern Canada for Influenza A/H1N1. At the same time, the intensity and velocity of the flows involved in the infectious flows were also critical factors involved in the spread of the diseases. The increased volume of passenger flow through airlines in the contemporary era increased the risk of infection, while the speed of air travel has meant that those infected can show no signs of the disease until after their arrival at a distant location, thereby contributing to the possibility of unwittingly spreading the disease across large distances (i.e., the development of a new type of vulnerability). The impact propensity can also be identified as a crucial consideration in infectious disease spread today, because the cross-over of the SARS Coronavirus from the civet cat, or Influenza A/H1N1 from the pig, to human beings may occur in some remote location that, in the past, would have only led to a localized outbreak in an isolated village – an event which ultimately burned itself out, thus preventing the disease from spreading beyond the confines of that locale. Today, because of globalized connectivity, such insularity or isolation no longer exists to the extent it once did, and a localized outbreak of a disease in a remote area is more likely to spread to more populated areas and therefore have a much greater potential for global impacts (for a detailed discussion of these aspects of globalization and SARS, see Ali & Keil, 2006, 2008). Such recent developments highlight how changing relationships of humans with the *natural* environment have implications for the spread of disease in a globalized world. Specifically, the cross-over phenomenon focuses attention on how increasing intervention in nature renders less and less of the world as "untouched natural habitats" (Kaplan & Kaplan, 1989). This has direct implications for disease spread in the contemporary era, as seen not only in the case of the SARS Coronavirsus but also for other diseases. For example, increased suburbanization has resulted in greater human encroachment in formerly untouched wilderness areas and this has led to increased exposure to the tick which causes Lyme disease. Another example relates to how development pressures that result in deforestation effectively open up corridors for mosquito vectors that transmit various diseases, most notably malaria, to humans.

If globalization is indeed a sort of underlying "master process" in today's world, then, just as rationalization has influenced other major societal developments such as capitalism, democracy, bureaucracy, and science, globalization would likewise be expected to have an influence on major developments in the contemporary era that qualitatively distinguish the character of life today from that of the past. I contend that this is indeed the case. In recent times, the collective effects of the types of transformations noted above may be seen in developments that characterize the contemporary epoch, including the advent of new forms of networks (e.g., virtual communities of interest based on Internet technologies); new institutional configurations such as those related to transnational social movements, subpolitics, and issues related to scalar politics (Keil & Ali, 2007); new forms of society-nature relations as the world reflexively confronts issues such as global transboundary environmental problems and emerging diseases; new vulnerabilities to threats, as complex interconnections and nonlinear developments resulting from the interplay of the technological, ecological, and social factors, lead to various types of disasters (Ali, 2008); and new forms of governance, particularly in relation to issues evolving from a crisis politics (Keil & Ali, 2007), such as those involving social control and security under the influence of capitalism (Ali & Keil, 2010; Hooker & Ali, 2009). Many of the effects of these transformations identified above may in turn be thought of in terms of other (constitutive) sub-processes that fall under the auspices of globalization, namely, urbanization, neoliberalization, and securitization. In what follows, I will discuss how the master process of urbanization and its associated sub-processes are relevant to understanding the origin, spread, and response to infectious disease threats in the contemporary era. For the sake of illustration, the case of SARS will be used as the primary empirical referent throughout these discussions.

Urbanization, the Built Environment, and Infectious Diseases

The United Nations (2007) has noted that, as of the year 2008, more than half of the world's population will live in cities. This development has important implications for the spread of infectious diseases. In order for a disease outbreak to be sustained, the host population of humans must be large enough, and dense enough, for the pathogen, especially a virus, to remain in circulation. That is, to survive, the virus must be able jump to another host before the first host succumbs to the disease. Without a living host, the virus cannot multiply, as it essentially requires the cellular machinery of the human host to reproduce itself. Cities, with a large number of people (i.e., hosts) and crowded conditions, are therefore well-suited to the survival needs of the virus. Further, the physical environment of the city itself, with its infrastructure and transportation patterns that serve to open up new avenues for spatial diffusion and proliferation, enhances the suitability of the city for the virus. Notably, cities serve as points of concentration for pathogens arriving via a variety of networks – food, water, air, waste distribution networks. In a sense, due to the very nature of the built infrastructure, cities are connected entities and it is these connections that serve to make human beings vulnerable to infectious diseases. This influence of the built environment was seen in several ways during the SARS outbreak. First, a community outbreak among 300 residents of the Amoy Gardens apartment complex in Hong Kong was found to have occurred because of transmission of the virus through a vertical sewage pipe that connected washrooms directly above one another (Ng, 2008). Second, the spread of SARS was found to be mostly nosocomial in nature, that is, limited to spread within the hospital system. The built environment of the hospital was therefore critical in channeling the disease spread in certain directions. Third, the international spread of SARS was clearly dependent on air travel (Bowen & Laroe, 2006), therefore implicating the built environment of airports as connecting hubs. Notably, the spread of SARS illustrated how the location of major hubs in certain types of cities – that is, those referred to as "global cities" – was instrumental in spreading the disease through the global city network of connections between Hong Kong, Toronto, and Singapore (Ali and Keil, 2006). Indeed, another dimension of "globalization" concerns the linkages between global cities, that is, those cities that act as the command and control posts of the global economy. Global cities integrate the world economy by connecting together various types of flows required to keep the world economy going, including information, communications, labor, capital, and commodities. Although global cities' research has recognized the significance of these types of flows, SARS has revealed that the spread of infectious diseases represents another type of connective flow that can arise in the network of global cities (see Ali & Keil, 2008).

Neoliberalization and Infectious Disease Flow

The process of neoliberalization is based on an ideology that emphasizes the minimization, or complete elimination of state intervention in the provision of goods and services (i.e., the economy), with the ultimate aim of expanding the free market and free trade with the concomitant protection of private property rights. As such, neoliberal policies and activities often promote privatization of state functions, including outsourcing and downloading of state functions to private agencies; and the elimination of environmental, industrial, trade, and health regulations that are viewed as costly barriers to privatized profit maximization. Although these are all matters of a political nature that seem distant from issues related to disease flows, neoliberalization in fact has had a direct impact on infectious disease flows by influencing those health and economic policies and regulations that ultimately have a direct impact on the channeling of these flows through the environment. Two examples will help illustrate this point.

Elsewhere I have described how an outbreak of *E. coli* O157:H7 in the year 2000 in the rural town of Walkerton, Ontario, was related to the neoliberal policy of the Ontario Government, known at the time as the "Common Sense Revolution" (Ali, 2004). The outbreak emerged as the outcome of several policy changes that dramatically influenced the flow of this particular bacterium. *E. coli* O157:H7 originates in the gastro-intestinal tract of livestock and persists in the fecal matter spread as fertilizer on agricultural land in the area. This was found to be origin of the bacteria that entered the town's drinking water supply, contaminating the water, and leading to the outbreak. The contamination event occurred during a time in which drinking water management in the province was being deregulated and privatized. In practice, this meant that the provincial government agencies and laboratories that were responsible for the testing of samples from municipal drinking water supplies were being downsized and faced heavy staff cuts. In fact, such provincial government services had just been privatized at the time of the outbreak, and rural municipalities, such as Walkerton, with a sparse population, and therefore a weak tax revenue base, simply could not afford regular privatized water testing (a monitoring function previously performed by the provincial government). In part, and as a consequence of these developments, the contamination of the drinking water was not detected in a timely fashion, and once detected, efforts to curb the flow of the pathogen were hindered by a lack of resources resulting from budget and staff cuts.

As part of the services provided by governments, public health is an important area that is subject to neoliberal policies. In several ways, the outbreak of SARS in Toronto illustrates how such policies influenced the trajectories the viral flows (Ali & Keil, 2009; Salehi & Ali, 2006; Sanford & Ali, 2005). Tracking the path of the virus through the process of contact tracing is the first step before public health actions, such as the social distancing methods of isolation and quarantine, can be taken. Contact tracing involves the process of identifying all those that are infected or possibly infected by a disease agent. This is accomplished by asking infected persons to list all of the people they had been in contact within a previous period of time. Through follow-up with those identified, it could be determined if the contacts were infected, and those found to be infected would be quarantined or isolated, to break the chain of transmission and contain the outbreak. In the case of Toronto, the ability to engage in contact tracing was hampered through budget cuts to public health, as transfer payments to the municipality by the province for public health had been dramatically reduced over the preceding years (Salehi & Ali, 2006). This had a number of consequences for the response to the SARS outbreak. First, the number of public health officials needed to engage in the laborious process of contact tracing was limited, because of reduced personnel in the Toronto Public Health agency (*ibid.*). The hospital personnel available to combat the disease was also severely limited, particularly in terms of the surge capacity required during the time of an outbreak – a significant problem, as many hospital workers had to self-quarantine as a safety precaution, because of the possibility of infection. Furthermore, reduced spending on public health had a direct impact on the technological infrastructure required to track the virus, particularly in relation to the technical capacity required for the collection, processing, and communication of epidemiological and clinical data across different locations. The province's mandatory information system for the surveillance of reportable disease was found to be very outdated, and the 14-year-old system was not able to support the quarantine management functions required for the management of SARS cases during the outbreak. Although a plan had been proposed by the provincial Public Health Branch in previous years to update the system, the proposal was not approved for funding by the government, forcing Toronto Public Health to develop its own software tools to track SARS cases and contacts (Naylor, 2003). Before this software tool was developed, Toronto Public Health was forced to maintain records on paper charts and maps, using color-coded Post-It notes (Basrur, Yaffe, & Henry, 2004). The city's chief medical officer of health remarked that this was like using 19th century tools to fight a 21st century disease (Naylor, 2003).

The limitations of and constraints on the ability to curb the flow of SARS due to the influence of neoliberal policies are perhaps most pronounced with reference to the weakening of the frontline field response, especially as it related to nursing work and surge capacity during the outbreak. To deal with the reduced amount of funds available to remunerate nurses, hospitals were forced to adopt certain strategies such as the casualization of health care work, whereby Ontario employers were not obliged to pay benefits to part-time (i.e., "casual") workers. In this context, the nursing labor force in Ontario has become highly "casual" (Nuttall-Smith, 2003; Perkel, 2002). Such circumstances increased the potential for inter-hospital spread of the SARS, because many nurses were forced to work at multiple jobs at multiple locations to earn a full-time wage equivalent. Fortunately, perhaps due to the directive that nurses only work at one site during SARS (Burcher, 2003), only one case of inter-institutional transmission was documented (Naylor, 2003). However, such restrictions further constricted the available pool of nurses (Hall et al., 2003). The casualization of nursing labor has several other implications for the ability to respond effectively to a disease outbreak. First, since casual workers are paid on an hourly basis, a latent incentive may be created for casual work nurses workers to continue to work while ill, thus increasing the potential for disease spread (Naylor, 2003). Second, casualization may lead to an attenuation of a sense of workplace community and a reduced awareness of infection control protocols, both essential for frontline workers faced with an outbreak situation (Naylor, 2003). Third, the nursing shortage meant that 4 to 5 patients were assigned to one nurse but it was soon recognized that this was a potentially dangerous ratio that could lead to enhanced disease transmission (Loutfy et al., 2004).

In sum, neoliberalization, as manifested in fewer regulations and enforcement activities aimed at containing disease spread, has indirectly yet significantly contributed to the ways in which infectious diseases may spread, although the adoption of this political program is by no means the only factor contributing to the nature of contemporary disease spread. Thus, the point being made here is that disease spread is a complex and multifaceted phenomena that involves both biophysical and socio-political factors, yet the latter tend to be neglected in conventional epidemiological accounts, thus producing an incomplete story of a very complicated process.

Securitization and Infectious Disease Flow

Social control has always been an important aspect of infectious disease and this is reflected by the fact that the state has adopted specific control strategies at various points in history to deal with various diseases. Citing the work of Michel Foucault, Sarasin (2008) relates, for example, that an *exclusion* orientation was adopted in the Middle Ages, when lepers were forced into colonies located at the periphery of the town. In contrast, an *inclusion*-oriented response to the plague was pursued by the state in the 17th century, whereby those infected were not expelled to outside the city but rather sequestered to an area *within* the city – a social control orientation based on a strategy that was in line with the newly influential liberal ideology of the time (i.e., respect for individual rights). Sarasin goes on to note that Foucault observed that a different strategy was adopted to deal with smallpox in the 18th century, namely, a security-orientation in which statistics were collected to differentiate areas of high risk from those of low risk, knowledge which, in turn, was used in conjunction with targeted immunization programs.

Clearly, social control measures still exist today to deal with infectious disease threats, as mentioned above for the contact tracing and social distancing methods used to deal with SARS. Social control, however, has also led to tension with other governance issues with which the state is involved – primarily, ensuring economic growth. In this context, there has long been a tension between public health and economic interests, because public health strategies based on the curtailment of human movement also prevent those activities necessary for trade and commerce. It was for this very reason that, for example, the imposition of quarantine on incoming ships in European city-state ports during the Middle Ages was met with resistance from the merchant class (Banta, 2001). Notably, such tension still persists today as evidenced by the fact that travel advisories issued by the World Health Organization (WHO), warning travelers not to travel to SARS-affected areas such as Toronto, were met with disapproval by Canadian officials who feared grave economic consequences to the hospitality, tourism, and other sectors of its domestic economy. Protection against public health threats has increasingly been recognized as a precondition for the stabilization of economic activity. In this light, security measures may be somewhat begrudgingly tolerated by state actors. This toleration of security measures may have intensified to an all-encompassing embrace since the events of September 11, 2001. This is seen, for instance, by a renewed emphasis on issues of surveillance, vigilance, and security that now play a prominent role in a state of affairs informed by a crisis politics referred to as the "new normal" (Hooker & Ali, 2009). Within the "new normal" worldview, public health and terrorism issues have become conflated, under the mantle of "national security." This may be illustrated by considering several examples in which public health has become securitized. The explicit recognition of public health as a security issue may be seen, for example, in a report developed for the US Central Intelligence Agency that framed the threat of new infectious diseases as a threat to the nation (CIA, 2003). A second illustration of this recognition is the publication of a special issue of the political affairs journal *Foreign Affairs* (2005: 84(4) July/August) dedicated to the issue of new and emerging diseases as threats to security.

Finally, a third indication of the significance of this issue is seen in the development of Canada's first national security policy: "Securing and Open Society," in which public health emergencies figure prominently (van Wagner, 2008). It has been noted that in a post-9/11 era of the "new normal," a focus on bioterror and infectious diseases as security threats has led to a renewed interest in traditional social control measures associated with national security, such as border control and intelligence capabilities (King, 2002, 2004). Examples of the adoption of such measures can be seen by considering the national and international response to the spread of SARS.

At the most fundamental level, securitization is based on the ability to differentiate the threat from the non-threat and to classify things accordingly. Securitization therefore explicitly involves a sorting process. Similar to contact tracing procedures, the surveillance techniques on which securitization is based are predicated upon sorting and monitoring practices, which today involve advanced digital technologies. Usually these surveillance technologies are located in places such as airports, where a homogeneous flow of people is to be differentiated into streams of possible threats versus non-threats. Since the airport represents a site of considerable sorting and re-sorting of disparate populations across national borders, it is no surprise that surveillance plays a critical role in contemporary airport operation. Clearly, in the post-9/11 era airport surveillance in the service of "security" has intensified, as indicated by the introduction of ever more pervasive surveillance technologies, including the increased use of detention centers, closed-circuit television cameras, global positioning systems, iris-recognition security, intermodal traffic interchanges, as well as through the embedding of data codes on airline tickets that digitally enscribe what the passenger is doing and predict other actions that the passenger may take (Dodge & Kitchin, 2004). Still further indications of the extent to which the post-9/11 "state of emergency," or the exception has become the "rule" are revealed by considering just a few of the many recent state initiatives aimed at "securing" airports. One example involves the "nationalization" of airport security, with the Department of Homeland Security taking charge of implementing standardized systems of person and baggage securitization. A second example is the US government's Total Information Awareness program that integrates and co-ordinates various types of data from private and public sources, including those from biometric technologies that are able to recognize humans at a distance, as well as the mapping of people's multiple connections across their social networks. A third example is provided by the passenger profiling systems known as Computer Assisted Passenger Pre-screening or CAPPS (now Secure Flight), that gathers comprehensive information about a passenger with the ostensible aim of enabling officials to make informed judgments and risk measurements about that passenger's propensity to become a threat (Bennett, 2004).

It was in this context of enhanced surveillance that the WHO "recommended" that airports in SARS-affected areas adopt certain monitoring practices, including temperature screening of departing and transiting passengers, the provision of information leaflets to travelers, exit questioning, and the completion of a mandatory health declaration form by passengers (Ali & Keil, 2010; Bowen & Laroe, 2006). Notably, the issuing of these recommendations violated an important principle of the long-held Westphalian political order, namely the principle of state sovereignty – that is, the view that the nation state holds the exclusive right to govern domestically in an autonomous manner free from external influence. Although these recommendations appeared to be voluntary, in effect they were not because, if the nation state did not adopt the WHO recommendations, then the travel advisories would remain in place, thus having a continued desultory economic impact on SARS-affected cities. The issue and persistence of the travel advisory represented ways of ensuring compliance of the nation state to WHO demands and, according to David Fidler (2004), illustrates a new form of public health governance (which he refers to as post-Westphalian). Yet, it should be noted that despite all this controversy over the imposition of such surveillance and monitoring initiatives at airports, it is questionable, or at least not clear, whether such measures actually succeeded in decreasing the spread of this infectious disease. Especially in light of the fact that almost no cases of SARS were identified through such airport screening strategies, despite the considerable amount of resources deployed in such initiatives (Naylor, 2003).

Concluding Remarks

Over the last quarter century, with intensified globalization, significant changes have occurred in the relationships not only between human beings, but between human beings and nature (including their relationship to physical space and time). Space-time compression, for instance, has important implications for the genesis, spread, and response to infectious diseases. As such, the increased speed of travel, a greater degree of human migration, intensified urbanization, and increasing human encroachment on untouched natural habitats have all enhanced the potential for pathogens to spread internationally in very short periods of time. Furthermore, interaction with other aspects of social change such as urbanization, neoliberalization, and securitization has collectively led to a situation of new vulnerability to disease spread, as new opportunities and pathways are opened up as unintended consequences of these interactions.

It is important to note that the spread of infectious diseases is one example in a set of problems that will probably become increasingly prominent as globalization processes continue to unfold in the future – specifically, I refer here to those developments that involve the crossing of geographic boundaries. Thus, in common with other transboundary problems, such as environmental pollution, international terrorism, and international trade in narcotic drugs, the management of infectious disease spread is challenging, because programs and policies aimed at disease containment necessarily implicate a myriad of other issues that have been discussed in this paper. These include issues related to the ability of sovereign states to govern autonomously, border regulation, the tension between ensuring free trade while protecting public health, and balancing the rights of the individual versus those of the collective. How such moral, political, economic, legal, and social issues are resolved will undoubtedly have significant implications for the battle against the threats of new and (re)emerging diseases – threats that show no sign of abating in the near future, due to increasingly profound human impacts on the environment.

Finally, in terms of the contribution that the approach I have developed here may have for the advancement of people-environment studies more generally, I believe that by focusing on the types of broader, socio-political issues that I have raised, and by situating them within the context of globalization, a more critical and less reductionist approach that more accurately captures the numerous nuances and subtleties of disease transmission in contemporary conditions may be developed.

References

Ali, S. H. (2004). A socio-ecological autopsy of the *E. coli* O157:H7 outbreak in Walkerton, Ontario, Canada. *Social Science and Medicine, 58*, 2601–2612.

Ali, S. H. (2008). SARS as an emergent complex: Toward a networked approach to urban infectious disease. In S. H. Ali & R. Keil (Eds.), *Networked disease: Emerging infections in the global city* (pp. 235–249). Oxford, UK: Wiley-Blackwell.

Ali, S. H., & Keil, R. (2006). Global cities and the spread of infectious disease: The case of Severe Acute Respiratory Syndrome (SARS) in Toronto, Canada. *Urban Studies, 43*, 491–509.

Ali, S. H., & Keil, R. (2008). *Networked disease: Emerging infections in the global city.* Oxford, UK: Wiley-Blackwell.

Ali, S. H., & Keil, R. (2009). Public health and the political economy of scale: Implications for understanding the response to the 2003 Severe Acute Respiratory Syndrome (SARS) outbreak in Toronto. In R. Keil & R. Mahon (Eds.), *Leviathan undone? Towards a political economy of scale* (pp. 195–208). Vancouver, BC: UBC Press.

Ali, S. H., & Keil, R. (2010). Securing network flows: Infectious disease and airports. In S. Graham & S. Marvin (Eds.), *Disrupted cities: When infrastructure fails* (pp. 97–110). New York, NY: Routledge.

Banta, J. E. (2001). Commentary: From international health to global health. *Journal of Community Health, 26*, 73–77.

Basrur, S., Yaffe, B., & Henry, B. (2004). SARS: A local public health perspective. *Canadian Journal of Public Health, 95*, 22–24.

Bennett, C. J. (2004). What happens when you book an airline ticket (revisited): The computer assisted passenger profiling system and the globalization of personal data. In E. Zureik & M. B. Salter (Eds.), *Global surveillance and policing: Borders, security, identity* (pp. 113–138). Cullompton, UK: Willan.

Bowen, J. T., & Laroe, C. (2006). Airline networks and the international diffusion of severe acute respiratory syndrome (SARS). *The Geographical Journal, 172*, 130–144.

Burcher, B. (2003, September 17). *SARS: The experience from nursing education*. Paper Presented at the University of Toronto, Faculty of Nursing, Toronto.

CIA – Office of Transnational Issues. (2003). *SARS: Lessons from the first epidemic of the 21st century: A collaborative analysis with outside experts (unclassified)*. Retrieved from http://www.pdhealth.mil/downloads/cia_sars.pdf

Davis, M. (2005). *The monster at our door: The global threat of avian flu*. New York, NY: The New Press.

Dodge, M., & Kitchin, R. (2004). Flying through code/space: The real virtuality of air travel. *Environment and Planning A, 36*, 195–211.

Drexler, M. (2003). *Secret Agents: The menace of emerging infections*. Toronto, ON: Penguin.

Fidler, D. (2004). *SARS, governance and the globalization of disease*. New York, NY: Palgrave Macmillan.

Garrett, L. (1994). *The coming plague: Newly emerging diseases in a world out of balance*. New York, NY: Penguin Books.

Hall, L., Angus, J., Peter, E., O'Brien-Pallas, L., Wynn, F., & Donne, G. (2003). Media portrayal of nurses' perspectives and concerns in the SARS crisis in Toronto. *Journal of Nursing Scholarship, 3*, 211–216.

Held, D., McGrew, A., Goldblatt, D., & Perraton, J. (2002). Rethinking globalization. In D. Held & A. McGrew (Eds.), *The global transformation reader: An introduction to the globalization debate* (2nd ed., pp. 67–74). Oxford, UK: Blackwell.

Hooker, C., & Ali, S. H. (2009). SARS and security: Health in the new normal. *Studies in Political Economy, 84*, 101–126.

Kaplan, R., & Kaplan, S. (1989). *The experience of nature: A psychological perspective*. New York, NY: Cambridge University Press.

Keil, R., & Ali, S. H. (2007). Governing the sick city: Urban governance in the age of emerging infectious disease. *Antipode, 40*, 846–871.

King, N. B. (2002). Security, disease, commerce: Ideologies of postcolonial global health. *Social Studies of Science, 32*, 763–780.

King, N. B. (2004). The scale politics of emerging diseases. *Osiris, 19*, 62–76.

Levy, E., & Fischetti, M. (2003). *The new killer diseases: How the alarming evolution of germs threatens us*. New York, NY: Three Rivers Press.

Loutfy, M. R., Wallington, T., Rutledge, T., Mederski, B., Rose, K., Kwolek, S., . . . Berall, G. (2004). Hospital Preparedness and SARS. *Emerging Infectious Diseases, 10*, 771–776.

Mayer, J. D. (2000). Geography, ecology and emerging infectious diseases. *Social Science and Medicine, 50*, 937–952.

McMichael, A. J. (2001). Human culture, ecological change and infectious disease: Are we experiencing history's fourth great transition. *Ecosystem Health, 7*, 107–115.

Merton, R. K. (2002). *Science, technology & society in seventeenth-century England*. New York, NY: Fertig.

Morse, S. S. (1993). Examining the origins of emerging viruses. In S. S. Morse (Ed.), *Emerging Viruses* (pp. 10–28). New York, NY: Oxford University Press.

Morse, S. S. (1995). Factors in the emergence of infectious diseases. *Emerging Infectious Diseases, 1*, 7–15.

Nadakavukaren, A. (2000). *Our global environment: A health perspective* (5th ed.). Prospect Heights, IL: Waveland Press.

Naylor, D. (2003, October). *Learning from SARS, renewal of public health in Canada: A report of the National Advisory Committee on SARS and public health*. Ottawa, ON: Health Canada.

Ng, M. K. (2008). Globalization and SARS and health governance in Hong Kong under "One Country, Two Systems". In S. H. Ali & R. Keil (Eds.), *Networked disease: Emerging infections in the global city* (pp. 70–85). Oxford, UK: Wiley-Blackwell.

Nikiforuk, A. (2006). *Pandemonium: Bird flu, mad cow disease and other biological plagues of the 21st century*. Toronto, ON: Viking Press.

Nuttall-Smith, C. (2003). Diary of a SARS victim. CanWest news service. Retrieved from http://www.canwest.com

Perkel, C. (2002). *Well of lies. The Walkerton water tragedy*. Toronto, ON: McClellan and Stewart.

Salehi, R., & Ali, S. H. (2006). The social and political context of disease outbreaks: The case of SARS in Toronto. *Canadian Public Policy/Analyse de Politiques, 32*, 373–385.

Sanford, S., & Ali, S. H. (2005). The new public health hegemony: Response to Severe Acute Respiratory Syndrome (SARS) in Toronto. *Social Theory and Health, 3*, 105–125.

Sarasin, P. (2008). Vapors, viruses, resistance(s): The trace of infection in the work of Michel Foucault. In S. H. Ali & R. Keil (Eds.), *Networked disease: Emerging infections in the global city* (pp. 214–228). Oxford, UK: Wiley-Blackwell.

Schlosser, E. (2002). *Fast food nation: The dark side of the all-American meal.* New York, NY: HarperCollins.

United Nations Population Fund. (2007). UNFPA state of World Population 2007: Unleashing the potential of urban growth. UNFPA state of World Population 2007: Unleashing the potential of urban growth. Retrieved from http://www.unfpa.org/swp/2007/english/introduction.html

van Wagner, E. (2008). The practice of biosecurity in Canada: Public health legal preparedness and Toronto's SARS crisis. *Environment and Planning A, 40*, 1647–1663.

Weber, M. (1978). *Economy and society.* In G. Ross & C. Wittich (Eds.), Los Angeles, CA: University of California Press.

Weber, M. (2003). *The protestant ethic and the spirit of capitalism* (T. Parsons, Trans.). New York, NY: Charles Scribner's Sons.

Urban Vulnerability Under Conditions of Global Environmental Change

Conceptual Reflections and Empirical Examples From Growing and Shrinking Cities

Christian Kuhlicke,[1] Sigrun Kabisch,[1] Kerstin Krellenberg,[1] and Annett Steinführer[2]

[1]Department of Urban and Environmental Sociology, Helmholtz Centre for Environmental Research – UFZ, Leipzig, Germany
[2]Rural Studies, Johann Heinrich von Thünen Institute (vTI), Braunschweig, Germany

---

Abstract

In this chapter, we develop a perspective on urban vulnerability by considering the impact of environmental hazards. We include various strands of research such as social vulnerability to negative impacts of environmental hazards, concepts from the climate change debate, and research situated within the field of urban studies. Our aim is to propose a social-science based conceptual framework of urban vulnerability encompassing the following three dimensions: Exposure, susceptibility, and coping and adaptive capacities. In particular, we focus on two distinct urban development paths: Growing and shrinking cities. Whereas in growing cities the extent and complexities of vulnerability increase, the situation in shrinking cities is quite different. They seem to offer the potential to reduce their vulnerability as fewer people and buildings are exposed. We conclude that urban vulnerability remains a field of research with many challenging and still unanswered questions.

Key words: urban vulnerability, environmental hazards, adaptive and coping capacities, growing and shrinking cities

---

Introduction

Global environmental change is characterized by a complex transition of the world and its human habitats, which is making them (even more) vulnerable to various social, economic, and environmental hazards. The risks associated with this change are ambiguous and hence controversial. The impacts of global environmental change differ according to the local conditions, not least because existing vulnerabilities vary between contexts and new vulnerabilities will arise in the future.

In this paper, we develop a perspective on urban vulnerability by considering, in particular, the impacts of global environmental change. We focus on the causes, conditions, and consequences that environmental hazards have on cities and their inhabitants. While the term "urban vulnerability" increasingly appears in policy documents and scientific discourses, its conceptualization remains rather weak. We therefore attempt to fill this gap by developing a social-science based understanding of the concept.

Our contribution is structured as follows: In the first section, we introduce and discuss our conceptual approach to urban vulnerability and some methodological implications. Then, urban growth (e.g., increasing number of inhabitants, urban land expansion) and its consequences for urban vulnerability will be dealt with. By way of contrast, the subsequent section addresses the interrelations of urban shrinkage (e.g., decreasing number of inhabitants, emerging urban brownfields) and vulnerability. Both cases make it apparent that context-related investigations in different types of cities are indispensable, in order to obtain a deeper understanding of the emerging research field of urban vulnerability.

Urban Vulnerability: Concept and Methodological Approach

Concept

While the concept of vulnerability is now well-established in scientific discourses (for a review see Fuchs, Kuhlicke, & Meyer, 2011), the discussion of urban vulnerability is still unfolding. The debate is currently dominated by case studies and a more descriptive conceptualization of what the term urban vulnerability might mean (Romero Lankao & Qin, 2011; Tipple, 2006). We propose an approach that considers the tightly coupled interdependency between various spheres of urban life and the environment and argue that it is too short-sighted to capture the essence of urban vulnerability simply by taking into account the sheer number or density of people or by focusing on a changing state of nature or the environment (O'Keefe, Westgate, & Wisner, 1976). It is, rather, cities' *complexity* and the various interconnections of both their socio-technical and socio-ecological systems that need to be considered as central features of urban vulnerability. We therefore agree with Pelling's notion of a city as "an engine for economic growth, an organism or integrated system linking consumption and production, a source of livelihoods, a stock of accumulated assets, a political and cultural arena" (Pelling, 2011, p. 550). For such systems, it is important that small but destructive hazards may result in cascading effects that ultimately will have major and disastrous consequences for people, infrastructures, and buildings. In this sense, we argue that cities' "complexity increases [their] vulnerability" (Perrow, 2008, p. 156).

Studies on urban vulnerability can be classified into research approaches that focus (a) on the various *impacts* of environmental hazards or the projected consequences of climate change on urban areas (Prasad et al., 2009) and (b) on the *contextual* characteristics defining the vulnerability of urban areas and their inhabitants (Cross, 2001; Romero Lankao & Qin, 2011). Impact studies most often consider vulnerability as the "end point" of a sequence of analyses (e.g., down-scaling climate scenarios, hydrological modeling of flood events, etc.) and usually focus on the changing *exposure* and *susceptibility* of different entities such as buildings, critical infrastructure, and/or population groups to different hazards (O'Brien, Eriksen, Nygaard, & Schjolden, 2007). Contextual studies, on the other hand, are based on the local context and focus on *exposure and susceptibility* as well as on actors' *coping and adaptive capacities*. Such studies are typically interested in understanding and explaining the social reasons for the production of hazardous situations and their consequences. They therefore place the social, institutional, economic, political, and cultural causes (and not the natural forces; O'Keefe et al., 1976) for the production of vulnerable conditions at the forefront of their analyses (Wisner, Blaikie, Cannon, & Davis, 2004).

In taking up this discussion, our conceptual approach to urban vulnerability includes three dimensions: (1) exposure, (2) susceptibility, and (3) coping and adaptive capacities.

Exposure describes the physical precondition to be affected (Fuchs et al., 2011). Susceptibility involves the preconditions to suffer harm, because a person or a group experiences some level of fragility or disadvantageous conditions. Coping and adaptive capacities refer to the ability to come to terms with stressing, threatening, or damaging events by coping with or adapting to them.

Thus, urban vulnerability, in our understanding, describes (a) the degree to which individuals, social groups, local communities, organizations (formal and informal), infrastructures, as well as physical assets in urban environments are exposed and susceptible to environmental hazards and (b) the capacities of these actors and assets to adapt to and cope with the negative impacts of such hazards (also see Mileti, 1999, p. 376 and the earlier concepts of the "hazardousness of place"; Hewitt & Burton, 1971). Our concept aims at deliberately adding a notion of "the urban" – as the predominant mode of the complex spatial organization of contemporary societies worldwide – to the existing (social) vulnerability discourse.

Methodological Approach

In order to approach urban vulnerability empirically (Figure 1), we argue for a research design that is based on both deductive and inductive methods (Kuhlicke, Scolobig, Tapsell, Steinführer, & De Marchi, 2011). Deductive methods mostly rely on indicators

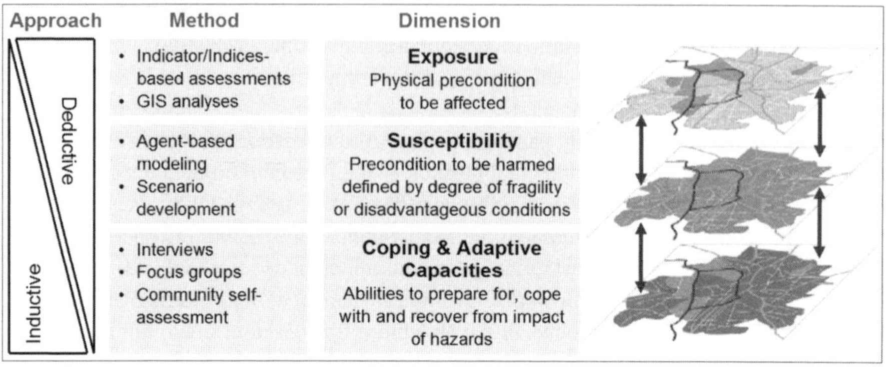

Figure 1. Conceptual and methodological framework for urban vulnerability analyses. (Source: the authors)

and indices. Such indicators or indices should be transferable and applicable to a variety of contexts, although they need to be adapted to the specific local conditions, to allow comparisons within and between urban areas (Borden, Schmidtlein, Emrich, Piegorsch, & Cutter, 2007). Quite often, these methods include spatial analysis, based on Geographic Information Systems (GIS) and remote sensing data, and focus on the exposure and susceptibility of actors, buildings, and infrastructure. Adaptive and coping capacities, when taken into account at all, are dealt with in a rather hypothetical manner, that is, they are presumed to be mirrored by high levels of certain measurable indicators (e.g., wealth, formal qualification, or homeownership). Inductive methods, in contrast, aim at developing an understanding of urban vulnerability that is based on empirical observation and applies some kind of direct interaction with the actors being the focus of the research. They usually rely on qualitative methods and are performed in a participatory manner. Typical tools are interviews, focus group discussions, and community self-assessments. Inductive methods facilitate an understanding of causes and effects but also allow the unraveling of implicit assumptions based on tacit knowledge, traditions, norms, and values, all of which are often hard to measure or cannot be measured at all by quantitative indicators. Agent-based modeling and scenario development are among the tools/approaches that offer an opportunity to integrate the results of both deductive as well as inductive methods. From our approach, we would like to stress that a comprehensive picture requires a reasonable combination of both methodological strands.

Growing Cities: Are They Especially Vulnerable?

On a global scale, urban growth is the dominant developmental trend of human settlements. Starting from more than 50% urban population today, it is expected that, in

2025, two-thirds of the global population will live in urban areas (Roy, 2009). Viewed globally, particularly very large cities (here: megacities with 10 million inhabitants and more) as well as intermediate cities (1–5 million inhabitants) have experienced rapid growth rates in the last decades (5.4% and 2.7% between 1975 and 2005, respectively; UN Population Division, 2008). Although urbanization is a global phenomenon, it does not occur as a homogeneous process across regions and countries, or even within one country (Heinrichs, Krellenberg, & Hansjürgens, 2012). For example, more than two-thirds of today's megacities are located in the so-called "developing" countries in the global south.

Unprecedented scale, rapid change, and complexity are the defining characteristics that transform growing cities, and especially megacities, into spaces of opportunity and risk in terms of social, economic, and environmental challenges (Heinrichs et al., 2012). Highlighting the opportunities, we point out that the concentration of people in cities harbors the potential to lower resource consumption through compact urban forms or mass transport systems. However, at the same time, population growth goes hand in hand with the expansion of urban areas, often in high risk environments threatened, for example, by floods or landslides.

In recent years, research on vulnerability in growing cities has increased and has focused on case studies from all over the world, often based on conflicting theories and paradigms. In the following, we will highlight the most visible interconnections between urban vulnerability and growing cities: First, an increasing exposure of people to environmental hazards as a consequence of urban growth; second, increasing socio-spatial fragmentation of the urban population; and third, poor governance.

The first process, the sheer enlargement of urban areas, is surely among the most decisive drivers of urban vulnerability in growing cities. It results in an overall increase in exposure due to (a) the expansion of the city into hazard-prone areas, (b) the emergence of new hazard-prone areas through the transformation of urban land use and natural systems as well as to changing climate conditions, and (c) the overall increase in the population that is potentially exposed.

In growing cities, where pressure on and conflicts about land are strong and land for new construction is becoming scarce, exposure to environmental hazards might be increasing generally among all socio-economic groups. Taking into account that climate change will further add to already existing vulnerabilities, it is presumed that, in growing cities, exposure of infrastructure and people will increase in the future (Hallegatte & Corfee-Morlot, 2011). The same holds true for heat stress, where increasing mean temperatures, larger sealed surfaces, and less evaporation result in higher heat exposure.

Ongoing research in Santiago de Chile, which provides access to a wide range of representative problems associated with megacities (Heinrichs et al., 2012), shows that changes in anthropogenic land use increase the exposure of residents to potentially hazardous flood events and aggravate the flood hazard by increasing the surface water runoff after precipitation events. Case study evidence for one selected municipality shows that not only poor but likewise affluent households are exposed to potentially hazardous events (Ebert, Welz, Heinrichs, Krellenberg, & Hansjürgens, 2010). However, this example should not be overestimated, because evidence from other cities (Brouwer, Akter, Brander, & Haque, 2007, Douglas et al., 2008) indicates that even though exposure overlaps with socio-spatial inequality, it does not directly predict higher vulnerability (Cross, 2001).

This relates to the second relevant process: Urbanization processes may lead to an increasing socio-spatial fragmentation, exemplified by gated communities of the more affluent and informal settlements of the poor population. This process is connected with new urban vulnerabilities, in particular for the poor. Informal settlements in many fast growing cities are often located in areas exposed to hazards (De Sherbinin, Schiller, & Pulsipher, 2007). Vulnerability however, depends very much on further factors, such as building materials of diverse origin and quality (Hardoy & Pandiella, 2011) that may show disadvantageous properties with respect to risky and dangerous living conditions. Furthermore, already existing inequalities can be reinforced, because the poor in many developing countries, women, the elderly, and children are considered to be particularly vulnerable (e.g., Kantor & Nair, 2005; Pelling, 2002).

In this context, the importance of differentiating between potential consequences becomes clear as various socio-economic groups tend to incur dissimilar damage, even though they share similar exposure. In other words, coping capacities may be a more important distinction between poor and better-off households than their sheer exposure. Nevertheless, coping is directly interlinked with differing preconditions that can be very diverse in cities. For example, lower income groups might suffer material damage due to fragile houses and a lack of protective measures, whilst better-off households have the possibility to protect themselves by constructing physical structures, such as walls or dams. In many megacities, but also in intermediate cities in the global south, protection schemes are still uncommon and rarely publicly financed. Moreover, at present, a private insurance market does not exist (Kunreuther, 2001). Thus, the enhancement of individual coping and adaptive capacities becomes fundamental for reducing vulnerabilities. This is closely linked to governance structures, the third issue we discuss.

Growing cities and especially megacities are complex with regard to governance structures. For example, the provision of basic services such as water, shelter, and mobility is organized by a variety of decentralized individual and informal institutions outside the formal set of rules (Webster & Lay, 2003), particularly when governments fail or are unable to ensure the provision of basic goods and services. This, in turn, results in a loss of formal governability and control (Kraas, 2007). Poor governance can amplify or even produce risks for and negative impacts on human security. On the other hand, good local governance and planning can make a tremendous contribution to decreasing the exposure of people and services to environmental hazards and to the enhancement of coping and adaptive capacities. In this context, adaptation in terms of a variety of measures such as forecasting, early warning systems, as well as targeted environmental education and information, is becoming a stronger focus of research.

Summing up, growing cities and particularly megacities show a high degree of vulnerability due to the "stress bundles" that are related to the three processes discussed above. High exposure, increasing socio-spatial fragmentation, and inappropriate governance make growing cities and their population vulnerable to environmental hazards.

Shrinking Cities: An Opportunity to Reduce Urban Vulnerabilities?

While cities in the global south are experiencing remarkable growth rates, in many parts of Europe, Northern America, and some parts of East Asia (e.g., Japan or South Korea) overall population growth is at a halt or is even declining. As a consequence, a polarized pattern of urban development is emerging. Besides metropolitan areas facing in-migration and natural population surplus, many cities and towns are losing population to a considerable extent (Oswalt & Rieniets, 2006). As the debate has one of its roots in Germany, the term "urban shrinkage" (*Schrumpfung*) is commonly applied to this urban trajectory (Kabisch, Bernt, & Peter, 2004; Mykhnenko & Turok, 2007; Rieniets, 2009). Shrinkage first and foremost relates to a significant decline in population within a short time span. As a rule, this process is caused or, at least interconnected with, a considerable decline in economic power, a cutback of the municipal budget, and a loss of urban functionality related to service and infrastructure offers.

Urban shrinkage confronts many cities with new problems and challenges, which might appear as less urgent, from a global perspective, than those triggered by enormous growth rates. However, as Rieniets (2009) states: "we know [. . .] less about [shrinking cities] and about possible solutions" (p. 232). While there is some knowledge about the causes and consequences of urban shrinkage (Lampen & Owzar, 2008; Schwarz & Rugare, 2009), the link to how this process relates to an increase or decrease of urban vulnerability to environmental hazards has not yet been made. Such a perspective would clearly go beyond the description of emerging research agendas of environmental mitigation and ecological restoration in terms of land use in shrinking cities (Blanco et al., 2009, pp. 228–229; see also Rink & Kabisch, 2009).

In the following, we focus on three interconnections between urban vulnerability and shrinkage. We first argue that disasters may trigger shrinkage. Secondly, we discuss the potentials of shrinking cities with regard to reducing their exposure and, thus, their overall urban vulnerability to environmental hazards. In a third step, we reflect on possibly decreasing coping and adaptive capacities in shrinking cities. At present, with the exception of the first line of argumentation, almost no empirical studies exist that shed light on the issues just mentioned.

The first case relates to disasters as catalysts of shrinkage. The occurrence of a major earthquake or a flood in densely settled areas quite often results (like wars) in the destruction of the built environment and triggers temporary or permanent out-migration and forced relocation of residents. At the very least, such disasters result in a reduced number of inhabitants and hence urban shrinkage. A recent and graphic example is the impact of hurricane Katrina on New Orleans. In 2010, the population of the city was one third of its 2005 total, when large segments of the residents had to relocate (Campanella, 2010). Equally important, the population structure of New Orleans has changed considerably (Robertson, 2011, based on findings by the sociologist Andrew Beveridge; Sastry, 2009). However, from a historical perspective, this development path is more the exception because, usually, the reconstruction of the *status quo ante* and rapid repopulation is the typical pattern, as

Thomas J. Campanella (2006) states: "the modern city is virtually indestructible" (p. 142). He bases this evaluation in particular not only on strong evidence of the social resilience of cities (particularly the willingness of the citizens to rebuild and reshape their environment) but also on the persistence of the urban hardware (concrete foundations and buried network utilities), the internal logic of private homeownership, and political steering mechanisms (see also the contributions in Vale & Campanella, 2005).

However, the Katrina example also leads to our second argument. Shrinking populations might contribute to a decrease in the overall vulnerability of a city. Ideally, the decline of the number of inhabitants can be utilized to also decrease the exposure of people (e.g., by moving them out of harm's way). There is, hence, a considerable potential for governing the very process of shrinkage towards reducing the vulnerability of cities.

Benefits and drawbacks of a reduced population of cities in general and of New Orleans in particular have been discussed controversially (Campanella, 2010; Wolf-Powers, 2007). One could argue that it would be a wise decision to use the disaster as a "window of opportunity" and transform the temporary evacuation of the city into a permanent one. New Orleans is largely located below sea level and, furthermore, many of its areas are facing local land subsidence and a projected increase of hurricane risks in the future (Wolf-Powers, 2007). Therefore, calls to transform the heavily destroyed parts of the city into open parks and nature preserves were formulated immediately after hurricane Katrina struck the city. Others argue that the "realities on the ground are considerably more complex" (Wolf-Powers, 2007, p. 87). If the hurricane had hit other, more affluent parts of the city, is seems rather unlikely that "anyone would be talking about not rebuilding" (James Dart from the New Jersey Institute of Technology, quoted in Wolf-Powers, 2007, p. 87). Therefore, the Katrina example can also be interpreted such that the disaster was utilized to "cleanse" New Orleans from its "undesirable residents" (i.e., in terms of socio-economic characteristics) and has hence "brought to light a 'shrinking city' agenda that looks more like a purge" (Wolf-Powers, 2007, p. 89).

The New Orleans example underlines the contested nature of transforming socio-spatial and land-use patterns in the aftermath of a disaster. Therefore, it seems more promising to develop strategies aimed at reducing vulnerabilities and risks *before* disasters strike. As stated above, shrinking cities seem to have considerable potential to transform former densely populated areas into open green spaces and to restore historical floodplains for retention purposes, to give just one example. This would not only contribute to reducing sealed surfaces but also to increasing retention of water in the landscape and would hence provide more space for the rivers (Kruse, 2010; Moss & Monstadt, 2008). In addition, such a strategy could also decrease exposure. However, from numerous examples of shrinking cities in Germany that are also prone to flooding, we know that flood risk management and urban restructuring (including relocation of neighborhoods at risk) are not pursued in an integrated way or, to put it differently: Decisions about large-scale demolitions of the housing stock (Kabisch et al., 2004) rarely take the results of risk assessments into account.

Shrinking urban habitats also face new challenges that might reveal new vulnerabilities; this leads to our third argument. If economic restructuring is the trigger for population decline, then urban shrinkage is related to selective out-migration, particularly of younger and better qualified age cohorts. In the medium term, the negative migration balance is typically followed by a negative natural balance: Fewer children are born because there are fewer women of reproductive age. Thus, the urban society is aging in absolute and relative terms and, in the long run, the proportion and number of the very old will increase. Yet, it must be stressed that not one, single variable leads to the vulnerability of a person. As numerous studies have demonstrated, it is not only the rather complex combination of different variables but also, and importantly, the respective societal context that influences the vulnerability of a certain group, such as the elderly (e.g., Klinenberg, 2002). Hence, there are indeed many "immaterial" aspects, such as local knowledge, culture, traditions, norms, and mores, which intervene in the process of vulnerability formation and are hard, if not impossible, to measure (Kuhlicke et al., 2011). In this sense, the simple assumption that a shrinking city inhabited by an aging population is automatically more vulnerable would be misleading, as there are many other intervening factors that would need to be taken into account (Steinführer et al., 2009, p. 82). Furthermore, such a general assumption is associated with a highly stereotyped picture of "the" elderly.

However, urban shrinkage is not restricted to the emergence of a new and more homogeneous socio-demographic profile of an urban area. It also represents a complex change in the mode of local governance, accompanied by severe budget constraints, declining levels of citizens' commitment, the closing down of public and private services of general interest, and the like. Accordingly, urban shrinkage may also challenge existing disasters management systems – which in many European countries rest, at least in part, upon volunteers (e.g., fire brigades in Austria, Germany, and Slovenia or the Civil Protection Service in Italy) – and other adaptive systems, because they are no longer able to provide their services, due to a diminishing number of actively participating members. Shrinking towns and cities might thus face increasing susceptibility to extreme events and declining individual and institutional capacities to adapt to and cope with them. Consequently, their dependency upon external support may increase.

However, as stated above, more thorough research is required in order to better understand the mentioned processes and their interdependencies with urban vulnerability.

Some Concluding Remarks and Outlook

The aim of this paper was to conceptualize urban vulnerability and to discuss whether distinct paths of urban development account for different dimensions and complexities of this type of vulnerability. Based on the examples, we argue for a context-specific evaluation of different development paths of cities worldwide, as provided by the examples of urban growth and urban shrinkage. A thorough understanding of urban vulnerability needs to include the changing patterns of exposure, susceptibility, and coping as well as adaptive capacities and how they interact with the interdependencies in various spheres of social, economic, ecological, and technological environments.

Furthermore, urban vulnerability should not be simply explained by changing patterns of environmental hazards (e.g., increasing rain intensity) or by the growing number of residents living in a certain area, resulting in higher exposure and thus, presumably, in higher vulnerability in the course of ongoing urbanization (Laukkonen et al., 2009, p. 288). Rather, the focus needs to be on revealing and understanding the various interconnections and interdependencies within urban areas, their actors, assets, and infrastructures. This is obviously a challenging task, because vulnerability analysis is (still) usually a rather "static" endeavor (e.g., when measuring the vulnerability of a community by using various standard social, economic, and demographic indicators). In order to capture some of the dynamics, we propose focusing specifically in future research on "second order" vulnerabilities, that is, on new vulnerabilities created unintentionally by trying to reduce existing ones. The notion of "second order" vulnerabilities tries to develop a dual and, to a certain extent, a paradoxical orientation: On the one hand, it relates to the fact that hazard and risk management efforts in many urban areas went at least hand in hand with, or even stimulated, urban growth and prosperity by providing a certain level of protection, and thus created the vulnerabilities that cities and residents are facing today. On the other hand, it acknowledges the so-called vulnerability paradox, because it captures the mostly unintended consequences of this development, namely that it contributes to perceiving the physical environment with its considerable variations as secured and controlled (De Marchi, Scolobig, Delli Zotti, & Del Zotto, 2007, pp. 26–31; Kuhlicke, 2010).

Besides the topics mentioned above, any analysis of urban vulnerability needs to account for various aspects and different types of urban development. Urban vulnerability differs not only between settlements facing population growth and those experiencing a decline in the number of residents, as discussed in this chapter, but also between megacities, cities, and small towns (Cross, 2001), as well as between urban habitats in various political settings. Thus, urban vulnerability seems to be a promising research agenda for the future that will need to address the many unanswered questions that could only be touched upon in this contribution.

References

Blanco, H., Alberti, M., Olshansky, R., Chang, S., Wheeler, S. M., Randolph, J., . . . Watson, V. (2009). Shaken, shrinking, hot, impoverished and informal: Emerging research agendas in planning. *Progress in Planning, 72*, 195–250.

Borden, K. A., Schmidtlein, M. C., Emrich, C. T., Piegorsch, W. W., & Cutter, S. L. (2007). Vulnerability of U.S. cities to environmental hazards. *Journal of Homeland Security and Emergency Management*, *4*(2), Article 5. Retrieved from http://www.bepress.com/jhsem/vol4/iss2/5

Brouwer, R., Akter, S., Brander, L., & Haque, E. (2007). Socioeconomic vulnerability and adaptation to environmental risk: A case study of climate change and flooding in Bangladesh. *Risk Analysis, 27*, 313–326.

Campanella, R. (2010). The geography of urbanization and the evolution of risk. In A. Dölemeyer, J. Zimmer, & G. Tetzlaff (Eds.), *Risk and planet earth: Vulnerability, natural hazards, integrated adaptation strategies* (pp. 73–80). Stuttgart, Germany: Schweizerbart.

Campanella, T. J. (2006). Urban resilience and the recovery of New Orleans. *Journal of the American Planning Association, 72*, 141–146.

Cross, J. A. (2001). Megacities and small towns: Different perspectives on hazard vulnerability. *Environmental Hazards, 3*, 63–80.

De Marchi, B., Scolobig, A., Delli Zotti, G., & Del Zotto, M. (2007). *Risk construction and social vulnerability in an Italian Alpine Region* (FLOODsite report T11–06-08). Gorizia, Italy: Institute for International Sociology.

De Sherbinin, A., Schiller, A., & Pulsipher, A. (2007). The vulnerability of global cities to climate hazards. *Environment & Urbanization, 19*, 39–64.

Douglas, I., Alam, K., Maghenda, M., McDonnell, Y., McLean, L., & Campbell, J. (2008). Unjust waters: Climate change, flooding and the urban poor in Africa. *Environment and Urbanization, 20*, 187–205.

Ebert, A., Welz, J., Heinrichs, D., Krellenberg, K., & Hansjürgens, B. (2010). Socio-environmental change and flood risks: The case of Santiago de Chile. *Erdkunde, 64*, 303–313.

Fuchs, S., Kuhlicke, C., & Meyer, V. (2011). Vulnerability to natural hazards – The challenge of integration [Editorial of Special Issue]. *Natural Hazards, 58*, 609–619. .

Hallegatte, S., & Corfee-Morlot, J. (2011). Understanding climate change impacts, vulnerability and adaptation at city scale: An introduction. *Climatic Change, 104*, 1–12.

Hardoy, J., & Pandiella, G. (2011). Urban poverty and vulnerability to climate change in Latin America. *Environment and Urbanization, 23*, 123–155.

Heinrichs, D., Krellenberg, K., & Hansjürgens, B. (2012). Introduction: Megacities in Latin America as risk habitat. In D. Heinrichs, K. Krellenberg, B. Hansjürgens, & F. Martinez (Eds.), *Risk habitat megacity* (pp. 3–17). Berlin, Heidelberg: Springer.

Hewitt, K., & Burton, I. (1971). *The hazardousness of a place. A regional ecology of damaging events* [Research Publication 6]. Toronto, ON: University of Toronto, Department of Geography .

Kabisch, S., Bernt, M., & Peter, A. (2004). *Stadtumbau unter Schrumpfungsbedingungen* [Urban restructuring confronting urban shrinkage]. Wiesbaden, Germany: VS Verlag für Sozialwissenschaften.

Kantor, P., & Nair, P. (2005). Vulnerability among slum dwellers in Lucknow, India: Implications for urban livelihood security. *International Development Planning Review, 27*, 381–406.

Klinenberg, E. (2002). *Heat wave: A social autopsy of disaster in Chicago*. Chicago, IL: University of Chicago Press.

Kraas, F. (2007). Megacities and global change: Key priorities. *The Geographical Journal, 137*, 79–82.

Kruse, S. (2010). *Vorsorgendes Hochwassermanagement im Wandel – Ein sozial-ökologisches Raumkonzept für den Umgang mit Hochwasser* [Changing preventive flood management. A socio-ecological concept of space concerning flooding]. Wiesbaden, Germany: VS Verlag für Sozialwissenschaften.

Kuhlicke, C. (2010). The dynamics of vulnerability: Some preliminary thoughts about the occurrence of 'radical surprises' and a case study on the 2002 flood (Germany). *Natural Hazards, 55*, 671–688.

Kuhlicke, C., Scolobig, A., Tapsell, S., Steinführer, A., & De Marchi, B. (2011). Contextualizing social vulnerability: Findings from case studies across Europe. *Natural Hazards, 58*, 789–810.

Kunreuther, H. (2001). Incentives for mitigation investment and more effective risk management: The need for public-private partnerships. *Journal of Hazardous Materials, 86*, 171–185.

Lampen, A., & Owzar, A. (Eds.). (2008). *Schrumpfende Städte zwischen Antike und Moderne* [Shrinking cities between ancient world and modernity]. Köln, Weimar, Wien: Böhlau.

Laukkonen, J., Kim Blanco, P., Lenhart, J., Keiner, M., Cavric, B., & Kinuthia-Njenga, C. (2009). Combining climate change adaptation and mitigation measures at the local level. *Habitat International, 33*, 287–292.

Mileti, D. S. (1999). *Disasters by design. A reassessment of natural hazards in the United States*. Brookfield, CT: Rothstein.

Moss, T., & Monstadt, J. (Eds.). (2008). *Restoring floodplains in Europe. Policy contexts and project experiences*. London, UK: IWA.

Mykhnenko, V., & Turok, I. (2007). *Shrinking cities: East European urban trajectories 1960–2005* (CPPR Working Paper 4). Glasgow, Scotland: Centre for Public Policy for Regions. Retrieved from http://www.cppr.ac.uk/media/media_51286_en.pdf

O'Brien, K., Eriksen, S., Nygaard, L., & Schjolden, A. (2007). Why different interpretations of vulnerability matter in climate change discourses. *Climate Policy, 7*, 73–88.

O'Keefe, P., Westgate, K., & Wisner, B. (1976). Taking the naturalness out of natural disasters. *Nature, 260*, 566–567.

Oswalt, P., & Rieniets, T. (Eds.). (2006). *Atlas of shrinking cities*. Ostfildern, Germany: Hatje Crantz.

Pelling, M. (2002). Assessing urban vulnerability and social adaptation to risk – Evidence from Santo Domingo. *International Development Planning Review, 24*, 59–76.

Pelling, M. (2011). The vulnerability of cities to disasters and climate change: A conceptual framework. In H. G. Brauch, Ú. Oswald Spring, C. Mesjasz, J. Grin, P. Kameri-Mbote, B. Chourou, P. Dunay, & J. Birkmann (Eds.), *Coping with global environmental change, disasters and security* (pp. 549–558). Heidelberg, New York: Springer.

Perrow, C. (2008). Complexity, catastrophe, and modularity. *Sociological Inquiry, 78*, 162–173.

Prasad, N., Rangieri, F., Shah, F., Trohanis, Z., Kessler, E., & Sinha, R. (2009). *Climate resilient cities. A primer on reducing vulnerabilities to disasters*. Washington, DC: The World Bank.

Rieniets, T. (2009). Shrinking cities: Causes and effects of urban population losses in the twentieth century. *Nature and Culture, 4*, 231–254.

Rink, D., & Kabisch, S. (2009). Introduction: The ecology of shrinkage. *Nature and Culture, 4*, 223–230.

Robertson, C. (2011, February 3). Smaller New Orleans after Katrina, census shows. *The New York Times*, p. A11.

Romero Lankao, P., & Qin, H. (2011). Conceptualizing urban vulnerability to global climate and environmental change. *Current Opinion in Environmental Sustainability, 3*, 142–149.

Roy, M. (2009). Planning for sustainable urbanization in fast growing cities: Mitigation and adaptation issues addressed in Dhaka, Bangladesh. *Habitat International, 33*, 276–286.

Sastry, N. (2009). Tracing the effects of hurricane Katrina on the population of New Orleans. The displaced New Orleans residents pilot study. *Sociological Methods & Research, 38*, 171–196.

Schwarz, T., & Rugare, S. (Eds.). (2009). *Pop up cities. Cities growing smaller*. Cleveland, OH: Kent State University.

Steinführer, A., Kuhlicke, C., De Marchi, B., Scolobig, A., Tapsell, S., & Tunstall, S. (2009). *Local communities at risk from flooding: Social vulnerability, resilience and recommendations for flood risk management in Europe*. Leipzig, Germany: Helmholtz Centre for Environmental Research – UFZ. Retrieved from http://www.ufz.de/data/Task11_Broschuere_7-0911060.pdf

Tipple, G. (2006). Housing, urban vulnerability and sustainability in rapidly-developing cities. *Built Environment, 32*, 387–399.

UN [United Nations] Population Division. (2008). *World urbanization prospects. The 2007 revision*. New York, NY: Author.

Vale, L. J., & Campanella, T. J. (Eds.). (2005). *The resilient city: How modern cities recover from disaster*. Oxford, UK: Oxford University Press.

Webster, Ch., & Lay, L. W. Ch. (2003). *Property rights, planning and markets. Managing spontaneous cities*. Cheltenham, Northampton, UK: Edward Elgar.

Wisner, B., Blaikie, P., Cannon, T., & Davies, I. (2004). *At risk: Natural hazards, people's vulnerability, and disaster* (2nd revised ed.). New York, NY: Routledge.

Wolf-Powers, L. (2007). Is New Orleans a shrinking city? *Places, 19*, 87–90.

Rethinking People-Environment Relations With Interdisciplinary and Transdisciplinary Contributions

Roderick J. Lawrence

Faculty of Social and Economic Sciences, Human Ecology Group, Institute of Environmental Sciences, University of Geneva, Switzerland

Abstract

People-environment relations are complex and difficult to understand at local, regional, and global levels. This chapter argues that the complexity of human ecosystems should be understood using interdisciplinary and transdisciplinary concepts and methods that can be applied to improve adaptation and resilience in precise localities. These approaches are pertinent because they address complexity, uncertainty, and unpredictable change. Often these concepts are considered only from a specific disciplinary perspective such as anthropology, biology, epidemiology, psychology, or sociology. In contrast, more integrated interpretations can provide innovative knowledge to deal with the multidimensional nature of complex environmental topics such as adaptation to predicted long-term climate change, or resilience to specific extreme weather conditions. This chapter shows how key principles of human ecology can be applied to reduce vulnerability. This more integrated approach illustrates that interdisciplinary and transdisciplinary approaches are necessary to deal with complex environmental topics.

Key words: adaptation, human ecology, interdisciplinarity, resilience, transdisciplinarity

Introduction

People-environment relations are multidimensional and specifically framed by cultural, ecological, geographical, and temporal variables (Lawrence, 2001). These relations are generally stable, but sometimes both predicted and unpredicted ecological or social events can challenge the sustainability of human habitats. Cyclones and hurricanes are extreme weather conditions that are predictable in many localities, especially those in tropical regions, including the Gulf of Mexico, during specific periods of the year. When Katrina hit the state of Louisiana during the last days of August 2005, the negative impacts on New Orleans surpassed all estimates, even though the strength of the hurricane had slightly diminished (Brunsma, Overfelt, & Picou, 2007). This catastrophe damaged natural ecosystems, agricultural production, tourism, buildings, and infrastructure, and the consequences for human life and well-being were dramatic (1,833 deaths and 250,000 displaced persons). These consequences illustrate the vulnerability of people-environment relations in that region: Meteorologists had predicted the intensity of the hurricane; doubts about the resilience of the dykes to retain the sea were documented; the risk of flooding was known to decision-makers. Five years after this catastrophe, daily life in New Orleans, especially for the poorest groups of the resident population, has still not returned to its former state. Hence, it is not unfair to claim that the resilience of New Orleans is low because adaptation was not implemented effectively, even though the risks were known, and remedial measures remain incomplete (Brunsma et al., 2007).

Environmental questions, such as the impacts of climate change over the long-term and extreme weather conditions on specific occasions, are difficult to deal with, owing to their complexity, to the compartmentalization of scientific and professional knowledge, to the sector-based division of responsibilities in contemporary society, and to the increasingly diverse nature of the societal contexts in which people live. In addition, the lack of effective collaboration between scientists, professionals, and policy decision-makers has led to the "applicability gap" in sectors that deal with both the natural and human-made environments (Lawrence, 1995). There is an urgent need for innovative approaches in many situations, such as the blatant failure of the wealthiest countries of the world to address ecological and social dilemmas, including an increasing number of environmental refugees who have been displaced by drought, earthquakes, floods, or landslides in recent decades. Migration flows of environmental refugees are predicted to increase during this century (Global Humanitarian Forum, 2009).

A different approach is required to deal with people-environment relations, especially at the local level. It could be one that not only includes the viewpoints of scientists, professionals, and decision-makers, but allows for groups and individuals in civil society to share their knowledge and know-how, especially in order to improve current understanding of complex environmental questions (Chambers, 2003). This chapter argues that future interpretations and representations formulated to describe both natural and human ecosystems should be elaborated by using both interdisciplinary and transdisciplinary contributions. The next section briefly explains some concepts and principles that help to define the local, regional, and global contexts of people-environment relations. Core concepts of adaptation and resilience are discussed to illustrate how various groups and societies define goals, projects, and strategies to deal with environmental issues that impact on their sustenance. Although these core concepts are usually interpreted by individual academic disciplines from precise perspectives, they ought to be reconsidered from a more integrated perspective. The following sections will explain how an interdisciplinary interpretation, based on the principles of human ecology, can be extended to apply transdisciplinary concepts and methods in precise localities to deal with adaptation and resilience in these localities.

Applying Interdisciplinary and Transdisciplinary Approaches to People-Environment Relations

At the beginning of the 21st century, it is important to rethink adaptation, resilience, and vulnerability with respect to core concepts and principles of people-environment relations. These have different interpretations which have a history in several scientific disciplines and professions. The majority of interpretations of people-environment relations in disciplines, including anthropology, biology, epidemiology, sociology, and psychology rarely adopt a holistic framework that includes the contributions from both the social and natural sciences (Lawrence, 2001).

In general, the environment has been considered as a determinant of human behaviour by those social scientists who adopt causal explanations borrowed from the natural sciences (Boyd & Richerson, 1985). In contrast, specific features of human settlements can be interpreted as adaptive responses to basic environmental constraints, including the availability of construction materials, the geology of the site, and the availability of building sites (Brunskill, 2000). The dykes in the Netherlands are one example of how the risk of flooding by the North Sea has been perceived and dealt with by scientists, professionals, policy decision-makers, and the general public over many generations (Van Baars, 2004). The dykes were constructed as a protective barrier to sustain human settlements constructed on sites that are vulnerable to flooding. Today, approximately 27% of the Netherlands is actually below sea level. This area accommodates about 60% of the country's population of 15.8 million people. Although the perceived risk of flooding is omnipresent, actual incidences have been rare during the last century, because Dutch society has applied adaptation measures to reduce the country's vulnerability. The case of Katrina and its impact on New Orleans show how societal responses to the same kind of risk can vary considerably.

Theoretical Framework: Concepts and Principles

There are certain conditions and limits overriding the sustenance of human groups and societies that are defined by some fundamental principles which will be summarized in this section. First, the biosphere and the Earth are finite (Boyden, 1987). Both natural and human ecosystems at all scales of the planet and its atmosphere are circumscribed by certain limits, such as the surface of land, its biomass and biodiversity, the water cycle, biochemical cycles, and thermodynamic principles related to the production and transformation of energy, including the accumulation and radiation of heat. Although these limits are fundamental, their relative importance has been interpreted in various, sometimes contradictory ways (even by scientists in the same discipline). These differences highlight the diversity and limitations of current knowledge in many scientific disciplines and a lack of coordination.

Second, human ecosystems are *not* closed, finite systems as Marten (2001) notes, because they are open to external influences from ecological (e.g., solar energy and earthquakes), biological, and also anthropological (e.g., disease and warfare) sources. This means that policies and programmes that deal with internal conditions and processes should also consider those external factors that impact on them. Some recent contributions on this subject include misconceptions about the autonomy of human habitats and the ability of modern technology to overcome ecological constraints. Human habitats (e.g., cities, towns, or villages) are not autonomous because, since the foundation of cities about 9,000 years ago, sedentary populations have been highly dependent on all kinds of imported goods and services (Bairoch, 1988).

Third, humans must create and transform energy by using materials, energy, and acquired knowledge to ensure their livelihood (Boyden, 1992). The increasing disparity between ecological and biological processes and products on the one hand, and the products and processes of urbanized societies, on the other hand, is largely attributed to the rapid growth of urban populations, the creation of many synthetic products that cannot be recycled into natural processes, plus increases in energy consumption based on the use of non-renewable and renewable resources (such as wood from forests) (Boyden, 1987). At the global level, the negative consequences of these trends include the depletion of the ozone layer, a reduction in biodiversity, the accumulation of wastes, and enhancement of the "greenhouse effect." At the local level, the frequency of extreme weather events, including avalanches, hurricanes, floods, and landslides has increased in recent decades and is expected to continue during this century.

Fourth, human beings can be distinguished from other organisms by the kinds of regulators they commonly use to define, modify, and control their living conditions (Moran, 1982). Individuals have several innate mechanisms that enable them to adjust to specific environmental conditions. These mechanisms include thermoregulation and circadian rhythms, which are used to ensure and maintain vital physiological functions (Laughlin & Brady, 1978). Collectively, human groups develop other socially acceptable mechanisms to sustain their livelihood; for example, nutrition is guaranteed not only by biological and physiological mechanisms, because cultural rules and practices (that vary between ethnic groups, across cultures, and within societies) are also used to produce and distribute edible products.

Adaptation is a set of interrelated processes that sustains all organisms and both natural and human ecosystems in the context of continual change (Marten, 2001; Smit & Wandel, 2006). Evolutionary adaptation refers to processes of natural selection and is only applicable to populations and it is transgenerational. Innate adaptation refers to physiological and behavioural changes that occur in individuals that are genetically determined and do not depend on learning. Cultural adaptation refers to adaptation by cultural processes that are not innate, such as rules and regulations, or changes in lifestyle, and therefore it includes institutional adaptations. Adger, Hughes, Folke, Carpenter, and Rockstrom (2005) show that the outcome of adaptation depends on a complex set of biological, ecological, cultural, societal, and individual human mechanisms.

Resilience is a concept that was first formulated in physics, then used in other disciplines including biology, ecology, psychology, and sociology (Adger, 2000, 2006; Berkes & Folke, 1998). It refers to the inherent property of objects, organisms, individuals, societies, and ecosystems that enables them to resist disturbances from an external source (Holling, 1986). The example of Katrina shows that New Orleans was fragile and unable to resist the impacts of a known risk, because decision-makers had not introduced regulatory measures to deal effectively with it and reduce vulnerability. The precise way that a human ecosystem responds to a planned or unexpected external disturbance depends on the nature of its impact and the internal properties of the ecosystem. If the ecosystem is elastic, accommodation processes will absorb change without modifying its initial state. This is illustrated by the case of nomads in their daily search for nutrition. If the ecosystem is plastic, assimilation processes will deal with change by altering the initial state of the ecosystem. This is precisely what Dutch society has done over a period of two centuries, in order to reduce its vulnerability.

Two interpretations of adaptation are commonly applied in ecological research (Moran, 1982). Several disciplinary contributions confirm that ecological systems include two types of adaptive processes that are meant to deal with change. The first type is dynamic equilibrium processes that operate in order to protect a system from rapid, disruptive change. The second type is resilience processes that are meant to sustain a system. In both cases, a beneficial adaptive process is one that contributes to the solution of a problem or to overcoming a stressful situation. These processes can only be understood in terms of the inherent complexity of ecological systems, and the nature and intensity of the agent instigating change. Both natural and human ecosystems have tensions between those components that resist change by negative feedback, and others that enable change by positive feedback (Marten, 2001).

Human responses to disturbances of ecological systems are varied and unpredictable, because they depend on the internal properties of the ecosystems as well as the type and intensity of the external impact (e.g., a small, single incremental disturbance in contrast to a large, enduring impact). These responses include short- and long-term change, with or without equilibrium states and internal transformations (Folke, 2006). In principle, ecological systems are not static but dynamic and change continually in terms of their composition, the interrelations between their components, and their equilibrium conditions. The dynamic nature of ecological systems is partly related to their diversity and their variability. Some changes to ecological systems stem from external sources such as unpredictable climatic conditions (e.g., frosts, hurricanes, or droughts). Ecological systems must adapt to these conditions in order to survive by self-regulation. These internal responses take into account not only the magnitude of the disturbance but also the degree of variability that a system has experienced historically (Holling, 1986).

In all human ecosystems, behaviours, concepts, ideas, and goals are used implicitly and explicitly to define and sustain the multiple relations between people and their immediate environment (Lawrence, 2001). These components of human culture not only influence the relations between people and natural ecosystems, but also the interrelations between various human groups. Some human ecologists refer to the "social environment," in order to distinguish it from the "natural environment." In order to understand the complexity of social sustainability, both the natural and social dimensions of human ecosystems need to be considered simultaneously.

Integrated Principles of Human Ecology

The following principles are part of a human ecology perspective that explicitly accounts for biological, cultural, ecological, societal, and individual human factors, which are essential components of human ecology and its applications (Lawrence, 2001). First, the interrelations between humans and their surroundings are manifested through a wide range of physiological, psychological, and cultural processes. All these processes include not only sensations and perceptions (which animals also share) but also beliefs, doctrines, ideas, and values, which are unique characteristics of the human intellect. Therefore, human ecology is also a subjective ecology and it should include human perception and cognition of the environment, such as those values and uses attributed to all kinds of resources (Berkes & Folke, 1998).

Second, the "human environment" can be contrasted with the environment of other organisms by the instrumental functions and the symbolic values that have been attributed to it (Marten, 2001). Human processes and products transform the constituents of the environment to meet prescribed aspirations, goals, and needs. In addition, human activities can provoke unintended consequences on abiotic and biological constituents of ecosystems which, in turn, have an impact on human health and well-being (Boyden, 1992). This is one reason why studies of the interrelations between the biological, ecological, and cultural characteristics of human ecosystems are crucial if an integrated perspective is to be effectively applied.

Third, unlike other systems of organisms, the interrelations between people and their surroundings are characterized by discursive and reflexive knowledge, including symbols and not just linguistic representations. Purposive human behaviour has been studied by social anthropologists, including Edward Hall, psychologists such as Irving Goffman, and sociologists, including Robert Merton (Lawrence, 2001). Their contributions consider the latent and manifest functions of human activities, as well as their intended and unintended consequences. Berkes and Folke (1998) show that the members of human groups and communities co-operate successfully to sustain themselves, because there is a social order founded on implicit or explicit conventions, customs, and norms that regulate human activities, including the sharing of information and tacit knowledge. These mechanisms can be illustrated by traditional methods of using resources, including food production, to sustain human groups.

Cultural and social regulatory mechanisms are transmitted by the tacit know-how of populations, including social rules and customs that are shared and respected, in order to ensure sustenance (Berkes & Folke, 1998). For example, famine is a recurrent event throughout human history in many regions of the world (Grigg, 1993). The mobility of Nomads should be considered as a conscious means of adjustment to seasonal and geographical variations in the quantity, variety, and distribution of edible matter and potable water. Sedentary communities employ one or more cultural dispositions to buffer environmental variations, such as drought, floods, or plagues of insects that can have a severe impact on the supply of food and water. One means that helps to counteract these kinds of temporal variations is the storage of surplus produce.

The preceding examples show that adaptability and resilience are fundamental characteristics of human culture that should be related to strategic action for the sustenance of human ecosystems (Laughlin & Brady, 1978). In principle, human groups may relocate or adapt their habitat in order to survive local environmental perturbations that may impact on the supply of food. Since the foundation of sedentary societies, human groups have primarily adapted to their environmental surroundings by modifying some constituents of their culture.

From Disciplinary to Interdisciplinary and Transdisciplinary Contributions

The preceding sections show that people-environment relations cannot be understood in a comprehensive way by concepts and methods from only one discipline, because environmental and social problems are not structured within traditional disciplinary and sector-based boundaries. Our capacity to deal with these complex topics is insufficient for several reasons, including their diversity and complexity; the difficulty of identifying and measuring the interrelations between them and their components; and the need to understand the relative importance of these components in precise localities (Thompson Klein et al., 2001). Therefore, it is necessary to shift from disciplinary and multidisciplinary to interdisciplinary and transdisciplinary contributions. In order to be effective, this shift should be founded on a clarification of definitions, goals, and methods.

What is Transdisciplinarity?

Transdisciplinarity is an ambiguous term that has been interpreted in various ways. Balsiger (2004) notes that there is no complete history of this term or concept. Like interdisciplinarity, there seems to be no consensus about its meaning. However, several shared aims of transdisciplinarity can be identified by an analysis of recent publications (Lawrence, 2010).

In principle, transdisciplinarity tackles complexity in science and it challenges the fragmentation of knowledge (Sommerville & Rapport, 2000). It deals with research problems and organizations that are defined from complex and heterogeneous domains such as climate change and its impacts on human health (Lawrence, 2004). As well as complexity and heterogeneity, this mode of knowledge production is also characterized by its hybrid nature, non-linearity, and reflexivity, transcending any academic disciplinary structure (Balsiger, 2004). Second, transdisciplinary research also accepts local contexts and uncertainty; it is a context-specific negotiation of knowledge (Thompson Klein, 2004). Third, transdisciplinarity implies intercommunicative action and it is the result of intersubjectivity (Després, Brais, & Avellan, 2004; Thompson Klein et al., 2001). It is a research process that combines the practical reasoning of individuals with the constraining and complex nature of social, organizational, and material contexts (Lawrence, 2004). Fourth, transdisciplinary research is often action-oriented (Després et al., 2004). It entails making linkages not only across disciplinary boundaries but also between theoretical development and professional practice (Lawrence, 2004).

Transdisciplinary contributions frequently deal with real-world topics and generate knowledge that not only addresses societal problems but also contributes to their solution (Pohl & Hirsch Hadhorn, 2007). A key objective is to understand the actual world and to bridge the gap between knowledge derived from research and from decision-making processes in society. However, transdisciplinary research should not be restricted to applied knowledge (Stokols, 2006). This common interpretation is too restrictive, because there is no inherent reason why theoretical development, especially the analytical description and interpretation of complex environmental concepts such as adaptation and resilience, cannot be achieved by transdisciplinary contributions.

Multidisciplinary, Interdisciplinary, and Transdisciplinary Approaches

Although the terms interdisciplinarity and transdisciplinarity have been used interchangeably by some authors, the author of this paper has stressed the difference between interdisciplinary and transdisciplinary contributions (Lawrence, 2010). In principle, the word interdisciplinary has been used consistently to denote scientific research that involves a number of disciplines. In contrast, the word transdisciplinary has not been restricted to scientific research, because it has been used since the 1970s in debates about teaching and professional practice (Sommerville & Rapport, 2000).

Bruce, Lyall, Tait, and Williams (2004) assert that, in multidisciplinary research, each discipline works in a self-contained manner and that in interdisciplinary research, an issue is approached from a range of disciplinary perspectives that are integrated to provide a systemic outcome. In transdisciplinary research, however, they argue that the focus is on the organization of knowledge around complex heterogeneous domains rather than around the disciplines and subjects into which knowledge is commonly organized. Després et al. (2004) propose that the difference between interdisciplinary and transdisciplinary contributions stems from the Latin prefix "trans," which denotes transgressing the boundaries defined by traditional disciplinary modes of inquiry. Lawrence (2004) compares interdisciplinary approaches to a mixing of disciplines, whereas trandisciplinary ones would have more to do with a fusion of disciplinary and other kinds of knowledge.

According to Thompson Klein (2004), transdisciplinarity requires an ingredient that some have called transcendence. This implies the giving up of sovereignty over knowledge, the willingness to share and understand disciplinary concepts, principles, and methods, the generation of new insights and knowledge by this kind of collaboration, while taking into account the know-how of professionals and tacit knowledge of lay people.

Interdisciplinary and transdisciplinary contributions should be grounded in the exchange of information, ideas, and knowledge between scientists, professionals, elected officials, and the local population. This means that information and data need to be collected and distributed to all actors and institutions in civil society (Ballard, 2005). This deliberative process is a prerequisite for applying participatory approaches to deal with complex environmental questions, such as the consequences of climate change and extreme weather conditions at the local level. Hall, Taplin, and Goldstein (2010) have applied this kind of approach to show how community action may impact on policy definition and implementation in Sydney, Australia. Public concern about this key environmental question has received broad media coverage in recent years in Australia. In Sydney, some locally-based groups of residents had been motivated by similar grassroots initiatives in the United Kingdom that had an influence on national legislation about climate change. According to Hall et al. (2010) this participatory process in suburban Sydney has not only developed public awareness but also influenced policy definition on climate change. Unfortunately, this kind of approach has never been applied in New Orleans, or many other cities that are vulnerable to extreme weather conditions.

Conclusion

The traditional Cartesian dichotomy of people-environment relations that prevailed during the last century has reinforced the compartmentalization and specialization of knowledge. This has hindered the development of a broad understanding of the contextual conditions of human habitats and broader ecological and environmental topics. At the beginning of the 21st century, it is unfortunate that the division between human and physical geography in both academic research and teaching still exists as does the division between physical/biological anthropology and cultural/social anthropology.

Common disciplinary interpretations of people-environment relations could be replaced by interdisciplinary and transdisciplinary contributions that study the mutual interrelations between humans, their ecosystem, and the biosphere as a dynamic, regulated network that can be studied as a holistic system, including its components and the interrelations between them. From this perspective, studies of people detached from their surroundings can be replaced by studies of processes and relations that occur between the nonhuman and human components of open, dynamic ecosystems.

Thompson Klein (2004) presents an optimistic review of the prospects for interdisciplinary and transdisciplinary contributions in the near future. Her optimism is based on the fact that increasing numbers of individuals and institutions are beginning to recognize the need to deal with sets of interrelated problems (such as sustainability, public health, and extreme weather conditions) that are not confined within conventional disciplinary boundaries or sector-based professional practice. She argues that academics, scientists, and governments have been challenged by civil society to provide knowledge that is relevant and applicable to current societal concerns about people-environment relations in precise localities.

References

Adger, W. (2000). Social and ecological resilience: Are they related? *Progress in Human Geography, 24*, 347–364.

Adger, W. (2006). Resilience, vulnerability, and adaptation: A cross-cutting theme of the international human dimensions programme on global environmental change. *Global Environmental Change, 16*, 268–281.

Adger, W., Hughes, T., Folke, C., Carpenter, S., & Rockstrom, J. (2005). Socio-ecological resilience to coastal disasters. *Science, 309*, 1036–1039.

Bairoch, P. (1988). *Cities and economic development: From the dawn of history to the present*. London, UK: Mansell.

Ballard, D. (2005). Using learning processes to promote change for sustainable development. *Action Research, 3*, 135–156.

Balsiger, P. (2004). Supradisciplinary research practices: History, objectives and rationale. *Futures, 36*, 407–421.

Berkes, F., & Folke, C. (Eds.). (1998). *Linking social and ecological systems: Management practices and social mechanisms for building resilience*. Cambridge, UK: Cambridge University Press.

Boyd, R., & Richerson, P. (1985). *Culture and the evolutionary process*. Chicago, IL: Chicago University Press.

Boyden, S. (1987). *Western civilisation in biological perspective: Patterns in biohistory*. Oxford, UK: Oxford University Press.

Boyden, S. (1992). *Biohistory: The interplay between human society and the biosphere, past and present*. Paris, France: UNESCO.

Bruce, A., Lyall, C., Tait, J., & Williams, R. (2004). Interdisciplinary integration in Europe: The case of the Fifth Framework programme. *Futures, 36*, 457–470.

Brunskill, R. (2000). *Vernacular architecture: An illustrated handbook*. London, UK: Faber & Faber.

Brunsma, D., Overfelt, D., & Picou, J. S. (2007). *The sociology of Katrina: Perspectives on a modern catastrophe*. Lanham, MD: Rowman & Littlefield.

Chambers, S. (2003). Deliberative democratic theory. *Annual Review of Political Science, 6*, 307–326.

Després, C., Brais, N., & Avellan, S. (2004). Collaborative planning for retrofitting suburbs: Transdisciplinarity and intersubjectivity in action. *Futures, 36*, 471–486.

Folke, C. (2006). Resilience: The emergence of a perspective for social-ecological systems analysis. *Global Environmental Change, 16*, 235–267.

Global Humanitarian Forum. (2009). *The anatomy of a silent crisis*. Geneva, Switzerland: Global Humanitarian Forum.

Grigg, D. (1993). *The World food problem* (2nd ed.). Oxford, UK: Blackwell.

Hall, N., Taplin, R., & Goldstein, W. (2010). Empowerment of individuals and realization of community agency: Applying action research to climate change responses in Australia. *Action Research, 8*, 71–91.

Holling, C. (1986). The resilience of terrestrial ecosystems: Local surprise and global change. In W. Clark & R. Munn (Eds.), *Sustainable Development of the Biosphere* (pp. 292–317). Cambridge, UK: Cambridge University press.

Laughlin, C. D., & Brady, I. A. (Eds.). (1978). *Extinction and survival in human populations*. New York, NY: Columbia University Press.

Lawrence, R. (1995). Meeting the challenge: Barriers to integrated cross-sectoral urban policies. In M. Rolén (Ed.), *Urban policies for an environmentally sustainable world* (pp. 9–37). Swedish Council for Planning and Co-ordination of Research: Stockholm, Sweden.

Lawrence R. (2001). Human ecology. In M. Tolba (Ed.), *Our fragile world: Challenges and opportunities for sustainable development* (Vol. 1, pp. 675–693). Oxford, UK: EOLSS.

Lawrence, R. (2004). Housing and health: From interdisciplinary principles to transdisciplinary research and practice. *Futures, 36*, 487–502.

Lawrence, R. (2010). Beyond disciplinary confinement to transdisciplinarity. In V. Brown, J. Harris, & J. Russell (Eds.), *Tackling wicked problems through the transdisciplinary imagination* (pp. 16–30). London, UK: Earthscan.

Marten, G. (2001). *Human ecology: Basic concepts for sustainable development*. London, UK: Earthscan.

Moran, E. (1982). *Human adaptability: An introduction to ecological anthropology*. Boulder, CO: Westview Press.

Pohl, C., & Hirsch Hadorn, G. (2007). *Principles for designing transdisciplinary research*. Munich, Germany: Oekom Verlag.

Smit, B., & Wandel, J. (2006). Adaptation, adaptive capacity and vulnerability. *Global Environmental Change, 16*, 282–292.

Sommerville, M., & Rapport, D. (Eds.). (2000). *Transdisciplinarity: Recreating integrated knowledge*. Oxford, UK: EOLSS.

Stokols, D. (2006). Toward a science of transdisciplinary action research. *American Journal of Community Psychology, 38*, 63–77.

Thompson Klein, J. (2004). Prospects for transdisciplinarity. *Futures, 36*, 515–526.

Thompson Klein, J., Grossenbacher-Mansuy, W., Häberli, R., Bill, A., Scholz, R., & Welti, M. (Eds.). (2001). *Transdisciplinarity: Joint problem solving among science technology and society, an effective way of managing complexity*. Basel, Switzerland: Birkhäuser.

Van Baars, S. (2004). Peat dike failure in the Netherlands. E-WAter, 3. Retrieved from http://www.ewaonline.de/journal/2004_03l.pdf

Coping with Climate Change and Natural Hazards

Adaptation to Climate Change by Coastal Communities

Amalia Fernandez-Bilbao

Principal Consultant, URS/Scott Wilson, London, UK

Abstract

Climate change is predicted to significantly affect coastal areas, due to sea level rise and an increase in storm intensity and wave height. Coastal communities in England and Wales are also experiencing a series of socio-economic challenges, including ageing populations and fragile economic conditions related to low incomes, seasonal employment, and pressure on services during the summer months. This paper examines issues that may affect the ability of coastal communities in England and Wales to adapt to the impacts of climate change. It is based on the results of research on community involvement in planning for climate change adaptation on the coast. The key findings related to the lack of agreement between communities and authorities on what constitutes "climate change adaptation," with authorities focusing on dealing with specific impacts of climate change (e.g., floods) rather than on reducing underlying vulnerability in their communities, as well as to communications and engagement gaps, lack of awareness of climate change impacts in communities, and distrust of residents in authorities. Communities do not appear to agree with the official message that there is a need to "adapt" or "change" because of climate change. The findings have relevance for policy and practice of climate change adaptation, particularly in the light of the new "localism agenda" in the UK, which seeks to give communities more of a say in managing local issues, including climate change.

Key words: awareness, climate change adaptation, coast, community engagement, localism, trust

Introduction

Published evidence on climate change suggests that coastal areas in the UK will be particularly affected by climate change impacts such as increase in the frequency of storms and flooding (Zsamboky, Fernández-Bilbao, Smith, Knight, & Allan, 2011). A rise in the sea level is also likely. In addition, some coastal communities face significant socio-economic issues that may reduce their ability to adapt to these impacts. Some coastal areas in England and Wales have disproportionately high numbers of elderly people, low education levels, low incomes, are geographically isolated and have poor transport networks.

This paper examines issues that may affect the ability of coastal communities in England and Wales to adapt to the impacts of climate change. It reanalyses the findings of a qualitative research project that was originally commissioned by the Department for the Environment, Food and Rural Affairs of England and Wales (Defra)[1] in 2009. The project produced a research report (Fernández-Bilbao et al., 2009) and a Guidance document (Woodin et al., 2009), to help local authorities engage their communities in planning for climate change and coastal change which includes modifications to the coastline due for instance to erosion.

The project came about as a response to the need, identified by the central governments in England and Wales, to involve communities in planning for adaptation to the climatic changes that they are likely to experience.

The emphasis on communities taking the lead in adaptation activities is particularly pertinent in the light of the emerging "localism agenda" in England and Wales. Localism aims to transfer planning and other responsibilities from central government and local authorities to communities. A series of white papers (e.g., Strong and Prosperous Communities White Paper, DCLG, 2006), legislation, and consultations published in the UK over the past decade have sought to transfer decision-making power from national to local government and also right down to individual citizens. This has culminated in the Localism Bill (2011) that aims to decentralise power as much as possible by devolving power and money to communities, civil society, and local authorities (CLG, 2010).

Under the localism agenda, local governments and communities should work together to deliver services, which is known as co-production. Co-production sees citizens as decision makers, thus blurring the distinction between service users and service providers. Those who are in favour of localism suggest that, if a community is more involved in delivering a service, it will have more control over it and therefore be more empowered. Local communities and authorities also stand to gain greater control over their services and budgets (National Council for Voluntary Organisations, 2010). The localism agenda applies to all sectors and this paper also raises questions about how this agenda can be implemented in practice with reference to adaptation to climate change.

The remainder of this paper provides background information on climate change adaptation and coastal communities in England and Wales, summarises the research aims and methodology, discusses the implications of the research for climate change adaptation and the localism agenda, and provides some initial conclusions and recommendations.

1 Note that the UK is divided into four countries: England, Wales, Scotland, and Northern Ireland. This research covered England and Wales only.

Background to Climate Change Adaptation

Climate change is one of the biggest challenges affecting Europe today. According to the European Commission (2010), the consequences of climate change can already be seen in Europe in the form or rising temperatures, changes in rainfall patterns, sea level rise, and an increase in extreme weather events that are causing floods and droughts. These impacts are expected to become more severe in the next few decades.

Within Europe, there are several regions that are particularly susceptible to the impacts of climate change such as southern Europe (due to heat and drought) and coastal zones, deltas, and floodplains (due to sea level rise, more intense rainfall, floods, and storms) (European Commission, 2010). Although climate impacts will be felt across the board, the effects will not be homogeneous within countries, with some regions (such as coasts), economic sectors, and age groups, particularly the elderly, expected to be more seriously affected (Tol, 2009).

There are currently two main types of responses to climate change: Mitigation and adaptation. Mitigation focuses on reducing carbon emissions in order to avoid climate change. Adaptation recognises that even if all emissions ceased today, a certain degree of climate change can still be expected and that society needs to change in order to adapt. The European Commission defines climate change adaptation as "taking action now so that we can be more resilient to our current climate, less susceptible to the impacts of future climate change and be in a position to take advantage of opportunities" (EC, n.d.).

Pettengell (2010) argues that although there is a degree of consensus that the warming of the climate is occurring, there is also uncertainty as to the specific effects at the local level, which can hinder conventional planning decisions. Therefore, she concludes that the approach to adaptation should manage uncertainty and foster adaptive capacity. In other words, it should comprise both reducing general vulnerability (e.g., by enabling people to make informed decisions about their lives and livelihoods in a changing climate) and also to prepare for specific hazards such as floods.

Consequently, adaptation should not be seen as preparing for or responding to the predicted impacts of climate change (e.g., flooding and heat waves) but rather as a process that also aims to address the underlying factors that determine vulnerability and adaptive capacity (Tschakert & Dietrich, 2010). Adaptive capacity is defined by the same authors as "the ability to undertake adaptations or system changes" (p. 10). Tschakert and Dietrich (2010) also argue that, too often, there is a "project focus" on climate change adaptation that favours clear and easily identifiable adaptation actions to be implemented before or after an event, rather than on an examination of vulnerability processes of a population or area.

There is uncertainty in climate change predictions but also about the outcomes of adaptation, which adds another layer of complexity. For instance, Moser and Ekstrom define adaptation as follows:

> Adaptation involves changes in social-ecological systems in response to actual and expected impacts of climate change in the context of interacting nonclimatic changes. Adaptation strategies and actions can range from short-term coping to longer-term, deeper transformations, aim to meet more than climate change goals alone, and may or may not succeed in moderating harm or exploiting beneficial opportunities. (Moser & Ekstrom, 2010, p. 22026)

Moser and Ekstrom (2010) state that adaptation has three main structural elements: The actors managing the issue, the governance context in which they act (e.g., national and local policies), and the larger context in which they act (e.g., the wider environmental and socio-economic situation). They conclude that barriers that can stop or delay the adaptation process can arise from all three components. Pettengell (2010) argues that empowering communities to manage risk and uncertainty requires both bottom-up and top-down processes. Making a change at the local level requires community-based action supported by high-level political will and devolved resources and decision-making.

The need to involve communities in adaptation planning has been recognised in other research. A report by CAG consultants (2009) on the social impacts of climate change in the UK highlighted that there are three types of strategic climate change adaptation responses: (i) policy, (ii) management and operational, and (iii) community-led adaptation. The report goes on to conclude that "action is needed at 3 levels within the UK: Nationally and regionally (e.g., by government, agencies, regional bodies, etc.), locally (including by local authorities) and, most importantly, by and with communities" (p. 31).

Characteristics of Coastal Areas in England and Wales

The coastline of England and Wales is rich in natural habitats and archaeological sites. It is an important tourist destination and many people want to visit, live on, or retire to the coast. Many key economic activities are located on the coast, including tourism, ports, fishing, and offshore energy generation. Coastal areas in the UK also experience important socio-economic challenges that may affect the ability of certain coastal communities to adapt to climate change (Zsamboky et al., 2010). In addition to ageing populations, some coastal areas are characterised by fragile economic conditions, including low incomes, seasonal employment, and pressure on services during the summer months (when there is a larger population). Coastal areas also experience high levels of youth out-migration. For instance, in the coastal district of East Lindsey (East England), for every two people aged 18-24 who move out of the area, three people aged over 60 move in (Centre for Rural Economy, University of Newcastle upon Tyne, 2006).

The UK population as a whole is ageing and this trend is particularly evident along Britain's coasts, which have traditionally been popular retirement destinations. Rural areas along the coast experienced an increase in the proportion of their population aged over 65

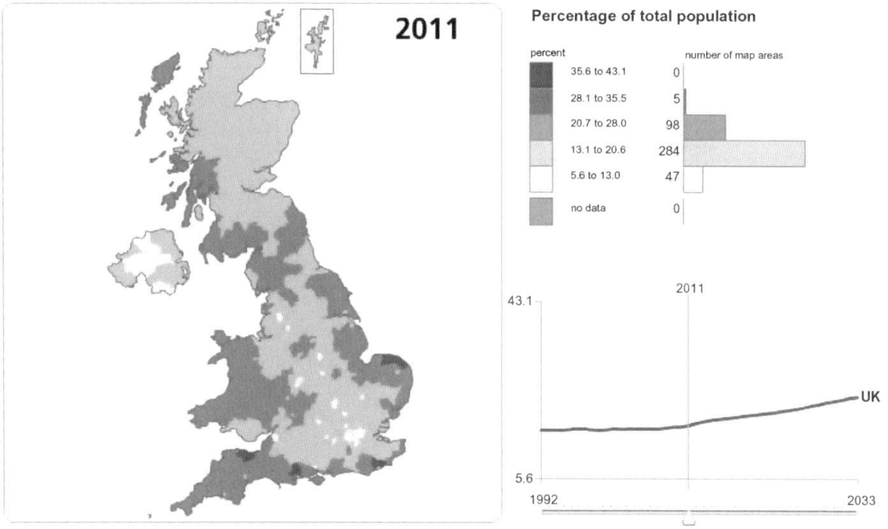

Source: Office for National Statistics, WAG, GROS, NISRA. Data: Population estimates 1992–2009; 2008-based National and Subnational projections 2010–2033. Graphic by ONS Data Visualisation Centre. © Crown copyright and database right 2010. All rights reserved. Ordnance survey Licence number ONS 100019153.

Figure 1. Population aged 65 and over in the UK (2011).

between 1981 and 2001. Rural coastal areas therefore have a disproportionate number of retired people (Centre for Rural Economy, University of Newcastle upon Tyne, 2006).

Population forecasts predict that the ageing of the population of the coast will become more acute. All but one of the British local authorities that are predicted to have the largest percentage of over 65 (34–41%) in 2033 are located on the coast (Office for National Statistics, 2011). Figure 1 illustrates the current concentration of older inhabitants in coastal local authorities. Figure 2 shows the situation predicted for 2033, which is characterised by a general ageing of the population in the UK but particularly in coastal areas.

The elderly are potentially more vulnerable to climate change impacts (the heat wave in 2003 caused a disproportionately high number of deaths amongst this group. Heat waves are expected to become more frequent with climate change). A review of the recent literature on flooding and social impacts also highlighted that the elderly are one of the groups that potentially suffer worse impacts during and after a flood (Fernandez-Bilbao & Twigger-Ross, 2009).

Not only are older age groups particularly vulnerable to climate change; they are also characterised by lower awareness of the issue. For instance, a study of elderly people's perceptions of heat wave risks suggests that this group does not perceive its vulnerability and therefore does little to prepare for and adapt to the risk (Wolf et al., 2009, cited in Adger et al., 2009). Those over 65 have also been reported in research to be less likely

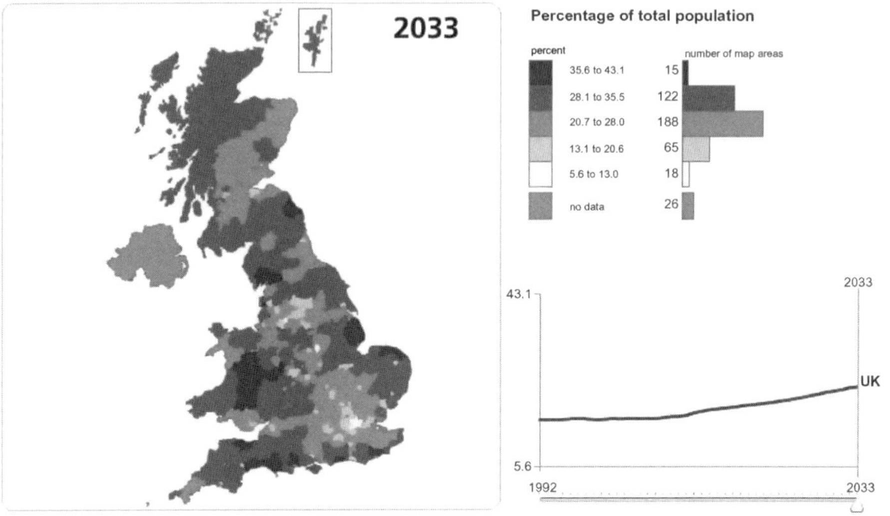

Source: Office for National Statistics, WAG, GROS, NISRA. Data: Population estimates 1992–2009; 2008-based National and Subnational projections 2010–2033. Graphic by ONS Data Visualisation Centre. © Crown copyright and database right 2010. All rights reserved. Ordnance survey Licence number ONS 100019153.

Figure 2. Population aged 65 and over in the UK (projection for 2033).

to think that they will be affected by climate change or will be able to take action to stop it (Haq, Brown, & Hards, 2010).

In addition to their population structure, some coastal resorts in England and Wales have high levels of economic and social deprivation caused by a combination of demographic issues (high proportion of retirees and benefits claimants), low wages, transitory populations (i.e., higher population in the summer than in the winter months, including an influx of seasonal workers and their families), and undiversified economic activities based mainly on tourism (Lincolnshire Research Observatory, n.d.). Although the poverty of coastal areas in England and Wales is relative, as they are still part of a western European country, poverty has been highlighted as a key factor that determines vulnerability to climate change and reduces adaptive capacity (Pettengell, 2010).

A significant number of coastal settlements in England and Wales are small in size, dispersed, and relatively isolated, which means that the transport infrastructure is often inadequate and these settlements are becoming increasingly isolated from other more prosperous areas. These settlements are also less likely to qualify for flood and coastal defence funding, as funding is currently based on a cost benefit analysis that favours larger places, that is, localities with larger populations and diversified economic activities. In addition, the dispersed nature of these settlements and the fact that they are on the coast means that they do not qualify for other types of European and national regeneration funding:

> If we were to roll up the English coast into its own region, the level of compressed deprivation would qualify the area for substantial European Union and national regeneration funding. As it is, many of the dispersed pockets of coastal deprivation fail to qualify for such support because of their size and their location. (Coastal Communities Alliance, 2010, pp. 33–34)

Climate change is likely to have significant impacts on coastal areas, due to sea level rise and an increase in storm intensity and wave height (Zsamboky et al., 2010). Currently, 46% of England's coastline is protected from flooding and erosion by hard defences (MCCIP, 2008).

Coastal defences protect properties, agricultural land, businesses, and other assets from flooding and erosion and have allowed development and economic activities to take place in areas at risk (Foresight, 2004). Maintaining current levels of protection on the coastline may not be possible everywhere, because of economic constraints and the increasing risk. Intergenerational issues could arise if certain decisions made now preclude adaptation in the future or cause problems further down the line. For instance, allowing certain kinds of development today may make it impossible to "roll back"[2] in the future, making those new communities vulnerable to the predicted increase in extreme events (O'Riordan, 2008).

In addition, sea level rise and the impacts of climate change will increase the challenges of protecting people and properties on the coast. Climate change and sea level rise are likely to have a severe impact on UK coasts by 2080. The total rise in sea levels off the coasts of the UK may exceed 1 m. The frequency of storms is expected to increase and, in combination with the rise in sea level, is expected to lead to increased flooding. Erosion is also expected to increase, partly because of sea level rise. Parts of the coast that are composed of low-lying and soft sediments will be most vulnerable (e.g., in the east of England) because they are most easily eroded. The most exposed locations and estuaries may be particularly vulnerable (Zsamboky et al., 2010).

The Research

The research described here aimed to identify on-going adaptation activities that were being led by communities. Potential barriers to and opportunities for engaging communities in adaptation planning were also investigated. The research was qualitative and exploratory and involved a range of stakeholders, from the national government to the communities affected by coastal change. Data was collected with the following methods:

[2] "Roll back" involves physical relocation of businesses, homes, and other assets to sites further inland, away from the threat of coastal erosion.

Policy and Context Review

The policy and context review included both peer reviewed and grey literature[3] as well as key policy documents and several organisations' websites. The review was undertaken in order to establish policy and governance related to coastal issues, to identify challenges to policy implementation, and to examine emergent thinking on community participation in adaptation planning. In this review, wider contextual issues such as predicted impacts of climate change, coastal erosion, and flood risk are explored. The emphasis was to find synergies or barriers for involving communities in adaptation planning. The review also informed the sampling and questionnaires for our stakeholder interviews and our selection criteria for case studies.

Stakeholder Interviews

Six semi-structured interviews were conducted with high-level stakeholders representing relevant organisations with a central role in coastal change or community engagement. The interview questions covered policy drivers and potential barriers for involving communities in adaptation, institutional frameworks, potential funding sources, organisational relationships, and community engagement practices. Interviewees' understanding of and vision for community adaptation planning were also explored. The findings of the interviews aided in the identification of case studies.

Case Studies

Five case studies on coastal communities were undertaken. The case studies had different foci in terms of scale, administrative level, specific adaptation issues, levels of community organisation, and geographical location. They were selected to obtain the greatest possible spread of issues and types of communities and to cover a range of examples:

- Range of coastal issues: Erosion, tidal flood risk, sea level rise;
- Range of communities: Variety of skills, deprivation, awareness levels, differing degrees of activism; and
- Range of administrative levels: Whole coast (county), local authority, parish council, town, village.

The five case studies were undertaken in Mablethorpe (Lincolnshire), Barrow (Cumbria), Suffolk Coast, Happisburgh (North Norfolk), and Shaldon (Devon). Information was gathered with methods that included desk research, site visits, meetings with local authorities and other agencies, and meetings with community groups and residents. The

3 Specialist/technical/research publications not always widely available.

case study areas are characterised by a high proportion of elderly residents and retirees and therefore these groups were over-represented in the community meetings.

Workshop

A stakeholder workshop was organised to validate the findings of our research with policy makers and other stakeholders who had participated in the project through interviews or case studies.

These methods generated a large amount of qualitative data which was coded and analysed for content and according to themes. Grounded theory formed the basis of the analytical approach. This facilitated an iterative rather than deductive approach to the analysis, which enabled the identification of key gaps and any emerging themes. The approach also toadied in drawing conclusions based on the data rather than from existing climate change adaptation theory.

Findings in Relation to Climate Change Adaptation

This section focuses on those issues identified through the research that have wider relevance for climate change adaptation, particularly in the light of the new localism agenda.

The community participants in the research reported that they were not currently involved in activities related to adaptation to climate change. Some research participants were involved in related activities but with a much narrower focus. For instance, they were involved in local flood groups. Some of the community groups we interviewed had been involved in consultations on local flood and coastal defence policies. In one of the case study locations, a rapidly eroding area where the current policy is to not provide coastal defences in the long term, the community had been actively lobbying for such defences for several years.

With regard to the consultations they had participated in, community participants did not feel that they had been meaningfully involved in decision-making. Several participants claimed that local communities are "consulted to death" on policies relating to coastal management and flood defences. However, they also felt that this was meaningless and that their contributions were not always taken into account in the final decisions. The documents they were consulted on were often highly technical and engineering-based, and the residents felt that the scope for influence was very small and that decisions were being taken by policy makers in central government agencies, without taking local views into account. In other words, they felt that national agencies were making decisions on their behalf.

Although there was knowledge about flooding and flood defences in the local communities that participated in the study, in general there was low awareness of climate change and the range of impacts (e.g., heat waves, storms, and sea level rise) and the expected consequences of these impacts for their areas. This seems to indicate that most of the authorities' efforts in relation to climate change adaptation in coastal areas are focused on addressing the risk of events such as flooding. Most of the concerns of the residents were also related to flood risk. However, they did not feel that this risk was related to or likely to increase with climate change. In some cases, they blamed authorities for a potential increase in flood risk, because it was perceived as stemming from a reduction in the available funding for flood defences.

The need to adapt or change, which was evident in the interviews with authorities, was not understood or agreed upon amongst the residents of the coastal areas that participated in the study. For these communities, adaptation means building and maintaining flood defences, in order to be able to live in the same locations and be protected from flooding and storms. Community members who participated in this research showed very low levels of trust in agencies and their messages about future climate change or sea level rise. Some mentioned the media and Google as their main sources of information on climate change. Participants highlighted that they did not trust the information provided by authorities in charge of coastal and flood protection. As well as lack of trust, participants cited the large number of agencies involved in managing the coast as a barrier to identifying who to go to for information on climate change impacts in their area.

The investigated areas have high levels of elderly and retired people and this age group was highly represented in our research. Older participants were particularly dismissive of the notion of climate change and they believed that it definitely would not happen in their lifetimes. This raises questions about the way climate change is currently communicated as something that will happen in the near or long-term (20–50 years) future.

Elderly people in some coastal areas are particularly vulnerable not only because of their age but also because they tend to live in bungalows or one-storey homes which, during a flood, would be more vulnerable because they lack an upstairs area for shelter. In the same locations, large numbers of people were living in caravans, which also increased vulnerability.

Local authorities and other agencies interviewed expressed concerns about the low levels of awareness of potential climate change impacts in the population. In their opinion, adaptation is needed, since some areas may no longer be protected from increased flooding and sea level rise by flood defences, partly because of limited funding, but also because the risk is too high for the residents behind the defences. For professionals, adaptation means change and awareness of climate impacts but the perception of the need to adapt is very low in the population. Authorities in one coastal community stated that, because of the characteristics of their residents, with large numbers of elderly citizens, low education levels, and low incomes, it was a particular challenge to increase awareness. In general, the local authorities interviewed felt that increasing the awareness of climate change impacts is very difficult.

Local authorities also mentioned that they need to manage the information they give to communities in order to avoid blight, that is, lack of investment due to future risk. This means that if they are too frank about the potential of flooding to increase significantly, this could cause an area to lose investments and lead to declining property prices.

The final research report (Fernandez-Bilbao et al., 2009) provides the full findings of this research.

Initial Conclusions and Implications for Climate Change Adaption

This paper provides only a snapshot of some potential issues and barriers for communities to adapt to climate change, particularly for coastal communities in England and Wales. It should be noted that the study was small and mainly exploratory. Therefore some of the findings may need to be further explored in future research.

The paper also challenges an opinion expressed by local and national authorities that certain coastal communities have low levels of awareness of climate change impacts. Although this is partly true and may be related to how climate change is communicated as a future risk, the reality seems to be more of a disagreement between communities and authorities on what "adaptation" to climate change should mean. The research also highlighted that current approaches to managing the risk in coastal areas seem to focus more on addressing specific impacts of climate change (e.g., floods) than on solving underlying problems that may be increasing the vulnerability of these coastal communities.

As mentioned in the introductory sections of this paper, Moser and Ekstrom (2010) highlighted that barriers to adaptation to climate change can emerge from any of its three structural elements. In our research, the main barriers seem related to the *actors* focusing only on certain impacts of climate change and increasing awareness of *their* definition of adaptation, as well as *policies* that are specific for flooding and coastal regions rather than more holistic ones that aim to decrease the vulnerability of these areas. The third element is the wider socio-economic *context* of coastal areas of the UK that may reduce their ability to adapt, in combination with the fact that coastal areas are likely to be severely affected by climate change.

Communities and authorities appeared to have different definitions of what climate change adaptation would mean in practice. The participant communities were reluctant to accept the "change" promoted by authorities and wanted to see the government investing in strengthening flood and coastal defences. Residents had yet not been involved in any adaptation or climate change related activities. Therefore, some communities may not be at the stage where they can be involved meaningfully in planning to adapt to climate change. More would need to be done in terms of increasing people's understanding of climate change impacts, as well as the consequences and constraints which authorities face (e.g., lack of funding, uncertainty), before certain communities can lead in planning for adaptation. In addition, it may be necessary to engage with communities on debates about what "adaptation" should mean rather than trying to impose an "official" view.

When authorities refer to a lack of awareness of climate change impacts and the need to adapt in communities, the real issue may be a lack of agreement with communities on those impacts and what adaptation means. It may be more useful to address this lack of agreement by more engagement around climate change issues and what the most appropriate responses should be. Having an initial agreement on what needs to be done will be essential if communities are going to meaningfully lead on adaptation efforts. In other words, if communities are to take the lead in planning for adaptation to climate change, they have to be involved in defining what the most appropriate response should be.

It is likely that coastal communities that have several socio-economic issues have an increased underlying vulnerability, therefore, this should be addressed as part of any adaptation response. Authorities' discourse is very much about "adaptation" but, in practice, they are talking about dealing with the impacts, rather than removing this underlying vulnerability or increasing the adaptive capacity of local communities. The low awareness in the population may also be leading to an increase in vulnerability, because people are less likely to prepare for the risks, for instance, by retrofitting their properties, taking precautions in the event of a flood or a heat wave, or obtaining insurance for flooding.

The lack of awareness of potential climate impacts was particularly high among the elderly, many of whom felt that climate change is not an issue they need to worry about, because the impacts are not expected to occur for 20 or 30 years. Climate change should be communicated as something that is already happening and not as a future risk, in order to avoid this type of reaction among the elderly.

Any consultation or involvement in climate change adaptation or flood-related activities needs to be meaningful, that is, the communities' views have to be taken into account and feedback on what is done with their comments needs to be provided, so that communities are able to trust authorities. This means that expectations need to be managed and, if the aim of the consultation is only to give information, this should be made clear. Communities also need to know what happens with the feedback and comments that they provide in these consultations. Lack of trust can hamper future involvement of communities in adaptation activities. It can also lead to communities not believing the messages that they receive from authorities about climate change.

Involving communities in adaptation planning under the localism agenda will require an agreement on what adaptation means. This is likely to require wider debates with local communities. There should also be a move from focusing on specific impacts, such as flooding, to a more holistic response that aims to reduce underlying vulnerability. This should include not just addressing the physical risk through defences, but also increasing understanding of the risks and uncertainty around climate change impacts and ensuring that poverty and social issues that affect coastal communities are addressed.

References

Adger, W. N., Dessai, S., Goulden, M., Hulme, M., Lorenzoni, M., Nelson, D. R., ... Wreford, A. (2009). Are there social limits to adaptation to climate change? *Climatic Change, 93*, 335–354.

CAG Consultants. (2009). *The differential social impacts of climate change in the UK. Final Report to Sniffer, UK.* Edinburgh, UK: Sniffer.

Centre for Rural Economy, University of Newcastle upon Tyne. (2006). *Ageing and coastal communities. Final report to the Coastal Action Zone, UK.* Newcastle, UK: University of

Newcastle Upon Tyne. Retrieved from http://www.coastalcommunities.co.uk/library/published_research/Ageing_Communities_Report.pdf

CLG (Communities and Local Government). (2010). *Decentralisation and the localism bill: An essential guide, UK*. London: Author. Retrieved from http://www.communities.gov.uk/documents/localgovernment/pdf/1793908.pdf

Coastal Communities Alliance. (2010). *Coastal regeneration handbook, UK*. Lincolnshire, UK: Author. Retrieved from http://www.coastalcommunities.co.uk/

DCLG (Department for Communities and Local Government). (2006). *Strong and prosperous communities – The local government white paper, UK*. Retrieved from http://www.communities.gov.uk/publications/localgovernment/strongprosperous

EC (European Commission). (n.d.). *Adaptation to climate change: An EU approach*. Retrieved from http://ec.europa.eu/environment/water/adaptation/pdf/adaptation.pdf

Fernández-Bilbao, A., & Twigger-Ross, C. (Eds.). (2009). *Improving response, recovery and resilience (WP2) – Science Report, Report to the Environment Agency, UK*. Bristol, UK: Environment Agency.

Fernández-Bilbao, A., Woodin, S., Richardson, J., Zsamboky, M., Bose, M., Orr, P., . . . Colbourne, L. (2009). *Understanding the processes for community adaptation planning on the coast* (Defra Technical Report FD2624, UK). London: Defra.

Foresight. (2004). *Future flooding: Phase 1 Technical Report, UK*. London: Office of Science and Technology.

Haq, G., Brown, D., & Hards, S. (2010). *Older people and climate change: The case for better engagement*. Stockholm: Stockholm Environment Institute.

Lincolnshire Research Observatory. (n.d.). *Statistics – The condition of the coast, UK*. Lincolnshire, UK: Author.

Moser, S. C., & Ekstrom, J. A. (2010). A framework to diagnose barriers to climate change adaptation, *PNAS, 107*, 22026–22031.

National Council for Voluntary Organisations. (2010). *Third Sector Foresight: Localism agenda*. Retrieved from http://www.3s4.org.uk/drivers/localism-agenda

MCCIP (Marine Climate Change Impacts Partnership). (2008). *Marine climate change impacts: Annual report card 2007–2008*. Retrieved from http://www.mccip.org.uk/arc/2007/default.htm

Office for National Statistics. (2011). Ageing in the UK: Interactive mapping tool. Retrieved from http://www.statistics.gov.uk/ageingintheuk/default.htm

O'Riordan, T. (2008). *How do we create a sustainable coastline?* [Summary of Tyndall Assembly Workshop on September 10th, 2008]. Norwich, UK: University of East Anglia.

Pettengell, C. (2010). *Climate change adaptation: Enabling people living in poverty to adapt* (Oxfam Research Report, UK). Oxford: Oxford International.

Tol, R. S. J. (2009). The economic effects of climate change. *Journal of Economic Perspectives, 23*, 29–51.

Tschakert, P., & Dietrich, K. A. (2010). Anticipatory learning for climate change adaptation and resilience. *Ecology and Society, 15*, 11–34.

Woodin, S., Fernández-Bilbao, A., Richardson, J., Zsamboky, M., Bose, S., Orr, P., . . . Colbourne, L. (2009). *Guidance for Community Adaptation Planning and Engagement (CAPE) on the coast, Defra, UK*. London: Defra. Retrieved from http://randd.defra.gov.uk/Document.aspx?Document=FD2624_8901_FRP.pdf

Zsamboky, M., Fernández-Bilbao, A., Smith, D., Knight, J., & Allan, J. (2010). *Impacts of climate change on disadvantaged UK coastal communities* (Report to the Joseph Rowntree Foundation. UK). York, UK: Joseph Rowntree Foundation.

Residents' Usage of, Adjustment to, and Evaluation of Donated Post-Disaster Housing

Rachma M. Syam and Ryuzo Ohno

Department of Built Environment, Tokyo Institute of Technology, Japan

Abstract

Natural disasters have physical, psychological, economic, and social impacts on survivors, especially if they are forced to relocate. Such drastic environmental transitions require adaptation and adjustment. To understand their influence on residents' lifestyle, as well as how residents use and evaluate donated post-disaster housing, a new post-disaster settlement was investigated by a comprehensive survey, using questionnaires, in-depth interviews, and observations. Items related to house conditions and usage, residents' evaluation, and participation in community activities were analyzed. The results suggest a discrepancy between design and actual usage. Moreover, evaluations indicate dissatisfaction with fixed aspects of the house design and outdoor spaces. Although certain behaviors are impeded, residents have adapted and adjusted, to retain their prior lifestyles. This study shows that facilitating social interactions and flexible, open-ended design are vital for allowing survivors to make the transition to a new environment.

Key words: adjustment, community participation, evaluation, post-disaster housing, usage

Introduction

Earthquakes often cause catastrophic damage, which requires immediate attention. To accelerate reconstruction and to improve disaster prevention performance, prefabricated materials and modern construction methods are sometimes necessary. However, donated post-disaster housing often originates from other countries, and it often differs culturally from the local dwellings. In such cases, the survivors' only choice is to accept donated housing, especially when they must relocate. Tercan (2001) has demonstrated that any attempt to remove people from their existing physical, social, and economic environments significantly impacts their lives. This is also true for the post-disaster housing settlement in New Ngelepen, Yogyakarta, and Indonesia. After the Java earthquake on May 27, 2006, original Ngelepen village was considered geographically unbuildable land, thus the residents were relocated to a new type of settlement and given dome post-disaster housing at New Ngelepen, which differed drastically from their Javanese vernacular dwellings.

Altman and Chemers (1980) have specifically addressed the interaction between behavior, environment, and culture that shapes a social system in an integrated manner. The physical environment influences people and their cultures and vice versa. Thus, it is necessary to consider both why environments should be culturally responsive and to whom they should be culturally responsive and, consequently, how (Rapoport, 1987). Ikaputra and Titisari's 2005 study found that imported culture inevitably influences the local identity (Ikaputra, 2008). Monolithic domes houses have been built because they are secure, low cost, disaster-resistant, low-maintenance, and durable, as Rick Crandall[1] has argued (Parker, 2008), but the construction of dome house settlements in New Ngelepen has raised many controversies about how it will influence the residents.

Several studies have investigated the impact of social behavior on Javanese dwellings. Some have focused on the influence of social interactions and their meaning for the morphology of traditional Javanese dwellings (Revianto, 1997), while others have examined the impact of social interactions on the presence of non-traditional dwellings, such as public housing (Subroto, 1995). Although cultural changes in a society are accompanied by changes in housing, social behavior apparently still influences Javanese dwellings. Consequently, the first hypothesis of this study assumed that social behavior is an important cultural factor in Javanese society that is reflected in the usage of dwellings and needs to be facilitated, even in post-disaster housing.

A built environment acts as a setting for human activities. It can facilitate or inhibit certain behaviors, cognitive processes, etc. Because inhibiting environments reduce competence, they have a greater effect than facilitating ones (Rapoport, 1969); therefore, flexibility is important. Environments that allow responses to cultural specifics (and their manifestations: Lifestyle, activity systems, social networks, etc.) are more satisfying and avoid obsolescence longer because they remain more congruent (Rapoport, 1990). Consequently, such settings are more supportive and become an active component of the culture (Rapoport, 1987). The second hypothesis of this study presumed that the limited condition of the donated dome houses is an inhibitive environment for some aspects of the residents' previous lifestyle. In order to fulfill their needs, adaptations and adjustments are necessary; thus, flexibility becomes essential.

Several studies have been conducted at New Ngelepen, including studies on the response to dome house (Ikaputra, 2008), as well as residents' perceptions about their new environment (Saraswati, 2007). However, studies about residents' dwelling conditions and usage before and after the earthquake, evaluation of their current dwelling

[1] Rick Crandall is an architect from Crandall Design Group. He was invited to join the Monolithic Dome Institute as the principal consulting architect.

conditions, and participation in community activities have received considerably less attention. Through questionnaires, in-depth interviews, and observations of residents' activities, this study strives to understand how residents use and evaluate their post-disaster housing, as well as to elucidate the influence of culturally different post-disaster housing on their lifestyles.

Overview of Survey Study

The case study area for this research is located at New Ngelepen, Sleman Regency, Yogyakarta special province, Indonesia. It is a new post-disaster settlement relocation area situated approximately one kilometer from the original village.

New Ngelepen introduced a house cluster site plan design, where 11 or 12 houses form a block with shared electricity, public toilet/washing areas, and approach pathways to the houses (Figure 1). Monolithic domes with a hemispherical roof and a circular floor plan were introduced, using a concrete cast as a single integral structure. The diameter of the

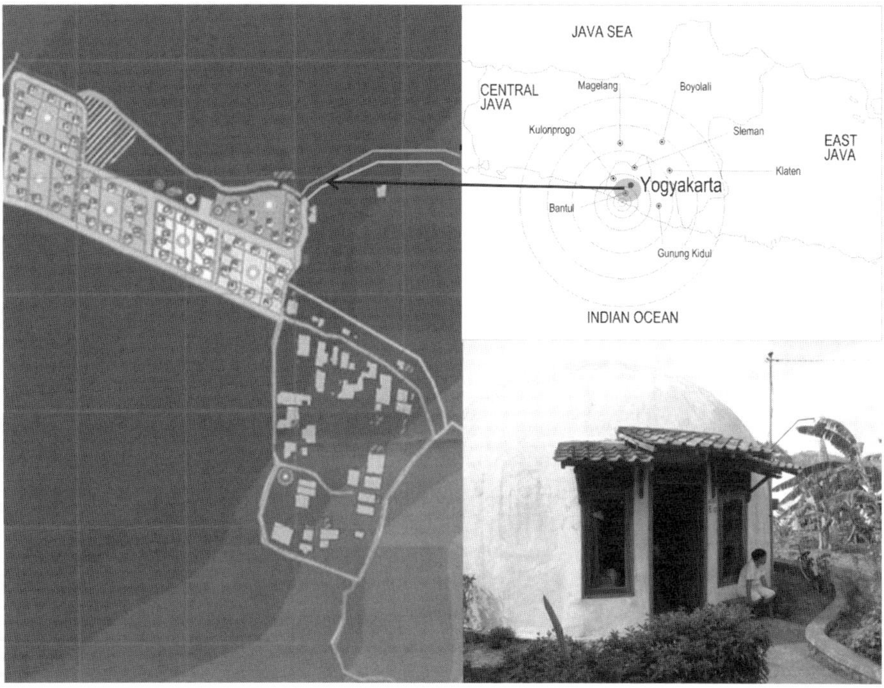

Figure 1. Site plan, cluster typology, and dome house. (Source: the authors)

two-story house is seven meters with a total area of about 38 m^2. Each house consists of a guest room, two bedrooms, and a kitchen on the first floor, and a family room on the second floor. The houses were built by the World Association of Non-Governmental Organizations and the Domes for the World Foundation. Construction began on October 10, 2006, and occupation began in late April 2007. The houses were built on land owned by the local government, Sleman's Regency, but they were donated free of charge. Survivors were allowed to live rent free for three years, and the donated dome houses were distributed via a lottery. Based on observations and confirmation from the village official in August 2009, 50 of the 71 dome houses built for and distributed to survivors were occupied.

Research Methods

A comprehensive fieldwork survey was conducted, using questionnaires, in-depth interviews, and observations of residents' activities. Additionally, the physical changes made to the original donated houses and site plan were documented for approximately one month (August 2009) with the assistance of volunteer architecture students from Gadjah Mada University, Indonesia. Thirty-four households (68%) from 50 occupied dome houses participated both in the questionnaire survey and in the in-depth interviews. The sample had follows:

- Age: < 30 years (40%), 30–60 years (60%);
- Education: College (6.5%), senior high school (25.8%), junior high school (32.3%), elementary school (12.9%), not educated (12.9%), unknown (9.7%);
- Income: No income (9%), < Rp. 500.000 ($US50) 39.4%, Rp. 500.000 – Rp. 1.000.000 ($US50-100) 45.5%, Rp. 1.000.000 – Rp. 2.000.000 ($US100-200) 6.1%.

Questionnaires and in-depth interviews contained the following items:

- *Dwelling conditions and usage:* Previous and current house conditions, living situation, etc., residents' recognition of room/space availability, future preferences to add to or alter the condition of the dome house, and changes residents have made to the original donated house, were assessed.
- *Residents' evaluation of house design and outdoor spaces:* Residents were asked about their capability to control their house design, including flexibility, personalization, and maintenance, their acceptance of situations related to the outdoor space of the house, neighborhood streets, and cluster facilities. For the evaluation questions, the items were derived from intensive field observations.
- *Participation in community activities:* Participation in community activities by heads of households and their spouses was assessed.

Additionally, place-centered behavior mapping, similar to Whyte (1980), was conducted by observations on random days (weekdays and holidays) between 8 am and 8 pm. Volunteers observed the residents' activities by walking a specific route at 15 minutes intervals and documenting the types of behaviors, age and sex, location, estimated duration, and number of people, on diagrams and maps.

The data in this paper was mainly derived from the questionnaire and in-depth interview data related to residents' background, prior and post disaster house condition and usage as well as residents' evaluation and participation in community activities. The behavior mapping and physical documentation were used as supporting data.

Results and Discussion

Dwelling Conditions and Usage

Most of the New Ngelepen residents are from the same destroyed original Ngelepen village (almost 50%), and the remainder are from neighboring villages. Based on the data achieved on residents' prior house condition, in previous Ngelepen village, residents lived in one-floor single detached houses with a non-circular floor plan and a gable or hip roof. There was a private washing, bathing, and toilet area, a stock house and a crop yard, as well as a private approach pathway to the house. Close neighbors usually had a familial relationship; one house sometimes consisted of multiple households, where more than one family lived together (e.g., parents lived with their child's family).

In New Ngelepen, the findings suggest that the residents use and recognize some rooms and areas within the house and its outdoor space differently than as intended in the architects' master plan. For instance, although the second floor in a dome house was designed as a family room, only 44% of the residents acknowledge the availability of a family room. This is understandable because the designers overlooked the latent meaning of a "family room." In a traditional Javanese dwelling, a family room is used as both a private family gathering space and an extension of a social interaction space. Consequently, continuity and accessibility with other rooms is important, which explains why a secluded second floor is not recognized or used as a family room.

Moreover, 67% of the yards are used as a crop yard, fish pond, or chicken coop, whereas the landscape master plan suggested that yards be esthetically clustered fruit tree and flower gardens. The real use indicates the need of the yard as part of the residents' lifestyle that provides additional income for the residents.

Additionally, the residents indicate that not all their needs have been met, because certain rooms are either unavailable in the original dome house or the available rooms are not flexible enough to change the intended function. Highest on the residents' lists for changing the conditions of the original dome house are adjustments to terraces, kitchens, and private bathrooms. More than 80% have an added terrace or canopy, to protect the original dome windows and doors, which are unsuited for a tropical environment, as well as to provide extension of social space.

Furthermore, more than 80% of the residents would prefer to add another kitchen where they can use cordwood cooking,[2] which is more affordable and the residents are more accustomed to, instead of the propane gas stove in the dome's kitchen. Residents' cooking customs consist of multitasking activities, which require a multifunctional kitchen that provides space to prepare, cook, store, eat, and to socialize with family and others. Because of this special cooking custom, they perceive kitchens as "dirty places." The high demand for another "dirty kitchen"[3] is understandable, as the "clean kitchen"[4] in the dome houses does not facilitate residents' cooking customs and differs from their perception of a "kitchen" (supported by Ikaputra, 2008). Additional private bathrooms are also necessary, because shared bathrooms in communal areas invade privacy. In the original settlement, some facilities, such as toilets or kitchens, were shared within neighboring family units, but in the dome settlement, residents share pathways to houses, public bathrooms, etc., with unrelated families.

Another concern is changes in living habits. According to data from the interview with a New Ngelepen village official, in the residents' original houses, 34% lived in multiple households, but only 6% of the dome houses are multiple households. The inability of dome houses to accommodate an extended family in one house indicates a situation where nuclear families are encouraged to each own a separate house. Consequently, some family members have returned to live in their original house, which is considered dangerous, or have moved to a nearby village.

Residents' Evaluation of House Design and Outdoor Spaces

The residents indicate that the house design has some limitations. For example, changing the function/layout of a room, enlarge space, and alter doors/windows are "very hard" (Figure 2a, b, c). Moreover, adding extra rooms to the original dome house is "neither easy nor hard" (Figure 2d), but adding a room compatible with the dome house is considered "hard" (Figure 2g).

As shown in Figure 2e, residents consider personalization to differentiated own dome house differently from others, using reused or recycled items from the previous house,

[2] Cordwood cooking is a traditional method using fire-place stove with cordwood as fuel, which produces ashes and smoke.

[3] The "dirty" kitchen is basically where all activities from storing raw materials, washing, to preparing and cooking food, etc., are conducted. The area is preferably hidden from outsiders, because it usually remains messy and dirty.

[4] The "clean" kitchen is rarely used as the actual area for cooking; rather, it is a place where cooked food is prepared and where food and clean tableware, etc., are stored. It is usually located near the dining area and sometimes open to visitors.

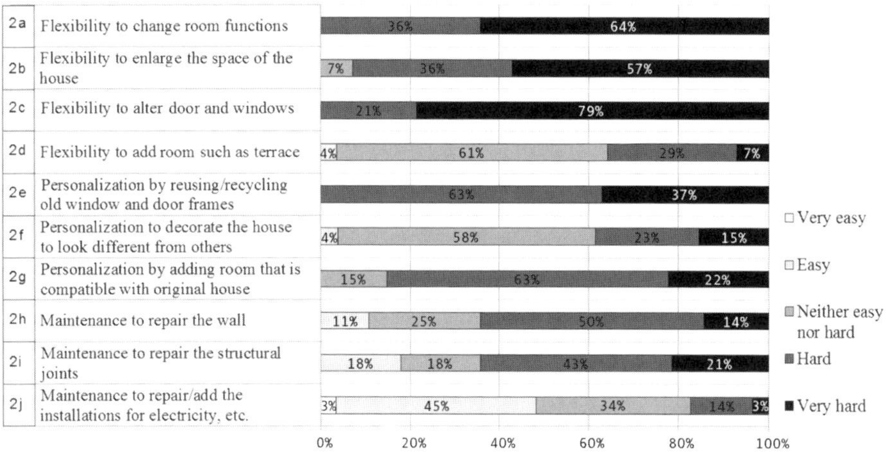

Figure 2. Residents' evaluation of the flexibility, personalization, and maintenance of the donated dome houses.

"hard or very hard." The only personalization considered "neither easy nor hard" is to decorate the house to make it look different, with paint or by adding new elements, without changing the house structure (58%). Maintenance related to fixing the damage to walls and structure joints is considered "hard" (Figure 2h, i). Only fixing or adding installations for electricity, etc., unrelated to structure, is considered "easy" (Figure 2j).

Residents express dissatisfaction with many aspects of houses' outdoor space, neighborhood streets, and cluster facilities. In particular, they indicate a need for private pathways (57%), additional social gathering spaces in the yard/house (61%), and a front terrace (50%) (Figure 3a, b, c).

The results suggest that a house's outdoor space plays an important role in household activities. Typical rural Javanese dwellings have multifunctional yards, which not only facilitate household activities such as doing laundry, harvesting crops, etc., but also serve as important socialization spaces, where housewives chat, children play, and social events such as weddings, family gatherings, etc., are hosted. The residents' agreement with the statement that "large front yards are better than large streets for social gatherings" is consistent with typical Javanese rural dwellings (Figure 3e).

Because it is impossible to conduct certain social gatherings inside dome houses, and also due to the house's limited outdoor space, residents have learned to use the neighborhood's streets for their activities. Therefore, the streets "act" as their yards and hang-out spaces. Although most consider the streets too dangerous for unsupervised play, the interviews and observations show that children play freely in the streets while their mothers chat on the sides (Figure 3d). The ground surface of the main street entrance is painted as a badminton field, where youths and men gather in the evening. Additionally, the streets

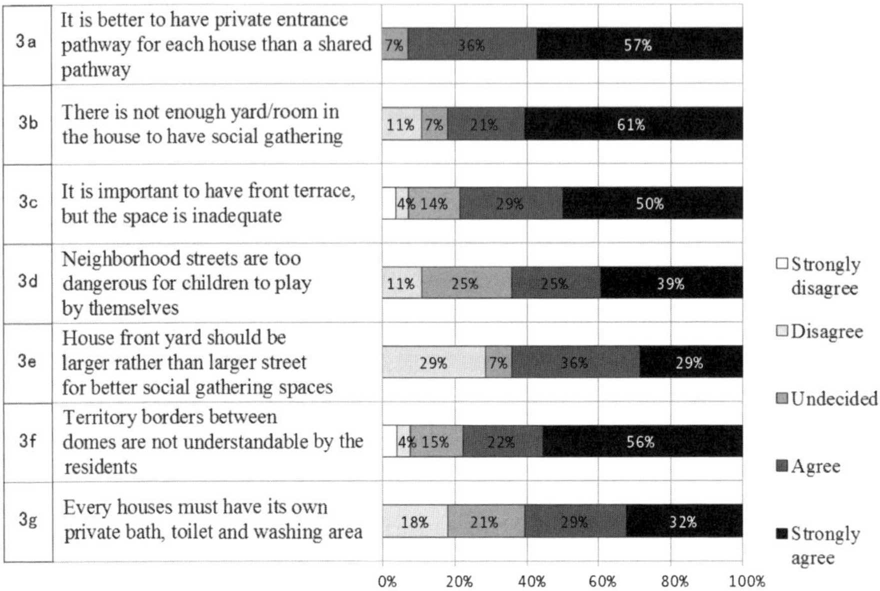

Figure 3. Residents' evaluation of outdoor space and neighborhood streets.

are closed to facilitate activities such as the weekly "ruwahan",[5] monthly meetings, and even a resident's wedding ceremony, because the house and its outdoor space are inadequate.

Unfortunately, residents' dissatisfaction does not only show their incapability to change the condition of the house design and outdoor spaces. Based on in-depth interviews, evidence has emerged that residents also worry about the land status, because their houses are built on land borrowed from the government. Thus, they are reluctant to make permanent changes. This may lead to other psychological problems. Because a house is the most fundamental basis for one's personal life, an alternative, long-term solution is needed to give residents a sense of ownership and self-identity, for example, they have a long-term payment scheme to own the land, instead of just renting it.

5 Ruwahan is a set of traditional ceremonies during the ten days before Ramadhan (Islam's sacred month), which consists of various community activities such as cleaning up the neighborhood, recitation from the Qur'an, etc.

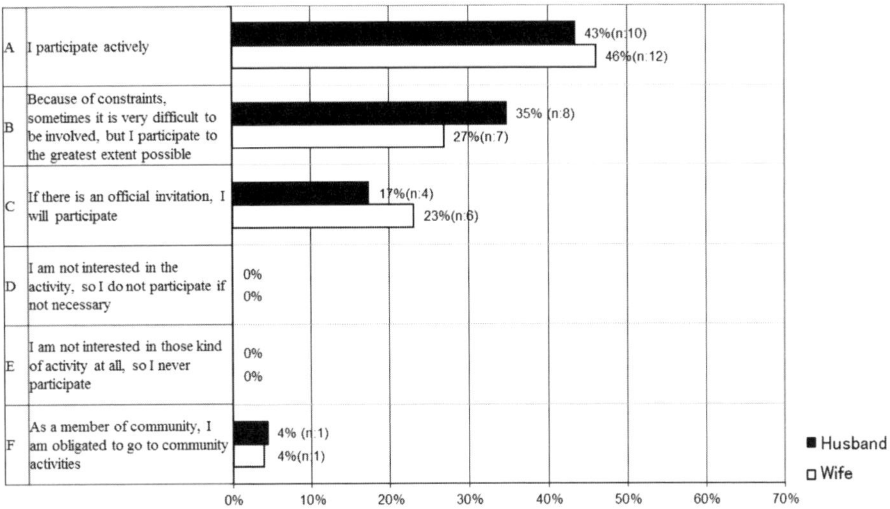

Figure 4. Residents' participation in community activities.

Participation in Community Activities

The results show that residents are highly involved and value their community activities. In the multiple-choice questions, responses A, B, and C are considered as "willingly participate," responses D and E are "do not participate" and response F is "participate due to obligation." Accordingly, more than 95% of heads of household and their spouses say that they willingly participate. Less than 5% participate due to obligation, and no one indicated that they do not participate (Figure 4). The high participation in community activities is supported by the residents' perception of the importance of community activities in their neighborhood. Approximately 65% of heads of household and 57% of spouses chose response A, "I feel that community activities are very important," while only 35% said that "they are neutral" (Figure 5).

Because most residents are from the same or nearby villages, social cohesion is still maintained in the post-disaster settlement. Previous weekly and monthly routines, as well as incidental events like "gotong royong"[6] still occur. However, community activities that used to take place inside the house now occur at different venues such as on the street, in the mosque, etc., due to limitations of the dome house. Additionally, in some cases,

[6] Gotong royong is a traditional practice of Javanese people's mutual assistance among community members.

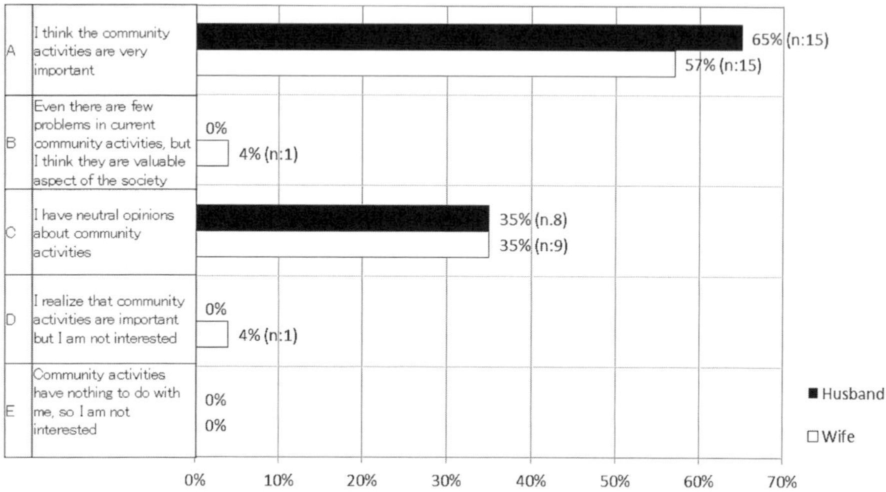

Figure 5. Residents' perceptions of community activities.

unavailability of the original setting means certain activities now occur less frequently or are suspended indefinitely.

The results confirm that social behavior is an important factor in Javanese society even in of post disaster housing, but the limited condition of the housing had enforced the residents to make adaption and adjustment in order to fulfill their lifestyle.

Summary and Conclusion

Although the implementation of a clean kitchen, shared bathrooms, large but secluded family room, esthetic landscape, shared pathway, etc., were designed to improve residents' quality of life, planners and designers failed to recognize or neglected the complex ways in which an environment acts as a setting for life, as clearly demonstrated by the discrepancies between the intended usages in the master plan and residents' actual usages and cognition. Consequently, the residents have had to adapt and adjust physical and behavioral aspects, to maintain their previous personal and social lifestyles. Understanding the latent function of services must be considered, so that the intended design does not become an environment inhibitive for residents' lifestyle, such as the presence of electric lightning (Rapoport, 1978) or efficient kitchens (Esber, 1972) that designed for better living quality but instead rejected by the residents. Hence, in order to create a settlement and house design for post-disaster housing that is close to the indigenous patterns, it is crucial that user lifestyle is fully investigated.

In New Ngelepen, the unchangeable house design and outdoor spaces have resulted in resident dissatisfaction. Residents' evaluation suggests that they are incapable of changing the condition of their donated post-disaster house design and outdoor space. The need for flexibility, personalization, and ease of maintenance are essential, because in the original settlement, residents routinely had social gatherings within the house, and used to repair or change the house on their own. Due to design limitations, dome houses and their outdoor spaces cannot facilitate prior activities such as larger gatherings. Thus, streets and public facilities now have a greater and more central role in facilitating social interactions.

Most Javanese dwellings consist of the core house itself, as well as the surrounding wells, streets, etc. These areas all serve as social spaces to accommodate various social interactions. The data shows that residents remain very involved and continue to value community activities (social interaction) very highly. Unfortunately, in post-disaster housing, these needs are sometimes neglected or cannot be facilitated, forcing residents to adjust by using neighborhood streets as gathering places and sport areas, going to a nearby village's mosque for large prayer gatherings, etc. In some cases, certain activities have been reduced or suspended indefinitely.

To aid in the post-disaster recovery process, post-disaster housing should not only be built for emergency situations, such as to accelerate reconstruction and to improve disaster prevention performance. Providers also need to understand survivors' previous lifestyle, activity systems, social networks, etc. in order to reconstruct a settlement that can accommodate these features, or is at least flexible enough for the residents to adapt and adjust. This could help them in their transition to the new environment and provide them with a better quality of life.

References

Altman, A., & Chemers, M. M. (1980). *Culture and environment.* Cambridge, MA: Cambridge University Press.

Esber, G. (1972). Indians, architect and anthropologists: A study of proxemic behavior in a Western Apache society. *Man-Environment Systems*, March, p. 58.

Ikaputra (2008, October). *People response to localize the imported culture. Study case: The dome house in the rural culture post Javanese earthquake 2006.* Paper presented at the 14th World Conference on Earthquake Engineering, Beijing, China. Retrieved from http://static.monolithic.com/pdfs/dftw/Ikaputra.pdf

Parker, F. (2008). *Architect Rick Crandall on the unique features of monolithic domes.* Retrieved from http://www.monolithic.com/stories/architect-rick-crandall-on-the-unique-features-of-monolithic-domes

Rapoport, A. (1969). *House form and culture.* New Jersey, NJ: Prentice Hall.

Rapoport, A. (1978). Culture and environment. *Ecologist Quarterly, 4*, 269–279.

Rapoport, A. (1987). On the cultural responsiveness of architecture. *Journal of Architecture Education, 41*, 10–15.

Rapoport, A. (1990). System of activities and system of settings. In S. Kent (Ed.), *Domestic architecture and use of space* (pp. 9–20). Cambridge: Cambridge, MA: University Press.

Revianto, S. (1997). *Omah: The construction of meanings in Javanese domestic settings* (Unpublished master's thesis). McGill University, Montréal, Canada.

Saraswati, T. (2007). Kontroversi rumah dome di Nglepen, Prambanan, D.I Yogyakarta [Dome house controversy in Nglepen, Prambanan, D.I Yogyakarta]. *Dimensi Teknik Arsitektur, 35*, 136–142.

Subroto, T. Y. W. (1995). *A study on the spatial linkage in urban settlement as an alternative tool for improving living environment in the cities of Java* (Unpublished doctoral dissertation). Osaka University, Japan.

Tercan, B. (2001). *Post-earthquake relocation process in Yalova* (Unpublished master thesis). METU, Ankara, Turkey.

Whyte, W. (1980). *The social life of small urban spaces.* Washington, DC: The Conservation Foundation.

Representations of Climate Change by Water Managers

Elisabeth Michel-Guillou

Centre de Recherches en Psychologie, Cognition et Communication (CRPCC), Université de Bretagne Occidentale, Brest, France

Abstract

This study explores the social representations of water managers with regard to climate change. These managers are local stakeholders who draw up and implement water management plans to collectively manage local water-related issues (water quality/quantity). The present study was conducted in Brittany (France) and was aimed at analyzing the significance attributed by water managers to climate change. This paper presents findings derived from semi-structured interviews with 18 water managers. These interviews were transcribed verbatim and then their contents were analyzed with a dedicated software, Alceste. The results highlighted a representation of climate change bound primarily to global warming. Moreover, they showed that climate change was perceived by the interviewees as being global and assessable over the long term, in contrast to local water problems, which are manageable in the short term. These characteristics lead managers to not consider climate change impacts in their management decisions.

Key words: climate change, environmental and social psychology, social representation, water manager

Climate change is at the forefront of public concerns today. It is a source of discussions and debates, as well as stakes that lead some people to campaign for far-reaching changes, given the gravity of the phenomenon. On the other hand, others consider it as a simple lever to exert more pressure, to compel changes in practices. This topic is thus a subject of conflict between groups of people, depending on their points of view, about the causes, and consequences of climate change. However, despite differences of opinion on specific points, there is a consensus within the scientific community about the rapid evolution of climate that has taken place during the second half of the 20th century and on its anthropogenic causes (Brechin, 2003).

In this context, we examined the meaning of the term "climate change," in particular for water managers in charge of the management of a natural resource (water) whose quality and/or volume probably depend on variations in climate.

Climate Change: A Social Construction

One's representation of the world shapes one's understanding of the environment, as well as the way one makes decisions and engages in certain practices. This representation is not individual, but social. In other words, it refers to collective models of thought, ways of apprehending the world collectively. Everyone, through affiliation to groups, thus builds a representation of reality that allows him/her to understand, analyze, master, and make sense of his/her environment. Since it is a source of both risk and stake whose seriousness and importance are overestimated by some groups, or underestimated by others, climate change cannot be considered as an "objective reality." Consequently, it has to be considered as a social construction of the reality specific to each group.

This issue interests researchers in social and environmental psychology for two reasons: (i) it is an object of social construction and (ii) this social construction shapes the relationship between people and their environment. In this context, taking into account the relationship between the environment and its shared and socio-cultural dimension seems essential to fully understand the various representations and their related behaviors (Michel-Guillou & Weiss, 2007).

This collective approach refers, in particular, to the idea that individuals are not isolated beings, but "social" persons, in other words, socially integrated people who interact within a given social context. This approach examines how regulatory systems (ideologies, beliefs, standards, etc.) justify social behaviors and relationships between groups. It thus seeks to understand how social knowledge is formed and, in particular, how science relates to common sense (Moscovici, 2001). This is an approach based on Moscovici's Social Representations Theory (1976).

Social representations correspond to specific form of knowledge (e.g., beliefs, values) conveyed by society and shared by a given social group. They are a form of social thought that allows individuals to apprehend their environment and provides them with a vision of the world. Consequently, social representations are a social construct of reality (Jodelet, 1999), aimed at making reality meaningful (Abric, 2001). They correspond to a group's mental representation of a given conceptual or concrete object, and they are closely linked to context. They are "social" insofar as they are (i) the product and the reflection of social processes (e.g., interactions, mass communication) and (ii) shared by members of the same group.

Various functions are attributed to social representations, such as the management of social relations, which allow one to analyze and explain the surrounding world, as well as to position oneself within an environment and to control it (Jodelet, 1999). Consequently, social representations shape our relationship with the external world. Finally, they not only allow groups to preserve their specificity and identity, but also allow individuals to justify their standpoints and behaviors (Abric, 2001; Doise, 1993, von Cranach, Doise, & Mygny, 1992).

Thus, with regard to climate change, there are potentially as many realities as there are groups or communities who feel concerned about this issue (Spence & Pidgeon, 2009). Some studies on social construction have described a moderate level of knowledge about climate change (Cabecinhas, Lázaro, & Carvalho, 2008). Furthermore, when this issue is compared to other environmental problems, it tends to be considered as less important (Brechin, 2003; Spence & Pidgeon, 2009).

Consequently, when individuals are faced with the issue of climate change, they will have their own representation of the phenomenon, according to their social association. In this context, the present research deals with the construction of this concept by water managers, their representation of climate change, and the way they handle it in the daily management of water resources.

Climate Change, Water Resources, and Water Managers

The Connection Between Climate Change and Water Resources

This study was focused on what water managers think about climate change because of the scientifically acknowledged link between changes in climate and variations of water resources (IPCC, 2007). Scientists have noticed a sea level rise as well as significant increases in precipitation in certain parts of the world and reductions in others. The IPCC has taken these elements into account to give some examples of potential impacts, with variable degrees of probability, of extreme climate and weather events on water resources.

The present study relates to a specific area of France: Brittany. At the regional level, long-term follow-ups have provided evidence of climate change in Brittany (Dupont, Dubreuil, & Planchon, 2000; Jouan, 2005). The impacts of climate evolution on water include: Hydrological events (e.g., flood), changes in water quality (e.g., nitrate levels), and ecosystems (e.g., distribution and abundance of fish). Such events and features provide evidence of the connection between climate change and water resources. These facts drove us to interview managers in charge of the water resource in Brittany.

Water Management Plans and Local Water Problems

The interviewed water managers are responsible for the definition and implementation of Water Management Plans called, in French, *SAGE*.[1] This scheme is an official document that defines strategies and directives to protect local water supplies. Thus, the objective of the water managers is also to identify local water-related problems and to propose

[1] *SAGE: Schéma d'Aménagement et de Gestion des Eaux.*

corrective solutions. This is done through meetings between the water managers within a Local Water Committee called, in French, *CLE*,[2] which is composed of local representatives (50%), water user representatives (25%), and local authority representatives (25%).

In Brittany, and more particularly in the *département* of Finistère where the present study was conducted, the main water resource issues are the quality of drinking water, the quality of coastal waters for swimming, water sports and fishing activities, the conservation of the wet zones, and floods.

Objective and Research Questions

These considerations and the perspective proposed in Social Representations Theory drove us to conduct this study to gain insights into what water managers know about climate change and how they understand this issue. Our goal was to identify their representation of climate change, including potential variations in representations, depending on their affiliation to a specific group.

To understand and analyze whether water managers agree or disagree with the existence of climate change, we asked the following questions: What approach do water managers have toward climate change? What knowledge do they have about climate change? Is there a significant recognition of climate evolution? In the event of an affirmative answer, which indicators, causes, and consequences are perceived? What do they see as potential direct or indirect impacts of climate change on water resources? Do they take climate change into account in debates about the development of future water plans? Their answers enabled us to gain further insights into how decision-makers in environmental management understand climate change, and their evaluation of this issue, that is, water resources, compared to other environmental problems.

Methodology

Survey Procedure

The survey was carried out among 3 *SAGE* from Finistère: (i) the Elorn *SAGE* is in its final stages of definition (pre-implementation), (ii) the Bas-Leon *SAGE* is in the early stages of definition, and (iii) the Odet *SAGE* is at the implementation stage. In this exploratory study, 18 semi-structured interviews were conducted with water managers as follows: 10 interviewees from the Elorn *SAGE*, 5 from the Bas-Leon *SAGE*, and 3 from the Odet *SAGE*. The interviewees consisted of 10 elected local representatives, 7 water user representatives, and 1 French government representative.

[2] *CLE*: *Commission Locale de l'Eau*.

The interview covered three main topics: (i) water problems managed by *SAGE*, spatio-temporal evaluation, causes, and consequences, (ii) climate change (its representation(s), connections with water resources, management of the phenomenon by *SAGE*), (iii) sustainable development (its representation(s), connection with *SAGE* and with climate change). The present study deals only with the climate change issue.

Data Analysis

The recorded interviews were transcribed verbatim for further qualitative analysis of data with Alceste software (Kronberger & Wagner, 2000; Reinert, 1990). This software uses a Hierarchical Descendant Classification to identify a limited number of word classes representative of the discourses. The classes are based on word co-occurrences. This exploratory analysis was supported by a manual analysis of thematic content.

Results

The results presented here are derived from the Alceste analysis about the representations of climate change by water managers: From the 61% of classified text, five thematic classes were identified (Figure 1).

These classes can be regrouped into three general topics: (i) the definition of climate change (37% of classified text); it refers to a description of climate change according to past seasons (class 1; 19%) and to changes on a global scale (class 2; 18%), (ii) climate change issues (26%) with the underlying question about their possible evolution (class 3; 15%), and how they are taken into account at a local level (class 4; 11%), and (iii) the connection between climate change and water resources (37%).

The first topic is thus a definition of climate change. In the opinion of most of the interviewees, the climate is changing. This change is described by increases in temperatures and sea level, as well as more frequent occurrences of storms and droughts. "There are examples of global warming: Snow melting, tornados, tsunamis, an increase in ever more violent climatic events" (local representative, interview 7). These phenomena are liable to affect aquatic life. Half of the interviewees referred to either past seasons, or to their childhood, to say, for example, "It is true that the wind (...) is much stronger, more violent; but that's just my opinion, it is not scientifically proven" (local representative, interview 18). However, although some interviewees have seen a change and consider the seasons to be less marked than before, others have seen no change. "The temperatures are about average. There are always some peaks, we often have a short memory" (user representative, interview 14). These changes have been identified as being more discernable at a global scale than at a local level: In other words, they have taken place elsewhere. Half of the interviewees have not perceived climate change at the local level. "Locally it is difficult to tell what the impact really is. Globally, we can see it" (local representative, interview 1).

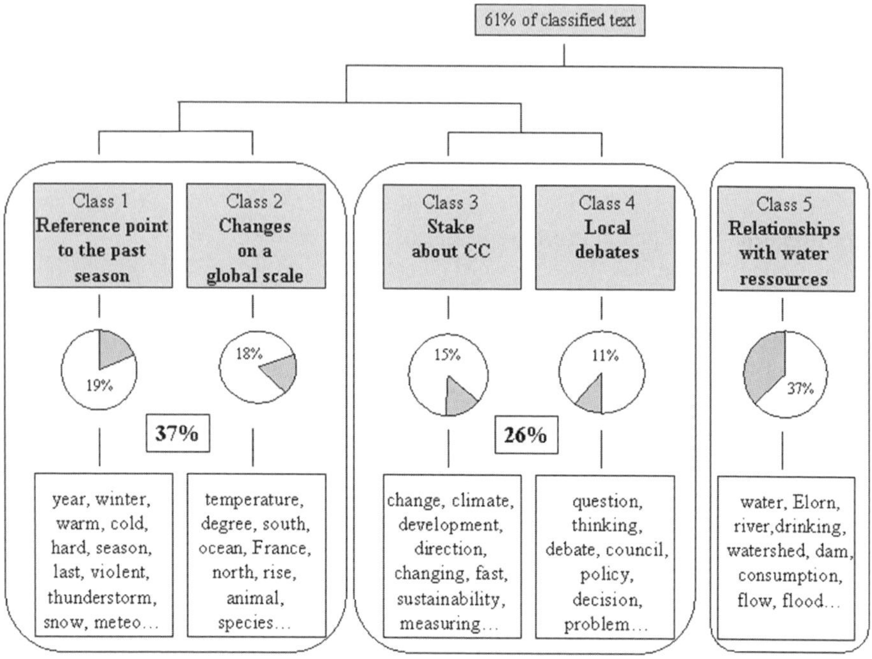

Figure 1. Interpretation of the topics from the classification provided by Alceste software. The percentage in each loss corresponds to proportion of each thematic class in the 61% of classified text.

Furthermore, when the problem is evoked at a local level, it is always in terms of probability and future.

> About the reality of the phenomenon, in my opinion, there can be no doubt. (. . .) Globally, you just have to read the press, even if it's unspecialized: There is a modification of the climate. Then locally, will we be rained on more or less, will there be more wind, less wind, more thunderstorms, less thunderstorms? I don't know. (user representative, interview 5)

Furthermore, the interviewees said that they trusted scientists.

> I also listen to what scientists say on the subject. I have no personal opinion on the subject, I could not, it would be stupid to have a personal opinion, I am not a scientist, it is necessary to take precautions. (...) And I think that on the human scale we cannot measure climate change, it can only be measured scientifically over decades. (local representative, interview 4)

But because there are a lot of uncertainties associated with the phenomenon (e.g., Where? When? How? Maybe, etc.), their discourse is clearly expressed as a feeling to evaluate climate change. This discourse is considered sometimes as inadmissible, because it is based, in part, on memory which is defined as "fallible." However, this feeling is the only possible reality for assessment of a phenomenon that is unperceivable in time and space. For almost half of the interviewees, the phenomenon is imperceptible at a human level. "At a human level, I do not think that it is discernible. We have time to learn to live on or under the water period" (user representative, interview 6).

Finally, most of the interviewees judged climate change to be obvious, despite the fact that it is not discernable at the local level. This contradiction is a source of discussion, and it forms the main theme of the second topic. For some people, rapid climate change is "worrying" and measures should be put in place to deal with the issue. "I don't know, it just worries me. I think that there will be a real problem in fifteen or twenty years" (local representative, interview 15). Others are doubtful about the existence of this phenomenon and question the need for far-reaching, long-term change. For example, they wonder about the climate cycle. "Is it climate change or is it rather an evolution or a cyclic rhythm? (...) I don't know, I am wary of this aspect" (interview 1). Regardless of the question posed, individuals' lifestyles and consumer behavior were always called into question. "It is especially due to our way of life" (interview 1). "Carbon dioxide" is mentioned in connection with human behavior, "We're all responsible for climate change because we all release carbon dioxide into the atmosphere" (interview 4).

Otherwise, because climate change is not discernible at the local level and because it is a source of uncertainty, consideration of this issue at the local scale seems to be impossible. Discussions about climate change are and have to be conducted on a global scale (e.g., Copenhagen), but the issue cannot be solved locally. Furthermore, blame for non-action is easily apportioned to others: Users blame the local representatives and vice versa. "At a certain point, economic models must be called into question. We have to stop believing that local communities and policies can do everything" (interview 15). Some interviewees highlighted financial problems, in particular about finding and developing alternative solutions that are sometimes more costly. At present, local climate change does not appear to be a pressing issue for water managers.

As climate change is not debated in such committees, it seems to be difficult for water managers to connect the problems associated with water resources and climate change. However, hypothetical connections(s) were evoked by some water managers through their concerns about water quantity and quality (Figure 1, class 5). Quality was mentioned in connection with events such as floods, sea level rise, as well as low water levels. Quality is affected by increasing water acidity, elevation of water temperature, and it also has an impact on biodiversity, especially for aquatic life (e.g., changes in the distribution of fish species, development of algal blooms). Nevertheless, as mentioned above, for almost half the managers, the connection between water quality and climate change is not really explicit, and they noticed the discrepancy between the water problems that they manage at the local level and the changes that occur on a global scale: "We are far removed from the tangible problems of a community with these problems of local pollution" (local representative, interview 3).

One-third of the interviewees expressed their feelings about the lack of connection: "Global warming? No, I do not think there is a connection" (interview 15).

Discussion

This study about the knowledge of and approaches developed by water managers toward climate change showed that most of the interviewees were aware of the reality of climate change elsewhere, not in Brittany, and could talk about it. The construction of this social reality proved to be largely influenced by mass media (Cabecinhas et al., 2008). The analysis of the interviews uncovered very little connection to the potential direct or indirect impacts of climate change on water resources and their daily management at the local scale: The problems associated with water resources are local, definite, empirical, and daily, whereas climate change is perceived as global, abstract, conceptual, or theoretical, and not perceivable at the individual personal level, in particular. These findings are consistent with existing evidence (García-Mira, Real, & Romay, 2005; Uzzell, 2000). Decision-makers committed to pro-environmental action (i.e., water managers), are aware of climate change, and some of them consider that this global environmental issue is serious. However, because water problems are immediate and salient, they are perceived as more serious than the issue of climate change. Alongside other issues, climate change is not rated as a priority (Brechin, 2003; Spence & Pidgeon, 2009). Decision-makers, especially water managers, do not believe that they can undertake pro-environmental actions, because they do not perceive ways to do so, a feeling enhanced by the spatial and temporal remoteness of the climate change issue. Furthermore, in their opinion, as this issue is a source of concern on a global scale, it must be thought about and managed collectively by competent institutions (e.g., governments). Finally, water managers said that they are unable to control it, in contrast to the local water problems for which they feel personally responsible. "When control is absent, internal psychological defenses, such as denial, can minimize fear" (Spence & Pidgeon, 2009, p. 11). This may also explain why people are more focused on the effects rather than on the causes of climate change (Cabecinhas et al., 2008), which the findings of this study corroborate.

Conclusion

The results of the present study showed no significant distinctions according to the affiliation to different groups of the interviewed decision-makers (local representatives, water user representatives, French government representatives). The local representatives, who deal with other environmental issues such as energy, talk more openly about climate change. In conclusion, the study showed a certain degree of knowledge about climate change, but illustrated that water managers do not consider it as a reality, because it is imperceptible, ambiguous, and not considered to be a priority. It remains to be seen whether stakeholders will take climate change into account when debating future plans for water management.

However, the study highlighted the lack of debates about this issue within the *SAGE* committees and the lack of intention to integrate it into the discussions in the near future.

References

Abric, J.-C. (2001). A structural approach to social representations. In K. Deaux & G. Philogène (Eds.), *Representations of the social: Bridging theoretical traditions* (pp. 42–47). Oxford, UK: Blackwell.

Brechin, S. R. (2003). Comparative public opinion and knowledge on global climatic change and the Kyoto Protocol: The US versus the World? *International Journal of Sociology and Social Policy, 23*, 106–134.

Cabecinhas, R., Lázaro, A., & Carvalho, A. (2008). Media uses and social representations of climate change. In A. Carvalho (Ed.), *Communicating climate change: Discourses, mediations and perceptions* (pp. 170–189). Braga, Portugal: Centro de Estudos de Comunicação e Sociedade, Universidade do Minho.

Doise, W. (1993). Debating social representations. In G. M. Breakwell & D. V. Canter (Eds.), *Empirical approaches to social representations* (pp. 157–170). Oxford, UK: Clarendon Press.

Dupont, N., Dubreuil, V., & Planchon, O. (2000). L'évolution récente des crues de la Vilaine: le rôle des précipitations et de l'occupation du sol du bassin versant [Recent evolution of the Vilaine floods: The role of both precipitations and river basin land use]. *Publications de l'Association Internationale de Climatologie, 13*, 91–99.

García-Mira, R., Real, J. E., & Romay, J. (2005). Temporal and spatial dimensions in the perception of environmental problems: An investigation of the concept of environmental hyperopia. *International Journal of Psychology, 40*, 5–10.

IPCC. (2007). *Climate Change 2007: Synthesis report. Contribution of Working Groups I, II and III to the fourth assessment report of the Intergovernmental Panel on Climate Change*. Geneva, Switzerland: IPCC.

Jodelet, D. (1999). Représentations sociales: un domaine en expansion [Social representations: An expanding domain]. In D. Jodelet (Ed.), *Les représentations sociales* (6th ed., pp. 47–78). Paris, France: Presses Universitaires de France.

Jouan, D. (2005). *Etude des conséquences de l'évolution du climat induit par l'effet de serre sur la fréquence et l'intensité des perturbations cycloniques et sur la distribution des valeurs extrêmes du vent au sol en Europe de l'Ouest* [Impacts of greenhouse effect-induced climate change on the frequency and intensity of cyclonic disturbances and the distribution of surface wind extreme values in Western Europe] (Unpublished doctoral dissertation). Université de Haute-Bretagne, Rennes, France.

Kronberger, N., & Wagner, W. (2000). Key words in context: Statistical analysis of text features. In M. W. Bauer & G. Gaskell (Eds.), *Qualitative researching with text, image and sound. A practical handbook* (pp. 299–317). London, UK: Sage.

Michel-Guillou, E., & Weiss, K. (2007). Representations and behaviours of farmers with regard to sustainable development: A psycho-environmental approach. In A. B. Larson (Ed.), *Sustainable development research advances* (pp. 207–221). New York, NY: Nova Science.

Moscovici, S. (1976). *La psychanalyse, son image et son public* [Psychoanalysis, its image, and its public]. Paris, France: Presses Universitaires de France.

Moscovici, S. (2001). Why a theory of social representations? In K. Deaux & G. Philogène (Eds.), *Representations of the social: Bridging theoretical traditions* (pp. 8–35). Oxford, UK: Blackwell.

Reinert, M. (1990). Alceste, une méthode d'analyse des données textuelles. Application au texte « Aurélia » de Gérard de Nerval [Alceste, a textual data analysis]. *Bulletin de Méthodologie Sociologique, 26*, 25–54.

Spence, A., & Pidgeon, N. (2009). Psychology, climate change, & sustainable behavior. *Environment, 51*, 9–18.

Uzzell, D. (2000). The psycho-spatial dimension of global environmental problems. *Journal of Environmental Psychology, 20*, 307–318.

Von Cranach, M., Doise, W., & Mygny, G. (1992). *Social representations and the social bases of knowledge*. Seattle, WA: Hogrefe.

Flood Hazards, Urban Waterfronts, and Cultural Heritage

Heiko Lieske

Faculty of Architecture, Technical University Dresden, Germany

Abstract

Riverine cities are specific risk habitats because of the concentration of people, material, and intangible values affected by flooding. Recent floods have focused attention on the vulnerability of settlements and on the urgent need for protection measures. This chapter reflects on the specific requirements for urban flood protection schemes in environments marked by cultural heritage, by distinct scenery, or by recreational or real estate value. Following a brief review of the history of hydraulic planning, a survey of the special conditions, the conflicts, and opportunities of inner-city flood protection today is presented. The findings are based on a comprehensive case study of recent German flood protection projects in historic townscapes. They are further illustrated by the case of the small Saxonian town Grimma, where the author's team is engaged as heritage consultants for a large-scale flood-proofing scheme.

Key words: cultural heritage, flood protection, integrative planning, risk habitats, urban waterfronts

Introduction

The August flood of 2002 in Saxony and Bohemia focused attention on the vulnerability of riverine settlements and on the urgent need to counter the risk of flooding with comprehensive protection schemes. Conventionally, these measures would be focused on the construction of dams, reservoirs, and flood walls. Recent disciplinary discussions, as well as public debate and political discourse, have placed emphasis on an additional aspect of flood control. The enhanced retention of rainfall within the catchment area of a river has become one of the main objectives of today's flood protection policy. Its effect, however, is frequently limited, because it is subject to a number of natural and anthropogenic conditions that often cannot easily be altered. Structural improvements therefore remain indispensable. Nevertheless, in many cases they have a severe impact on urban waterfronts and on cultural landscapes as human habitats. Issues of urban development and of heritage conservation, as well as social issues, are frequently affected by these side effects. To date, problems of this kind have rarely been taken into consideration, neither in the planning and construction of protection measures nor in scientific research on risk management. By what means can these challenges be met? How can we make sure that large-scale flood-proofing structures truly enhance urban living conditions, instead of screening off or even damaging the very qualities they are intended to protect?

After a brief description of the approaches and methods on which this chapter is based, the discussion turns to critical reflections on the traditions of hydraulic engineering. The body of the chapter deals with the special conditions, the conflicts, and opportunities of inner-city flood protection today. It refers to the author's experiences as a member of an interdisciplinary team at the Technische Universität Dresden that conducts planning of projects and research in the field of heritage conservation and its confrontation with flood protection.

Methodological Design

The historic review of hydraulic planning referred to in the following part of the chapter, as well as the recommendations given in the final section, are based on findings of a recent comparative empirical study that focused on heritage conservation and urban development confronted with large-scale flood protection schemes in Germany (Lieske, Schmidt, & Will, in press). Its results have been derived from the application of a set of specifically-tailored methods. The historic review involved a comprehensive search for and analysis of historic and recent information. Hydraulic, heritage, architectural, and landscape architectural literature, as well as historic maps and archive material, were studied at this stage of the research process.

The recommendations reflect the results of a nationwide case study. A survey of recent and historic hydraulic projects was conducted, involving comprehensive literature and archive research, as well as interviews and *in situ* analyses. The research focused on two questions:

A) Which procedures have been implemented for the planning and construction of the flood protection schemes referred to in the study? To what extent has the public been involved in these processes? What effects have these procedures and processes had on the cultural heritage and urban qualities?
B) Which hydraulic and architectural constructions have been executed in the various projects; what were the design principles? What effects have these construction and design solutions had on the cultural heritage and urban qualities?

An initial symposium was held to reflect upon experiences collected in the field by planners and hydraulic engineering authorities. In order to identify locations and flood protection schemes within the scope of the research project, the following selection criteria were chosen:

1. completion within the last 20 years;
2. cultural heritage affected; and
3. availability of information, especially archive material and interviewees.

Using these criteria, more than 50 schemes were identified, of which the 22 most significant were then studied in further detail. *In situ* investigations covered spatial and functional architectural analyses, examinations of the local heritage stock and photo documentations. Guided interviews were conducted with local and state authorities, as well as with planners and experts. All the information was analyzed, interpreted, and displayed, in order to allow for comparability between all the individual cases. The results are presented as detailed case studies, in comparative statements, and in recommendations for authorities, planners and politicians.

The middle section of this chapter is a work-in-progress report on a flood protection scheme in Grimma/Saxony, which the author's team is contributing to, as experts for heritage conservation, urban and landscape design. Initially, comprehensive analyses to describe and visualize the topographical, spatial, environmental, esthetic, and functional qualities of the city-river relationship were undertaken. The process was also supported by meetings with stakeholders and experts. Stakeholder sessions were held in 2006, based on the design of a consensus conference (Arbter, Handler, Purker, Tappeiner, & Trattnigg, 2005), to find out about the image of the city as perceived by its inhabitants, about their shared values and future conceptions, and about their concerns related to and hopes for flood protection. Programming methods (Henn, 2001) were applied at this stage. Town hall meetings, along with a questionnaire, provided deeper insights into the opinions of the interested public. Numerous meetings with the city council had been held from the beginning, to coordinate the scheme with the city's development plans. All these contributions were considered to be just as important for the integrated planning as the expertise of hydraulic engineering professionals. The comprehensive analysis was followed by the production of a number of alternative design proposals for characteristic river front sections and focal points. Exposure of these proposals to professional and to public scrutiny led several times to either rejection or refinement and, eventually, to general agreement on solutions tailored to the special needs of Grimma.

The Background to Integrative Hydraulic Planning in Germany

Nineteenth-century Germany witnessed a number of fundamental hydraulic projects, when virtually all major rivers were rectified and configured to meet the demands of shipping and flood protection. In the last quarter of the century, hydraulic engineers – called *hydrotects* in those days – were facing an emerging new culture of public debate addressing environmental issues. After the 1882/1883 flood of the Rhine river, both the public and politicians accused the engineers of causing the disaster, as an observer noted: "The people on the Rhine very soon made up their minds and concluded unanimously: This is the work of the hydrotects with their bricks, they increased the river's speed too much." (Stenographische Berichte, 1883, transl. HL) The hydraulic engineers responded with an unbroken line of scientific argumentation that was published instantly, and they succeeded in identifying exceptional meteorological circumstances as the cause of the disastrous event. This instance marks a new stage of debate and public scrutiny in large-scale water engineering schemes. On the one hand, the emergence of professional media in 19th century allowed for instant provision and distribution of information. This, in principle, meant that an independent source of news and background knowledge became available to the public. For the planning authorities it created an effective platform for engaging in public relations. On the other hand, the increased public awareness and scrutiny called for extensive involvement of the personally affected and the wider public in the planning process. In addition to public exposure, hydraulic engineers soon found that they also had to cope with a number of competing issues that were advocated by planners from various disciplines. The integration of agricultural irrigation and drainage, urban development, fresh water supply, esthetic qualities of the cultural landscape, and tourism, to name just a few issues, into flood protection confronted engineers with new challenges. An early example of a collaboration that proved to be successful was the incorporation of major hydraulic schemes into the plans for urban renewal in central Vienna, undertaken from 1892 to 1915, and directed by the architect Otto Wagner.

Whereas the consideration and integration of additional requirements and interests continued to be taken into account after the political transition from the Weimar Republic to the Third Reich, approaches to democratic planning procedures for hydraulic projects in Germany came to an end. In West Germany, the 1950s, and especially the 1960s, became a time of radical restructuring of inner-city areas as well as of the countryside, fostered by the incredible economic boom of the time. Massive infrastructure and urban renewal programs were founded on the euphoria about economic growth and technical feasibility shared by many planners and politicians of the time.

In Ilzstadt, a historic district of Passau located at the confluence of the Danube and Ilz rivers, between 1957 and 1975 a large-scale joint scheme of road construction and flood protection was implemented, virtually without any consideration of social structure, of every day functional relationships, of townscape qualities, of listed buildings or of the special quality of a life on a river. The historic residential, retail, and public buildings of an entire neighborhood were demolished to make way for the new scheme. Two hundred fifty households, which equals half of the local population, were forced to move and resettle in other parts of the city. The waterfront was raised by up to 4.5 meters. On top of this new embankment, a new major road was laid out that completely cut off the remaining parts of the settlement from the rivers, both in terms of view as well as of accessibility.

In view of the social, economic, and environmental problems that accompanied projects like the Ilzstadt scheme, a process of reconsideration in flood protection planning gradually emerged during the 1970s. Especially in Bavaria during the 1980s, a number of successful projects integrating historic preservation and urban quality into flood protection was carried out, for instance in Wasserburg, Kelheim, Weilheim, Marktzeuln, and Würzburg. (Knauer, 1987; Kraus, 1987; Kupfer & Grieblinger, 1987; Ullmann & Grosse, 1987). Wolfgang Heyenbrock stated: "The particular protection measures with no doubt become more and more technically intricate. They frequently require enormous funding, while all too often they become the target of a debate concerning competing interests and demands of hydraulic perfectionism, town planning, historic preservation, agriculture, nature and landscape protection, to name just a few. It is the task of a very open planning procedure to produce balanced solutions." (Heyenbrock, 1987, transl. HL)

Surprisingly, the publications listed above apparently were, for a long time, the only scientific contributions to the subject. Considering the importance of the Bavarian experience, it is regrettable that these approaches were not widely publicized. As literature on the subject still is very rare, many of the subsequent hydraulic projects in Germany show little effort to encourage public participation or to integrate external issues and schemes from other planning sectors. However, along with the recent, rapidly growing political and research interest in risk management, issues of environmental side effects of flood protection are increasingly being dealt with. And, occasionally, questions about the compatibility of flood protection and heritage conservation are also being raised (Lieske & Will, 2007; Will & Lieske, in press; Will, Meier, & Petzet, 2008).

What lessons can be learned from this brief excursion into the history of integrative hydraulic planning? Flood protection and its planning culture follow certain trends that cannot easily be traced back to definite social or political circumstances. Currently, the subject is undergoing profound shifts in terms of its opportunities, demands, and public appreciation. The *a priori* assertion given here is threefold.

A) Due to the consequences of global climate change, with its risk of increased flooding in Central Europe on the one hand, and ground-breaking advancements in hydraulic sciences and technology, like hydraulic 3D-modeling and temporary protection structures, on the other, hydraulic engineering and inner-city flood protection are facing entirely new and challenging circumstances, which mean new risks and new opportunities alike.
B) New conditions produce new areas of conflict. They may involve the requirements and interests of various stakeholders, planning departments, and authorities. They may also induce socio-spatial conflicts, and last but not least, by compromising the urban fabric and cultural heritage, they can cause a loss of local identity and a long term impairment of location factors, thus leading to economic downturn.
C) Planning authorities in charge of flood protection need to cooperate intensively with other planners of urban development and with the public. A policy of unrestricted information access therefore is as crucial to the success of large-scale flood protection schemes as the application of integrative trans-sectoral planning procedures and comprehensive stakeholder participation.

Flood Protection Under Urban Conditions

As hydraulic expertise increases and new technologies become available, the physical opportunities for integrative planning today are greater than ever before. Tools for large-scale planning procedures, including public participation and paying attention to the requirements of public agencies or planning departments, seem to cover all imaginable aspects of life in riverine cities. Yet their success is subject to the provision of comprehensive information that needs to be professionally edited and made suitable for public and political evaluation and balancing interests.

Flood protection today consists of three main areas of action: Prevention of flood formation by retention of rainfall in catchment areas, structural flood defense by construction measures in the affected locations, and non-structural measures, such as disaster management plans, designation of flood plains excluded from development, early warning systems, and the adaptation of housing and residential habits at ground floor level by individuals. The advancement of a non-structural protection approach has recently gained wide popularity. However relevant to the overall issue all three of these areas may be, for reasons of focus, this chapter only covers considerations of structural measures under urban circumstances. Besides the concentration of lives and goods, and the spatial limitations for harmlessly containing and channeling off water, a number of additional urban characteristics call for a specific approach in settled areas:

1. The density and permanence of buildings do not allow for spacious structures like embankments;
2. In urban settings, a large number of functions support or interfere with one another;
3. Settlements are places of concentrated history and meaning. Listed buildings, places of interest, and livability, as well as consolidated socio-spatial relations make for an urban living environment that cannot easily be subjected to profound structural alterations;
4. Within the last few years, bodies of water and waterfronts in urban settings have undergone a substantial change in public recognition. Once places for shipping, sewage disposal, trade, and industry during the 19th century, then subject to neglect over much of 20th century, they now face an upsurge in public appreciation.

The Grimma Case

During the disastrous August 2002 floods, the old town of Grimma was one of the worst affected places in the German state of Saxony. The river Mulde reached an unprecedented level and submerged the old town up to a height of 3.5 meters. The steep gradient of the riverbed caused a riptide, damaging up to 700 houses, sometimes up to the second floor level, or even totally destroying them. To protect Grimma from further damage, the Dam Authority, responsible for flood control in Saxony, proposed an initial plan that entailed a rather one-sided approach, based solely on hydraulic requirements. The proposal called for a 1,200 meters long concrete wall up to 3 meters high, blocking the river from the city (Figures 1 and 2). It was presented to the permit authority and to the city council. The plan was rejected promptly by both bodies.

Grimma is a small town of approx. 18,000 inhabitants, situated on a terrace in the Mulde River valley, in a rural setting not far from the major city of Leipzig. What distinguishes it from many other towns is its unusually well-preserved setting in the river landscape, with the medieval city wall still bordering the banks. Historic urban and landscape ensembles of great beauty, mostly situated along the river, make the city attractive to both inhabitants

Figure 1. Grimma, Mulde River, the waterfront and castle as seen from Pöppelmann bridge, *status quo.* (Photograph: courtesy of AG Will, TU Dresden)

Figure 2. Grimma, Mulde River, the waterfront and castle as seen from Pöppelmann bridge, retouched flood protection wall according to preliminary planning proposal 2004. (Photomontage: courtesy of AG Will, TU Dresden)

and visitors. With the preliminary planning proposal presented in 2004, Grimma citizens saw themselves confronted with the prospect of being separated from the river by a monolithic concrete wall. It was obvious that the realization of this proposal would inevitably involve severe and irreversible damage, not only to the river landscape, but also to the functional and esthetic qualities of the city, as well as to the historic fabric and the visual experience of the architectural heritage. The ensuing controversial discussions led to the general agreement that, for a place like Grimma, flood control planning based only on hydraulic and monetary parameters was insufficient, if not counterproductive, since it is was likely to screen off, to damage, or even to destroy those elements and features that it was supposed to protect.

Alternatives were needed, in order to devise flood control structures that would meet the hydraulic requirements without destroying or unduly damaging the cultural, spatial, and landscape values of the Grimma riverside. In 2005, our own team of architects and landscape architects was called in for consultation, to help generate new proposals (Will & Lieske, 2008).

Besides the revised proposal for the above-mentioned wall (Figure 3), one further example of these proposals will be given here. Situated in a small park linking the river

Figure 3. Grimma, Mulde River, the waterfront and castle as seen from Pöppelmann bridge, flood protection wall according to revised planning proposal 2010. (Photomontage: courtesy of AG Will, TU Dresden)

Figure 4. Historic District Administration Building, integrated brackets for stop logs 2008. (Photograph: courtesy of AG Will, TU Dresden)

to the system of open spaces that surround the town where the moat used to be, the historic District Administration Building constitutes a unique sight in the old town. Preliminary planning not only threatened to block the view to this historic building, but it also would have cut the connection between the green belt and the impressive river landscape.

Setting the alignment of the flood protection structure back from the waterfront saved the view to the historic site and building, as well as the prospect from the property to the river. Investigations into the structure of the monumental historic edifice itself revealed that its outer walls were capable of becoming an integral component of the flood protection structure. Consequently, the protection system that was proposed initially – a high monolithic wall – now is detectable only at very close sight, where brackets are installed to accommodate stop logs in case of a flood (Figure 4). With the help of integrative planning, spatial and esthetic losses could be minimized effectively.

Because construction in other sections is under way, there still are controversial discussions in Grimma, but now the issues have moved from *protecting* the image of the town and river landscape to *enhancing* it, to making it an even more attractive and vital part of the urban fabric than it used to be. Stakeholder participation is being supported by a policy of providing comprehensive, up-to-date information, in the form of exhibitions, public tours on site, and the publication of brochures and leaflets (Landestalsperrenverwaltung des Freistaates Sachsen [LTV], 2007, Will, Lieske, Schinker, Fuchs, & Werner, 2010). As recent findings show, in addition to the structural flood-proofing conducted by the authority in charge, non-structural individual measures are also being adopted in Grimma (Siedschlag, 2010).

Demands, Conflicts, Chances – Recommendations From a Nationwide Comparative Study

Recently, our research team was given the opportunity to conduct a nationwide comparative study of projects similar to the Grimma case (Lieske et al., in press). By reviewing flood protection schemes in more than 50 locations, we were able to identify certain lines of conflict that frequently occur in large-scale hydraulic projects. The controversies we detected fall into three main categories, namely, socio-spatial conflicts, discrepancies between various planning sectors, and problems with heritage conservation.

The setting of goals, the discussion of alternatives, the balanced consideration of competing issues, as well as decision-making prove to be fiercely contested stages of a political process that roots in the call for the protection of life and limb and material goods, yet the results of which can produce social and political tensions. Conflicts frequently occur between flood-affected residents and engineers, authorities and politicians in charge. Tensions can also emerge between residents living next to the waterfront and others having no direct access to the waterfront. Frequently, residents living next to the river, although even more exposed to flooding than others, opt against any flood-proofing structures, because these are likely to alter the aspects and prospects of the river landscape they are familiar with and fond of. Other discrepancies arise between parties committed to the protection of an ancient city and those demanding maximized safety from flooding. Last but not least, tensions can occur between any two places receiving unequal financial support.

Besides these conflicts at the level of the citizenry, there are competing demands about various political and planning issues that touch not only on matters of social justice, but also sustainability in urban development or on local identity. For example, flood protection constructions usually are associated with considerable impacts on the structure of monuments, on the cultural landscape, on the natural environment, or on the urban fabric. They can cause severe and lasting damage to the waterfront aspect and to the cultural heritage, and thereby easily spoil the livability and attractiveness of a special place. Thus, they can jeopardize future development opportunities, because the image and quality of urban public space has become a strong criterion for the location of residential, trade and industry, service, or tourist developments.

In this sense, flood protection can be the cause of serious problems far removed from conventional hydraulic engineering issues. On the other hand, flood protection can create opportunities to convert abandoned industrial sites into highly valued neighborhoods and urban open spaces, to rediscover, redesign and revaluate urban waterfronts for public access and benefit. Thus, places near the water will not only be protected from flooding, they will also benefit from urban and environmental improvements. Finally, property values are very likely to increase considerably. Given the importance of flood protection for the safeguarding of the future of a riverine community and its inhabitants, and considering all its side effects, the extra effort of integrative planning is an investment that more than pays off.

Conclusion

Inner-city flood protection proves to be a highly complex political and planning endeavor involving a great number of issues and interests. Its success today can only be ensured if unrestricted information about planning procedures and constructional solutions is available, if extensive public participation is exerted, and if innovative integrative planning cultures are adopted. By meeting these requirements, the livability and functionality of human habitats exposed to flood risks can be ensured under the conditions of global climate change. However, flood control planning based solely on hydraulic and monetary considerations is insufficient, if not counterproductive, since it is likely to screen off or to damage those elements and features that (in addition to its primary task of safeguarding the population) it is supposed to protect.

References

Arbter, K., Handler, M., Purker, E., Tappeiner, G., & Trattnigg, R. (2005). *Das Handbuch Öffentlichkeitsbeteiligung. Die Zukunft gemeinsam gestalten* [Participation Manual. Shaping Our Common Future]. Vienna, Austria: Austrian Society for Environment and Technology.

Henn, G. (2001). Programming. Form Follows Flow. In G. Henn (Ed.), *Prozess Architektur* (pp. 8–17). Dresden, Germany: Technical University Dresden.

Heyenbrock, W. (1987). Hochwasserschutz in Bayern [Flood Protection in Bavaria]. *Bau Intern, 7*, 134.

Knauer, J. (1987). Marktzeuln schützt sich vor den Hochwassern der Rodach [Marktzeuln Protection against Flooding of the River Rodach]. *Bau Intern, 7*, 142–144.

Kraus, W. (1987). Der Hochwasserschutz von Wasserburg a. Inn [Flood Protection in Wasserburg/Inn]. *Bau Intern, 7*, 135–137.

Kupfer, A., & Grieblinger, H. (1987). Weilheim und die Ammer [Weilheim and the River Ammer]. *Bau Intern, 7*, 138–139.

Landestalsperrenverwaltung des Freistaates Sachsen (LTV). (2007). *Hochwasserschutz Grimma* [Flood Protection Grimma]. Pirna, Germany: Author.

Lieske, H., Schmidt, E., & Will, Th. (in press). *Hochwasserschutz und Denkmalpflege. Fallbeispiele und Empfehlungen für die Praxis* [Flood Protection and Heritage Conservation. Case Studies and Recommendations for Practice]. Stuttgart, Germany: Fraunhofer IRB.

Lieske, H., & Will, Th. (2007). Flood Protection vs. Heritage Conservation? An integrated urban and landscape design approach for Grimma, Saxony. In A. Schumann & M. Pahlow (Eds.), *Reducing the vulnerability of societies to water related risks at the basin scale* (Vol. 317, pp. 441–446). Wallingford, UK: IAHS.

Siedschlag, D. (2010). *Hochwasser & Eigenvorsorge – Untersuchung von Einflussfaktoren persönlicher Schutzmaßnahmen* [Flood Protection and Individual Precautions] (Report 03/2010). Leipzig, Germany: Helmholtz Centre for Environmental Research.

Stenographische Berichte über die Verhandlungen des Reichstags [Stenographic Reports on Reichstag Meetings]. V. Legislaturperiode II. Session 1882/83, Bd. 72, 83. Sitzung am 9. Mai 1883, pp. 2431–2432.

Ullmann, F., & Grosse, A. (1987). Hochwasserschutz und Städtebau am Beispiel von Würzburg [Flood Protection and Urban Development: The Example of Würzburg]. *Bau Intern, 7*, 145–147.

Will, Th., & Lieske, H. (in press). Grimma (Germany) – A small-river flood in a small walled town. In M. Drdacky, L. Binda, C. Hennen, C. Köpp, L. G. Lanza, & Z. R. Helmerich (Eds.), *CHEF cultural heritage protection against flooding*.

Will, Th., & Lieske, H. (2008). Grimma an der Mulde. Hochwasserschutz und Denkmalschutz. [Grimma by the River Mulde. Flood Protection and Heritage Conservation]. In H. Lieske, E. Schmidt, & Th. Will (Eds.), *Bauliche Integration von Hochwasser-Schutzanlagen in historische Stadtbereiche* (pp. 7–17) [Constructional Integration of Flood Protection Measures into Historic Urban Fabrics]. Dresden, Germany: Technical University Dresden.

Will, Th., Lieske, H., Schinker, N., Fuchs, A., & Werner, Th. (2010). *Grimma auf dem Weg zum Hochwasserschutz* [Grimma Heading for Flood Protection]. Dresden, Germany: Technical University Dresden.

Will, Th., Meier, H.-R., & Petzet, M. (Eds.). (2008). *Cultural Heritage and Natural Disasters. Risk Preparedness and the Limits of Prevention* [ICOMOS-Series: Heritage@Risk, Special 2007]. Dresden, Germany: TUDpress.

Social Dimensions of Vulnerability and Risks

The Social Dimension of Forest Fires

Risk, Responsibility, and Participation

Ricardo García-Mira,[1] Carmela García González,[2] Adina Dumitru,[3] and Xosé Luis Barreiro Rivas[4]

[1]Faculty of Educational Sciences, University of A Coruña, Spain
[2]IES Val Miñor de Nigrán, Pontevedra, Spain
[3]Faculty of Sociology and Psychology, West University of Timisoara, Romania
[4]Faculty of Political and Social Sciences, Santiago de Compostela University, Spain

Abstract

Forest fires, and the risk they generate for the ecosystem and for human beings, are not only technical problems that can be solved solely by technical experts and their evaluations. Fires possess a social dimension, and addressing this type of disaster involves environmental understanding, the active commitment of citizens, and their participation in land management. This chapter aims to analyse and evaluate the value of citizens' subjective perception of forest fires and explore the dimensions that explain that perception. It also includes a discussion on how to make public policies in the fight against fires more efficient by including a number of technical and objective criteria, together with other more subjective aspects. A number of aspects is addressed, given their importance as lessons with implications for the field of risk policy and management: a) the attribution of responsibility and perceived causes; b) trust, in relation to communication and information management processes; and c) improving task coordination and social participation.

Key words: communication, forest fires, responsibility, risk, social participation, social perception

Introduction

The social dimension is considered today to be an essential element in both the analysis and introduction of environmental programmes and lines of action, and the implementation of risk prevention policies. Environmental problems in general, and forest fires in particular, together with their associated risk management, are not only technical problems requiring a solution provided by experts, with whose advice the corresponding public policies and proposals are produced, but are also social issues that require a higher level of understanding, a system of public participation and, at the same time, the adoption of a civic commitment to the environmental practices and actions that need to be put in place.

The technical analysis of risk moves between the threat or danger of a fire and other dangerous events on the one hand, and the vulnerability and exposure to such a risk on the other. It attempts to evaluate the most appropriate measures and decisions for a preventive management of such events, focusing on the potential danger from the standpoint of an analysis of its probability, its causes, and the intensity or severity it may eventually reach. By defining the areas and moments of risk, we can produce risk maps, but what we are interested in highlighting here, from the social dimension that concerns us, is the concept of vulnerability, or degree of susceptibility and propensity to damage and loss, whether of lives or of economic or environmental assets. This is a concept that enables us to think about the responses, actions and measures intended to reduce the level of risk, allowing us to consider the importance of the degree of awareness and knowledge of the population exposed to the risk in question. It is therefore a concept with a distinct social content that incorporates the idea of protection and the response measures we are willing to adopt, in order to reduce the possibility of suffering damage. This approach allows us to introduce concepts of social responsibility in risk prevention and to distinguish between spheres of action in the social and natural environment.

Social Perception and Communication

It is impossible to plan an adequate solution strategy for forest fires without taking citizens into account, but also without abandoning the technological approaches that characterised the way in which environmental crises were dealt with in the past. The study of social perception brings the perspectives of citizens into the equation. Perception, however, is not enough, since it is not the same as knowledge, nor does it involve a person in any action. The *risk management* of forest fires inevitably demands a social education and communication strategy that promotes *participation* and favours a change of attitudes and values, bringing them more in line with preventive practices.

The literature on risk in general has stressed the necessity to involve people from the moment of problem formulation, in order to ensure effective participation, and also to identify their needs adequately, in order for preventive practices to work (Drew et al., 2003). We are thus faced with the paradox of a society that has never made so much use of expert knowledge, that has never been in such need of technical advice when it comes to defining public agendas and policies, one in which knowledge has never been so relevant, and yet what we see is that the environmental problems we face cannot be solved on an exclusively technical or scientific basis. The policies we have to put in place to deal with environmental crises require consensus and social participation, without which it is impossible to promote change in behaviours, activities, and practices.

The study of social perception is a way of accessing the values, interests, affects, and beliefs of the general population, which are, simultaneously, elements in the melting-pot of the risk situations that intersect with the processes of communication and education of the population. Education and communication are always put forward as tools for improving risk prevention, and thus, in the case of fires, and given the high degree of deliberate intent cited as the cause of such disasters, they become a top-priority objective.

From this perspective, the direction in which we orient the communication and education process can be complemented by the analysis of public opinion and people's habits with the intention, amongst others, of detecting resistance to change. We can therefore highlight note the interest in obtaining a deeper analysis of public opinion for the following reasons: a) to clearly define the images, the difficulties and deficiencies, and the gaps that exist between knowledge, intention, and action; b) to create public opinion observatories as a means of systemising the monitoring of people's behaviour and the information that is generated, which is vital for the design of strategies for communication and the prevention of deliberate intent; c) to obtain relevant information and propose thought out avenues of communication; and d) to better define the global framework of the problem, namely spatial planning and appropriate development.

We must not forget that more knowledge does not necessarily mean more and/or better action, nor does it, on many occasions, produce an immediate change in people's behaviour. Forest fires in particular, as environmental disasters, display a complex and multidimensional structure with an impact on the territory and on the community that is explained by multiple factors of a technological, ideological, demographic, social, and economic nature (Oliver-Smith, 2004). There is also a series of intersecting psychological and social mediations that deserves greater attention. Changes in awareness are not immediately reflected in social or behavioural changes. Hence, the importance of ensuring that participation processes integrate not only the knowledge provided by experts, the opinions of the lay public, and those in positions of management responsibility must be recognised. They also promote a more active commitment to changes in practices.

Risk and Responsibility

For a better understanding of the framework in which a forest fire, seen as an environmental crisis, is contextualised, we can call on one of the most suggestive characterisations that has been put forward by the world of sociology: The risk society. Risk is seen as a characteristic feature of post-modern society, or "risk society" as Beck (1992) puts it. Although it might be going a little too far to consider risk as the element that best defines our society, it is true that risk is central in a large number of political debates relating to environmental, social, or health crises.

"Risk" is not the opposite of "safety" (Bechmann, 2004); rather, it should be defined in contrast to the concept of "danger." Danger is what a potential receiver of harm has to deal with, a receiver who is entirely lacking in both responsibility and decision-making capacity with regard to the origins and causes of the harm in question. Risk, therefore, is related to responsibility. Actions are multiple, at various levels and performed by a variety of bodies, and impossible to relate to a single point or moment; at the same time there is a collective responsibility for their outcome, or what has also been referred to as an "organised irresponsibility" (Beck, 1992). The framework of risk is therefore a question of the attribution of responsibility to the numerous prior human actions that generate consequences that can, at times, be neither imagined nor initially foreseen. This is the framework within which we should understand the changes in social perception detected in the population after the forest fires that affected extensive areas of Galicia (NW Spain) in 2006.

Social Participation

A key issue to take into account when analysing these changes is to determine the way in which public participation should form part of the context of collaborative risk management to achieve the required objectives of reducing the threat. This participation should be seen as a way of structuring society by integrating proposals put forward by its members through their commitment to and involvement in the decision-making process. Informing and educating the population is a necessary prior step to ensure participation and involvement in addressing social and environmental problems. Encouraging debate about problems affecting the territory and promoting a culture of appropriate forestry management not only serves to reduce the risk of forest fires, but also leads to an obvious benefit when it comes to tackling other problems such as floods, town-planning issues, or the sustainable revitalisation of rural economies.

In the literature dealing with the social management of risks and the relationships between science, technology, and society (Frewer, 1999; Rowe & Frewer, 2000), a number of methodologies have been used to encourage social participation. They take the form of deliberative procedures, with different degrees of linkage, which can either be decision-making or merely consultative (public hearings, surveys, negotiated management, citizen and expert panels, consensus meetings, citizens' advisory committees, amongst others). Through these procedures, people can obtain information, debate and express their opinions, recommendations, and considerations, and reach consensuses and joint decisions, according to the degree of independent identity of the procedure chosen. The initial degree of education and information required of the public varies, but there can be no doubt that all of these methodologies provide and favour access to new information, reflection, and the rigorous assessment of proposals.

Amongst the most widely used procedures are the consensus meeting and the citizens panel, both with roots in the Danish experience, which enable citizens to familiarise themselves with issues arising from the application of technical and scientific innovations. They have brought the problems surrounding the applications of technological and scientific advances closer to the general public, and have endowed socio-political participation with a totally new concept (López Cerezo, Méndez Sanz, & Todt, 1998). However, if there is a single procedure that has found widespread favour and thus acquired a certain value as the basis for decision-making, this has to be the opinion survey, used by Eco-barometers as a participative instrument and in studies for obtaining information about environmental perceptions, attitudes, and behaviours. The study of public perception thus becomes an extremely valuable tool, not only for orienting education, but also for designing effective communication strategies that will help to improve the social prevention of forest fires (Barreiro Rivas, 2007; García-Mira, García González, & Barreiro Rivas, 2008).

A Survey on the Social Perception of Fires

Within this framework, and from the perspective, mentioned above, of analysing public perception of the environmental problems caused by the wave of forest fires that swept over Northwest Spain (Galicia) in 2006, we designed a study to answer the following research questions: a) To whom do people attribute the causes and origin of forest fires? b) What characterised the situation that arose after the fires, and what damage did people perceive, both from an individual and social point of view? c) What was the extent of the response provided by the authorities, public institutions, and social agents? and d) What degree of trust was shown by the general public in institutions and the perceived efficiency of the relevant legislation?

Meetings were arranged with other research groups that were investigating the disaster from other standpoints. Participants in these meetings were drawn from a variety of fields and specialist areas, as well as politicians and stakeholders, and debate focused on a range of aspects relating to the damage caused, the perceived causes, and the measures adopted. The risk analysis performed from this technical, social, cultural, and educational perspective generated greater knowledge not only about the impact of the disaster or the characteristics and danger of the threat, but also about the vulnerability of the territory and the measures and decisions taken to face the latter.

The information generated in this way was used as the basis for producing a tool, in the form of a questionnaire, that was administered to a representative sample of the population, consisting of 651 persons over the age of 18 and living in Galicia in the north-west of Spain (see García-Mira et al., 2008), with a view to obtaining data that would be relevant to the proposed research questions.

Responsibility and Crime

Until very recently, the debate over how to tackle the problem of forest fires has focused more on the scarcity of fire-fighting resources, whereas discussions of the most appropriate ways of achieving more effective control of vulnerable areas, or scrutinising their management more efficiently, were of secondary importance. As we shall see below, the perceived origin of the cause of the fire is basically the hand of man, caused wilfully for financial reasons or to force a change in land use, which indicates that investing funds in analysing

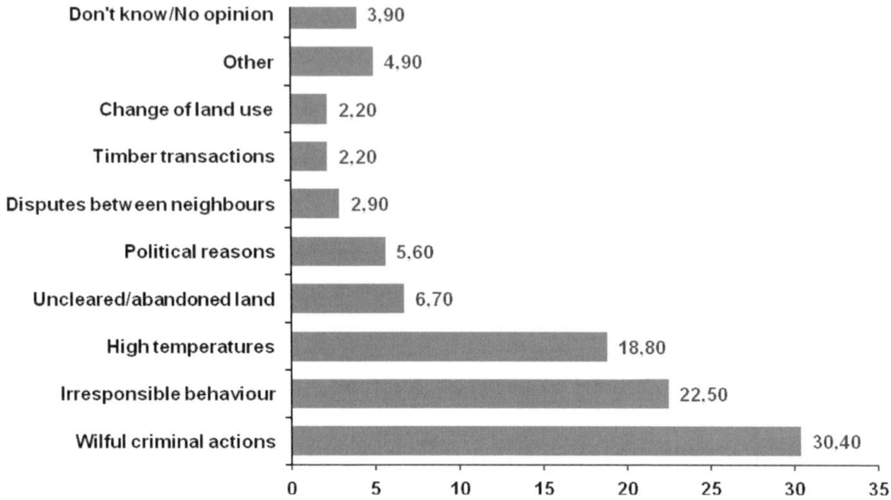

Figure 1. Attributed responsibility of forest fires in percentages of respondents. (Reprinted from R. García-Mira, C. García González, & X. L. Barreiro Rivas, 2008, p. 192, © 2012 Nova, Reproduced with permission)

the circumstances favouring fires, in an effort to prevent them, is much more productive than using the money to extinguish them. In this regard, educational and public prevention campaigns have proved to be quite effective. This notwithstanding, the control of criminal actions continues to be essential.

An initial analysis of the results of a previous study (see García-Mira et al., 2008) indicates that the main causes attributed by the general population to forest fires were wilful criminal actions (30.4%) or irresponsible behaviour (22.5%) (see Figure 1). Other factors identified by the Spanish public in 2004 as being the main cause of forest fires, such as a change of land use for speculative reasons or timber transactions, did not achieve this status in our more recent study. Other studies have also included financial interests as one of the major factors underlying forest fires (see Varela Mallou, 2005). Land transactions and financial interests linked to property development, together with timber transactions, were formerly considered to be significant causal factors. Nevertheless, there has been a growing trend to attribute forest fires to wilful criminal actions because, as Doiz Reus (2005) points out, these attributions are mutually compatible.

At the same time, when asked to choose who they considered to be mainly responsible for forest fires from a list of official bodies, groups, or social agents, the first choice of those surveyed was the regional government, with 20.9% of subjects opting for this attribution (see Figure 2).

Our survey also revealed an evident sensitivity to issues linking land management and risk, which would make it easier to steer public debate on spatial planning and changes of

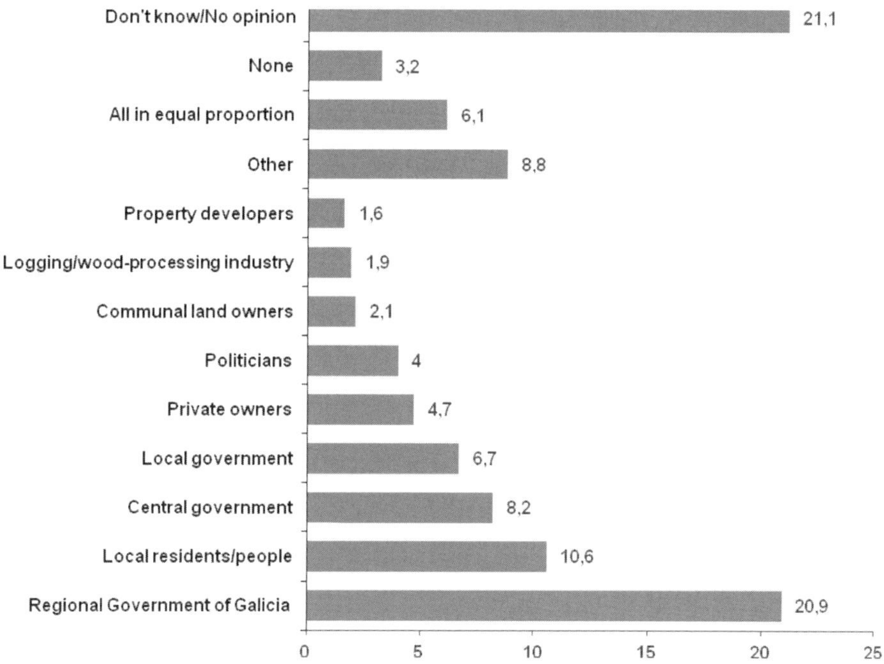

Figure 2. Perceived responsibility for the situation experienced after forest fires in percentages of respondents. (Reprinted from R. García-Mira, C. García González, & X. L. Barreiro Rivas, 2008, p. 194, © 2012 Nova, Reproduced with permission)

land use, thus reducing combustibility and the wilful starting of fires. A further aspect reflected in the findings of this survey was a better understanding by the population at large of the complex network of practices and situations regarding forested land. These are structural aspects that impinge on the issue and were previously not apparent.

Proximity of Fires to Residential Areas: A Lived Experience

Recent urban development has had, as a first implication, a reduction of the wooded surface area, which in turn leads to an increase in the risk of fires and the difficulty of extinguishing them, with the resulting harm to people and property. The proximity of residential environments to forests has thus increased their own very vulnerability as well as that of wooded land, making management even more difficult. Furthermore, such residential areas lack self-protection measures, which again greatly increases their vulnerability to forest fires.

This real increase in vulnerability for the population was apparent in the results of our study, and it marked a turning point in the subjective perception of associated risk, not only because of the number of fires that had occurred or the acreage of woodland destroyed by fire, but precisely as a result of the proximity of fires to residential neighbourhoods. Most of those interviewed indicated that one of the most striking features of the situation was the greater exposure of people and property to fire. Their responses were either the consequence of personally experiencing the real proximity of the flames or of news media coverage. The result was a greater sensitivity in the minds of the general population, even though the surface area destroyed by fire in 2006 was less than in other years.

From a territorial standpoint, the perceived damage appeared to affect the ecosystem above all, although harm done to the lives of those living near the fire-hit areas was also seen as one of the main reason why people felt very severely affected. This trend, namely that of relating the damage done by fires not only to the ecosystem, but also to the persons living within it, appears to be progressively increasing. An anthropocentric concern over the impact of forest fires must thus be added to the existing ecocentric one (see García-Mira, Santos, Gómez-Durán, Romay, & Fernández, 1998).

From a psychological standpoint, the impact of forest fires has been studied in relation to the emotions they arouse and the stress they activate in those potentially affected. In its more clinically oriented version, this impact has been studied in relation to the harm experienced by victims, to the interaction between victims and those involved in fire-fighting operations and, in general, in relation to the resulting traumatic and post-traumatic experience, which may lead to a comprehensive strategy for individual or collective action.

From the point of view of the individual, and of the perceived psychological impact, the feelings most commonly referred to by those interviewed were sadness (almost 30%) and indignation (again, almost 30%), followed by nervousness (18%), and fear (10.8%). Superficially, no significant differences in psychological perception were observed between those living in rural, semi-urban, or urban environments, with varying levels of education and relationships to forested land, in terms of percentages attained when valuing the impact, incidence, and causality of forest fires.

The Response of the Authorities and Other Social Agents: Information and Trust Processes

In this dimension, confidence and credibility are two essential factors for understanding the perspectives of the population with regard to the environmental risk posed by forest fires (see García-Mira & Lema Blanco, 2007; García-Mira, Real, Uzzell, Blanco, & Losada, 2005; García-Mira, Real, Uzzell, San Juan, & Pol, 2006). Confidence depends on cultural and economic circumstances, experiences, and needs. Subjective factors (personal experiences, personal involvement) and circumstances such as living in neighbourhoods threatened by fires, where the population lives in a situation of real environmental risk, can make some residents extremely distrustful of authorities. Nevertheless, studies

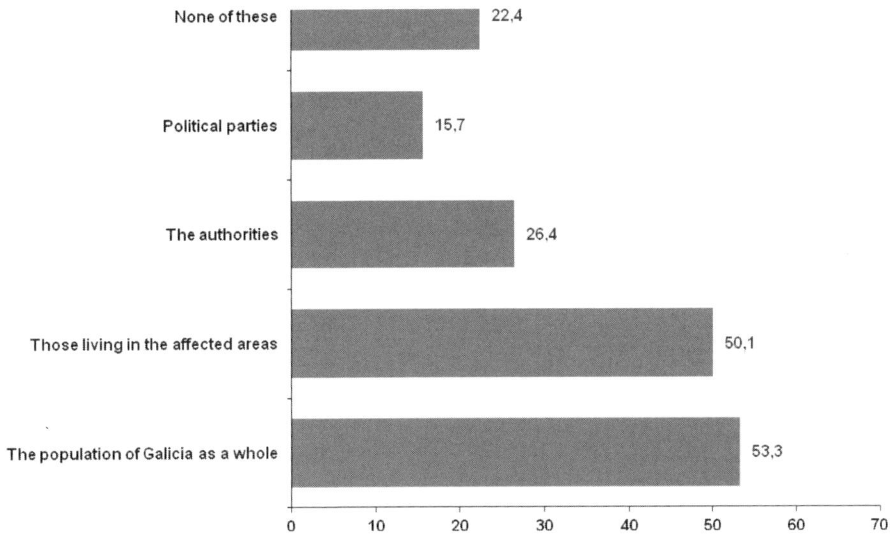

Figure 3. Adequacy of the response given by different actors in percentages of respondents. (Reprinted from R. García-Mira, C. García González, & X. L. Barreiro Rivas, 2008, p. 198, © 2012 Nova, Reproduced with permission)

such as that of Williams, Brown, and Greenberg (1999) indicate that citizens show great trust in the ability of scientists to resolve dangerous or risky situations, which suggests that they strongly support using scientific competence in resolving conflicts relating to risk resulting from fires.

In the study reported here, interviewees considered the response of local residents and the general population as more adequate than that of the authorities (see Figure 3), which reveals the widespread low level of trust in public policies or actions taken by the authorities. This positive assessment of popular response may be linked to the growing commitment to volunteering and social mobilisation that has developed hand in hand with recent environmental risks and crises.

The low level of trust in authorities is confirmed by the respondents when asked about their level of trust in different groups and institutions. They manifest that the degree of trust in local residents is higher than in institutions at national, regional, and local levels (see Table 1). A high level of trust is also placed in ecologists, undoubtedly due to the greater sensitivity and awareness induced by the campaigning done by these groups, which promote a specific value of solidarity and trust.

Our study also revealed that citizens award a low rating to the effectiveness of, and thus a low degree of trust in, legislation in the fight against forest fires. Further work is therefore needed to determine exactly at which aspects of law-making this negative assessment is directed.

Table 1. Degree of trust in different organisations and groups

	No trust	Some degree of trust	Quite a high degree of trust	A lot of trust	Don't know/ No opinion
The Spanish government	31.3	42.4	15.7	2.5	8.1
The Galician government	24.6	45.2	21.8	3.1	5.4
Local councils	27.6	45.2	20	2.8	4.5
Local residents, the population in general	8.3	31.2	**44.4**	**12.7**	3.4
Ecologists	16.3	36.6	**31.5**	**10.9**	4.8
Private land owners	16.7	45.5	26.6	5.7	5.5
Communal land owners	16.3	45	23.8	3.7	11.2

(Reprinted from R. García-Mira, C. García González, & X. L. Barreiro Rivas, 2008, p. 199, © 2012 Nova, Reproduced with permission).

As far as communication is concerned, this study, demonstrated that television was the main source of information about the fires (47.7%), followed by the press (33.9%). The quality of the information received was rated favourably, with 48.4% of those interviewed considering it to be "good," although, when it came to rating the information dealing with specific aspects, only 30% gave it this rating.

The way in which events are perceived is linked to how the mass media present them. In this regard, the greater sensitivity of the population to environmental issues, particularly those that imply a degree of risk, is related to the presence of such issues on TV and in the press. Visualisation of catastrophes is as important as the debate and discussions that raise public awareness of the complexity of the issues surrounding risk.

Finally, an invitation to think about the kind of measures that should be adopted, from a citizen's perspective, revealed that the largest number of respondents (27.8%) demands greater surveillance and police involvement; this is consistent with the attribution of responsibility for forest fires to criminal actions (see Figure 4). A similar percentage of interviewees call for greater prevention (27.5%), whilst a very limited role (7.2%) is given to education and information, which one would have thought to be essential for bringing about a change in the processes by which trust is built and conferred.

Discussion

In the light of the results presented, there are a number of aspects worth stressing because they constitute important lessons with implications for the field of risk policy and risk

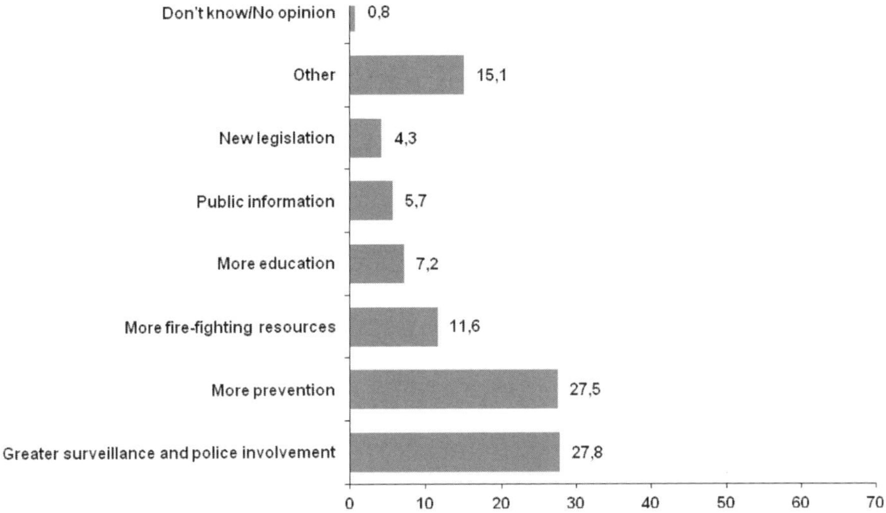

Figure 4. Relevant measures to prevent damage caused by forest fires, from a citizen's standpoint in percentages of respondents. (Reprinted from R. García-Mira, C. García González, & X. L. Barreiro Rivas, 2008, p. 203, © 2012 Nova, Reproduced with permission)

management. One of them has to do with a process of singular importance in crisis management: Trust related to communication and information management processes. A second aspect has to do with the attribution and assumption of responsibility, not only from the technical perspective of emergency management, but also from a socio-educational perspective. Finally, some recommendations are made on how to improve task coordination and social participation.

Social Perception, Trust, and Communication

Citizens are becoming increasingly critical and demanding of the authorities. With regard to the degree of public trust in the way forested land is managed, the results of our analysis revealed a major lack of confidence, above all in the authorities. Respondents were less critical and demanding when it came to the roles of local residents and land owners in the response to forest fires. Trust is one of the most important aspects in the environmental management of crises caused by forest fires. Given its dynamic nature, trust has a wide range of determinants, derived not only from perceived experience and knowledge, but also from the degree of information made public, honesty, openness, and the awareness of the issue shown by the media (see Peters, Covello, & McCalumm, 1997; Williams et al., 1999). In our study, even though the perceived quality of the information contained in the emitted crisis messages was generally considered to be good or adequate, the degree of trust was nevertheless affected. Let us briefly recapitulate the requirements of crisis communication theory. Apparently, the crisis messages were honest and transparent, the response was immediate, and there appeared to be no loss of control over the messages, in the sense that there were no contradictory sources. So, why was trust affected? The building up of public trust in institutions at times of crisis is important, because it is linked to the very perception of risk, which determines how people react to forest fires. Trust in institutions and their products decreases the perceived risk (Frewer, 2003), and is of particular importance in those circumstances in which people feel they have little control over any potential damage (e.g., fires that are fanned by high temperatures or wind).

Peters et al. (1997) hold that the role of institutions and the information received by the public with regard to risk are major determinants of perceived trust. Many people who belong to the disadvantaged socio-economic groups do not perceive those responsible for managing environmental risks as people who work to protect their interests (Petts, Horlick-Jones, & Murdock, 2001), and this finding can be generalised to include forest fires.

As we pointed out previously (García-Mira et al., 2008), a key factor for achieving a greater trust on the part of society is the communications strategy adopted by the institutions and organisations managing any crisis. People prefer to have information that they can use to determine the personal relevance of a given risk (Frewer, 2004). Regardless of the strategy adopted, if trust is to be established, the person or persons in charge of communication must respect three basic principles that apply to any communications strategy: 1) *Credibility*: Since we are dealing with official information, there must be a relationship of trust between the source of the information and the target (e.g., the general public) receiving it, 2) *Context*: The characteristics of the environments in which the information is transferred must be respected, 3) *Continuity*: Public information should not be an isolated or occasional event. Permanent, ongoing links need to be set up between the authorities and the general public or the target audience.

Furthermore, when designing communication campaigns, one of the factors that needs to be taken into account is time, that is, when communication with the public should be established. In this respect, communication initiatives can largely be categorised as being reactive or proactive (McComas, 2004). *Reactive communication* occurs when the public or the mass media focus their attention on a specific topic, for instance, in response to a sudden outbreak of fire. *Proactive communication* calls attention to a potential or existing risk and proposes an agenda for discussion, makes recommendations and suggests a format for the exchange of information. In addition, proactive communication can also alert the public to a risk of which people were unaware, which may be a disadvantage, particularly in the case of risks that pose no significant threat to human health or the environment.

Responsibility, Prevention, and Education

The finger of accusation in the case of the Galician fires of 2006 points to the hand of human beings, basically to criminal actions or carelessness, with the responsibility for prevention then being seen to rest with the authorities. There is a demand for further surveillance and policing measures, in order to combat criminal behaviour, and further preventive policies are seen as necessary. However, in the case of the fight against forest fires, there seems to be little appreciation of the responsibility of society as a whole and the importance of raising not only the awareness, but also the level of commitment of the population in order to initiate the changes required to achieve risk prevention.

When people identify possible causes, such as willful criminal actions or negligence, they focus their attention on what seem to be immediate causes of forest fires. As we have seen, less attention is given to the long-term causes, which can include change of land use, timber transactions or the fate of uncleared/abandoned land. Also, under the category of "others," respondents did not name any of the elements considered as part of the causes by environmental associations, such as the lack of self-protective measures or the inflammable nature of the majority of species of plants used in landscaping, both of which make houses more vulnerable to the impact of fire. These are more distant causes, belonging to a complex pattern of causality; various factors are involved in generating forest conditions that make forest fires more likely, or at least enhance the impact of willful criminal acts, by contributing to more disastrous fires.

Social psychology has taught us that, unless thoughts about complex causality are elicited through coordinated and coherent structures of social participation, people tend to think more economically and spend less cognitive energy in conceptualising issues. This is a possible explanation for the lack of understanding of the complex causes of forest fires, considering that the majority of the sample has close personal experience with both forests and fires, as land-owners (arable and forested), and has also have been exposed to many fires. Also, causes such as the change of land use or the lack of care for the land, that create the conditions that make fires more likely, refer to processes that have developed over time and require an understanding of both the cumulative effect of such conditions, and the complex interactions that generate a situation of fire vulnerability.

In spite of the apparent lack of understanding of these complex causes and of the interactions between them, we also see that, when asked about the relationship between the manner in which forested land is exploited and fires, a high percentage of respondents considers that these are related. This points to the possibility that people actually understand these complex causes, but they have not had a chance to actually construct an appropriate mental model, in spite of their exposure to them.

In other words, it is possible that they know what the causes are, and how they interact, and they activate this knowledge when asked by the interviewer, but they did not have a chance to actually fully articulate this knowledge before. This is consistent with the lack of effective mechanisms for social participation.

Finally, when talking about the measures needed to prevent fires in the future, very few respondents consider that there is a need for education or for further information. This is consistent with the fact that people might not have an activated understanding of the complex causes and of the interaction between natural and human factors in generating the conditions for rapidly spreading and disastrous fires.

Improving Coordination and Participation

Communication in crisis situations requires a high degree of flexibility and swiftness of response in order to respond to the demands of the media, as well as a particular sensitivity towards individuals or groups negatively affected by the situation. Whenever a serious and negative event or unforeseen situation arises, there is a tendency to make certain mistakes in institutional communication, resulting in the loss of trust and credibility on the part of the society concerned.

In this regard, although in the initial stages of the crisis situation there is little information or quantifiable data available, information should always be provided about what is known and about the decisions made. Care should be taken to avoid speculating with undocumented hypotheses; what is needed is a response that should be swift, but considered, especially when rumours or erroneous information abound.

An oft-repeated mistake in these cases is to apply a policy of "news blackout." If the authorities do not provide the necessary information, the mass media will try to obtain it from other sources, with a consequent loss of control over the message by the authorities. This, however, was clearly not the case for the way in which information was handled in the forest fire crisis of 2006, when the efforts of the authorities, mindful perhaps of how previous crises had been managed, may well have exceeded the level of attention and information that any citizen could reasonably demand.

The question arises, in the present case, about how public participation in the process of managing the risks deriving from forest fires could be structured. If the associations of communal owners of forested land, interest groups in the forestry sector, and experts already participate regularly in discussion forums with the authorities, then it would seem to make sense to include the opinions of other social groups and members of the general public concerned about the future of forested land, by making them participants in the process.

Conclusion

While there is a tendency to approach disasters solely from a technical perspective, requiring the interventions of experts and public officials in charge of policy formulation, forest fires, and the risk they generate for the region's ecosystem and for human beings, also possess a social dimension. Addressing this type of disaster involves environmental understanding, active commitment of citizens and their participation in land management. The study we have described here aimed at analysing and evaluating the subjective perception of the citizens of Galicia (Northwest region of Spain) with regard to forest fires, and to explore the dimensions that explain that perception. Because there are significant differences between the perspectives of experts and those of lay people when it comes to forest fires (Slovic, 2000), bringing these two perspectives together could shed more light on the strengths and weaknesses of policies directed at forested land and could potentially contribute to the design of better policies to mitigate risk. As there are only a few studies that aim to investigate the social perception of forest fires, this study was designed to obtain a better understanding of the public perception of the consequences of fires and the damage they cause, of the attribution of responsibility for and causes of the disaster, of the current level of trust in public institutions, and of measures to be taken for prevention of future occurrences.

The educational system is responsible for establishing the framework in which members of the school community can gain awareness of the immediate problems of their environment. Schools have to enable a global approach to local problems, and in the specific case with which this chapter is concerned, this means an approach to forest fires in which preventive measures are placed within the context of sustainable management of the environment.

The educational approach to the problem of forest fires should take into account, amongst others, the following aspects (García-Mira et al., 2008): a) Encouraging reflection on territorial planning and organisation and risk prevention, with attention to all the facets involved: Legislation, awareness-raising, and changes in people's customs and habits, structural measures such as the organisation, planning of forested land and limitations on certain land uses; b) Promoting a culture of appropriate forestry and land management that will guarantee a reduction in the wilful starting of forest fires and the combustibility of forested land; and c) Covering all the elements that constitute risk, both technical and social.

Finally, in the framework of democratic societies, any measure taken to drive forward the processes of citizens' participation not only possesses the instrumental goal of reducing resistance and making it easier for the authorities to achieve their objectives, but also has the value of being a formula that enhances the cohesion of society and cements ideas of citizenship by means of a commitment to making decisions that affect the future of the population as a whole and of the environment.

Acknowledgments

The authors would like to express their gratitude to the Galician Committee for Culture (CCG), and to the Caixa Galicia Foundation. The work reported here is based on the results of a wider study on forest fires in the Autonomous Region of Galicia, promoted by the CCG and carried out under the direction of F. Díaz Fierros, X. Balboa, and X. L. Barreiro Rivas (2007).

References

Barreiro Rivas, X. L. (2007). O papel da comunicación na xestión de catástrofes: o caso Prestige [The role of communication in disaster management]. In J. M. Garrido Fernández & J. M. Lema Rodicio (Eds.), *¿Qué aprendimos de la catástrofe del Prestige?* (pp. 67–89). Santiago de Compostela: UIMP and "Arao" Foundation.

Bechmann, G. (2004). Riesgo y sociedad postmoderna [Risk and post-modern society]. In J. L. Luján & J. Echeverría (Eds.), *Gobernar los Riesgos. Ciencia y Valores en la Sociedad del Riesgo* (pp. 17–34). Madrid: OEI-Biblioteca Nueva.

Beck, U. (1992). *Risk society*. London, UK: Sage.

Díaz-Fierros, F., Balboa, X., & Barreiro Rivas, X. L. (Eds.). (2007). *Por unha Nova Cultura Forestal fronte aos Incendios. Informes e Conclusións* [Towards a New Culture of Forestry Management as a defence against fires. Reports and Conclusions]. Santiago de Compostela, Spain: Consello da Cultura Galega.

Doiz Reus, M. L. (2005, August). Percepción de la sociedad ante el problema de los incendios forestales [Society's perception of the problem of forest fires]. Paper presented at the summer school "Paisaje, Sociedad e Defensa contra Incendios." Universidad Internacional de Andalucia, Spain.

Drew, C. H., Grace, D. A., Silbernagel, S. M., Hemmings, E. S., Smith, A., Griffith, W. C., . . . Faustman, E. M. (2003). Nuclear waste transportation: Case studies of identifying stakeholder risk information needs. *Environmental Health Perspectives, 111*, 263–272.

Frewer, L. J. (1999). Risk perception, social trust, and public participation in strategic decision-making: Implications for emerging technologies. *Ambio, 28*, 569–574.

Frewer, L. J. (2003). Trust, transparency and social context: Implications for social amplification of risk. in N. Pidgeon, R. E. Kasperson, & P. Slovic (Eds.), *The social amplification of risk* (pp. 123–138). Cambridge, UK: Cambridge University Press.

Frewer, L. J. (2004). The public and effective risk communication. *Toxicology letters, 149*, 391–397.

García-Mira, R., & Lema Blanco, I. (2007). The role of information and trust in the process of risk perception. In E. Edgerton, O. Romice, & C. Spencer (Eds.), *Environmental psychology: Putting research into practice* (pp. 74–90). Newcastle upon Tyne, UK: Cambridge Scholars.

García-Mira, R., García González, C., & Barreiro Rivas, X. L. (2008). The social perception of the 2006 Forest Fires in the Northwest Region of Spain: Information, trust, education and participation. In K. H. Kiefer (Ed.), *Applied Psychology Research Trends* (pp. 183–210). New York, NY: Nova Science.

García-Mira, R., Real, J. E., Uzzell, D., Blanco, G., & Losada, M. D. (2005). Exploring cognitive representations of citizens in areas affected by the Prestige disaster. In B. Martens & A. Keul (Eds.), *Exploring social innovation: Planning, building, evaluation* (pp. 137–145). Göttingen, Germany: Hogrefe.

García-Mira, R., Real, J. E., Uzzell, D., San Juan, C., & Pol, E. (2006). Coping with a threat to quality of life: The case of the "Prestige" disaster. *Revue Européenne de Psychologie Appliquée, 56*, 53–60.

García-Mira, R., Santos, M. C., Gómez-Durán, B., Romay, J., & Fernández, M. A. (1998). Evaluación de las actitudes proambientales en trabajadores de grandes organizaciones [An assessment of environmentally-friendly attitudes among employees of large organisations]. *Revista de Psicología Social, 13*, 435–444.

López Cerezo, J. A., Méndez Sanz, J. A., & Todt, O. (1998, March). Participación pública en política tecnológica. Problemas y perspectivas [Public participation in technology policy. Problems and perspectives]. *Arbor*, CLIX, 627, 279–308.

McComas, K. (2004). When even the "best-laid" plans go wrong: Strategic risk communication for new and emerging risks. *European Molecular Biology Organization, 5*, 61–64.

Oliver-Smith, A. (2004). Theorizing vulnerability in a globalized world: A political ecological perspective. In G. Bankoff, G. Frerks, & D. Hilhorst (Eds.), *Mapping vulnerability: Disasters, development and people* (pp. 10–24). London: Earthscan.

Peters, R., Covello, V. T., & McCalumm, D. B. (1997). The determinants of trust and credibility in environmental risk communication: An empirical study. *Risk Analysis, 17*, 43–54.

Petts, J., Horlick-Jones, T., & Murdock, G. (2001). *Social amplification of risk. The media and the public* (Research report 329/01). London, UK: Health and Safety Executive.

Rowe, G., & Frewer, L. J. (2000). Public participation methods: A framework for evaluation. *Science Technology and Human Values, 25*, 3–29.

Slovic, P. (2000). *The perception of risk*. London and Sterling, VA: Earthscan.

Varela Mallou, J. (2005). *Percepción social dos incendios forestais* [The social perception of forest fires]. Unpublished study, Department of Methodology and Social Psychology, University of Santiago de Compostela, Spain.

Williams, B. L., Brown, S., & Greenberg, M. (1999). Determinants of trust perceptions among residents surrounding the Savannah River Nuclear Weapons Site. *Environment & Behavior, 31*, 354–371.

Housing Development and Vulnerability Reduction in Informal Railway Settlements in Bangkok

Rittirong Chutapruttikorn[1] and Peter Kellett[2]

[1]Faculty of Fine and Applied Arts, Bangkok University, Thailand
[2]Global Urban Research Unit, School of Architecture Planning and Landscape, University of Newcastle upon Tyne, United Kingdom

Abstract

Conditions in informal settlements located along the railway tracks in Bangkok present the residents with a variety of risks. This paper focuses on the changing situation of two squatter communities that have registered as part of a large scale, official re-housing programme called Baan Mankong, which has led to significant improvement in their housing conditions. This, in turn, has reduced their vulnerability to physical dangers, poor health and even social exclusion. However, the costs of the upgrading programme have also increased the long-term debt burden, and the complexities of the housing design processes have created other vulnerabilities at both household and community levels. This paper draws on detailed qualitative data gained from close interaction with the communities and charts the changes throughout the key stages of the re-housing process. It examines the participatory processes of design and planning, identifies the main challenges and risks associated with the programme, and concludes with the implications and lessons of these complex physical and social processes.

Key words: collective actions, informal settlements, participatory upgrading, risk reduction, vulnerability

Introduction

Large numbers of low-income residents in Bangkok have created their own illegal, informal communities, but many of them are now faced with eviction and at least 200,000 households are affected (CODI, 2008). The majority of these are squatter settlements, many of which are located precariously along the railway lines and on adjacent land that belong to the State Railways of Thailand (SRT), which is seeking to maximise its property assets (Chiengtong & Witayasomboon, 2009). Despite their limited resources, these communities are well organised and have formed a city-wide network to negotiate with the landowners and to work with the official Community Organisation Development Institute (CODI). Together, they are implementing a government-initiated housing programme (Baan Mankong – "Secure Housing"), to achieve improved housing on sites which are usually close to the existing settlements. Although the programme is designed to be participatory, it is a complex process in which the demands, knowledge and expectations of the squatter communities do not easily map onto the resources and procedures of the official agencies.

The underlying causes of vulnerability in urban area are linked to the unequal distribution of resources as well as to exclusion from urban planned and services, which forces people to live on unsafe sites, often in overcrowded low-quality housing (Hamza & Zetter, 1998). Vulnerability is hence directly related to "complex systems of socio-political, institutional and economic processes which lie at the heart of urban" risk (Wamsler, 2006). The precarious situation of many existing railway communities makes them vulnerable to a range of risks – the most obvious are the physical dangers associated with the trains. In addition, their current illegal status means that infrastructure services are limited, which leads in turn to greater health risks. Many residents are therefore positive about opportunities which should lead to improved material conditions and hopeful that such changes will also improve their low social status.

In an urban studies approach, vulnerability is defined as "insecurity and sensitivity in the well-being of individuals, households and communities in the face of a changing environment" (Moser, 1998, p. 3), which means that people's responsiveness and resilience can be challenged during the process of change. To achieve the new housing conditions, the informal residents must engage with complex bureaucratic procedures and negotiations as well as financial management and housing design participation. The financial aspects of the programme are difficult for residents to understand and involve substantial loans that may be beyond the economic capacity of many beneficiaries. This can lead to problems of long-term debt and increasing risk from the additional stresses of repaying the loans. Furthermore, they are especially vulnerable when it comes to the details of housing designs, as they are unfamiliar with technical jargon and unable to read architectural drawings. These constraints mean that their participation can be compromised.

To explore these issues, the paper draws on a range of qualitative data obtained through intensive fieldwork in several railway communities over a 6-month period and over 2 months the following year. The first author adopted a collaborative "Action Research approach" (Stringer, 1999) and worked closely with the participants as an architect and technical advisor. Fifty interviews and 8 focus group discussions with residents and officials were recorded, plans of existing and planned dwellings were drawn and key documentation obtained from various NGOs and official organisations. The paper identifies the key elements of the programme and offers a critique of the participatory approach adopted. It also examines how design decisions were made and analyses the resulting architectural and planning solutions from a social and economic perspective. Moreover, the paper aims to examine how the mechanisms of the housing programme impact on vulnerability and risk reduction, and seeks to identify lessons for improving the situation of vulnerable communities. These can be expressed as questions: To what extent does the new housing help to reduce vulnerabilities and risks? Does the re-housing programme create other vulnerabilities for residents, and how can such risks be mitigated? The paper firstly explains the participatory upgrading programme and briefly introduces the case study settlements. It then examines the range of risks associated with housing, including physical and health risks; issues of vulnerability associated with low social status, the increasing financial risk of higher cost housing and, finally, explores the relationship between vulnerability and the housing design process.

A Participatory Upgrading Programme

Bangramard and Chonglom are two railway line communities where residents live illegally on land which belongs to the State Railways (SRT), which makes them vulnerable to eviction. Land values are increasing and the SRT is planning to reduce its high debt burden by creating development projects as well as leasing land to private companies (Thansettakit, 2008). Such projects inevitably affect a large number of residents. For example, 159 Chonglom households have been evicted to make way for a proposed city pumping station, and the 32 households of Bangramard have been moved to create a parking area for tourists.

Over ten years ago, both settlements gathered together with other low-income communities to demand their right to adequate living conditions, and successfully managed to negotiate a 30-year lease of land close to their original settlements. Throughout their history of fighting, acquiring temporary house registration and security of tenure have been their crucial successes. The house registration ensures that their children can enrol at school and the community can begin to instal proper infrastructure (Chantarapa, 2009). Security of tenure not only confirms their right to live in the city (Durand-Lasserve, 2006), but also protects them from the vulnerability of summary eviction, which can be violent and very destructive.

Since the United Nations promoted increased community participation and local control in settlement upgrading (UN-Habitat, 2003), CODI have adopted a participatory approach to project implementation. The ambitious Baan Mankong programme aims to strengthen community organisations through increasing their autonomy and reducing their dependency on outside resources. The key to this is social capital. Many scholars have indicated that to build community in neighbourhoods, social interaction among community members should be developed, community resources strengthened, and collective and individual assets that can be identified as social capital should be promoted (Chaskin, 2002; Hyman, 2002; Saegert, 2004). In the programme, the inhabitants are encouraged to take a key role in project management, design, and decision-making. They collectively engage in collaborative activities such as savings groups, participating in design and

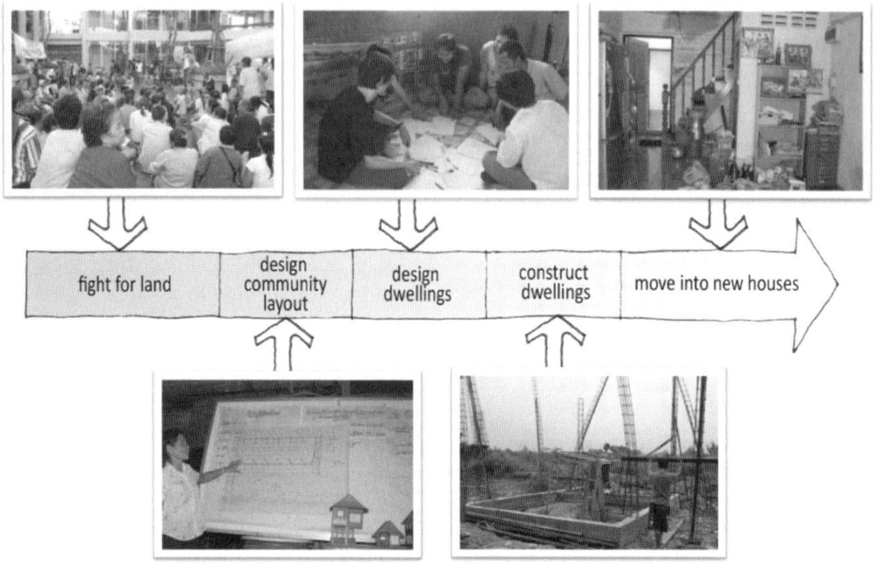

Figure 1. Process of housing development and design. (Source: the authors)

construction, and establishing community cooperatives. Such an approach is aimed at reducing "feelings of dependency [and building] *self-reliance, self confidence, and responsibility*" among the participants (McNeely, 1999, p. 742).

In addition, the residents are stimulated to use their own energy and creativity to build their future environments and participate in every stage of the improvement, beginning with a survey of property conditions and residents' requirements, then planning the community layout, designing the dwellings, and requesting building permission. They are usually supported throughout the process by facilitators/architects employed by CODI and sometimes by NGOs. The process is illustrated in Figure 1.

Housing Development and Risk Reduction

Informal housing is usually located on land considered unsuitable for formal development (Tipple, 2005), for example, areas prone to flooding or close to railway tracks, like at Bangramard and Chonglom. Tipple (2005) acknowledges that apart from their vulnerability to abusive evictions, the residents who live in such demanding living conditions also confront various risks and vulnerabilities to their physical and mental health. However, such vulnerabilities can be decreased when their housing is upgraded. These issues will now be discussed in turn.

Reducing Physical and Health Risks

Living close to the railway tracks places the dwellers in Bangramard and Chonglom at risk from train accidents. Additionally, both communities are settled in swampy areas. Bangramard was an old orchard where a large number of ditches had been excavated, whereas Chonglom is located on the edge of a canal which regularly floods at high tides. The contaminated water directly affects the residents' health because of increased risk of infection through water-borne diseases. Moreover, dwellers must keep a careful watch on their young children when they play near such areas. Several children in Chonglom have died from drowning.

Fortunately, such precarious conditions are changed when housing is redeveloped. The swampy land in Bangramard was levelled and the Chonglom community was relocated further from the canal. These are key steps in reducing the physical dangers. Likewise, an informant in another study investigating the Baan Mankong programme mentioned: "I am proud to have my own house. My children enjoy better living conditions. The environment is better. I don't worry that my children will be hit by trains or cars anymore" (Benchasup, 2005, p. 27).

The lack of good quality infrastructure (e.g., poor sanitation and unsafe water supply) is also harmful to the residents' health (Berner, 2000). In Bangkok, one third of all urban households rely on water from wells, while many of the urban poor are supplied through collection of rainwater or by polluted water in the canals (Daniere & Takahashi, 1999). Such poor water quality unavoidably impacts on all aspects of life (e.g., cooking, drinking, and bathing) and leads to many life-threatening infectious diseases, which gradually compromise individual hygiene and the domestic environment (Unger & Riley, 2007).

In Chonglom and Bangramard, the dwellers could not directly obtain formal infrastructure services because they had built their informal houses without permission and registration. However, a few families who had lived longer in the settlement managed to connect their utilities to the formal systems from neighbouring communities. Those who arrive later frequently connect to these electricity and water supplies. Although this informal option helps them to access safe water and other infrastructure services, it is very costly.

The condition of infrastructure is a key indicator of the outcome of low-income upgrading projects (Viratkapan & Perera, 2006). It was found that facilities supplied by public utility companies not only provide standard services and reduce vulnerability to poor health, but also decrease residents' excessive expenditure on informal infrastructures. One of the residents in Chonglom explained:

> I was thinking, now we pay far too much for water, for electricity - a thousand something (Baht) per month. If we were to pay the proper rate, it would probably cost only three hundred Baht. We will have lots money left over. We can use that money to pay for our housing loan. (Na pe in focus group, 18 September 2008)

Improving Vulnerable Social Status

To construct their informal dwellings, the residents frequently used substandard materials which they found or bought very cheaply. Such materials are often flammable and directly associated with accidental injuries such as collapses and burns (Unger & Riley, 2007). Moreover, the disorderly layouts and conditions can present access difficulties for ambulances and emergency vehicles.

Besides risks such as fire hazard, the physical characteristics of these environments can make their informal dwellings appear improvised, chaotic, and substandard. Thus, it is common for outsiders to view them pejoratively and as inferior, and the residents are frequently stigmatised with negative assumptions. Although conscious of their low social status and lack of respect, and even when they have the resources, dwellers lack the motivation to invest their money in physical assets because of the threat of eviction (Rhabibhadana, 2009). Similar to informal settlements in other countries, these communities are typically considered "shameful and unhealthy – not only for the inhabitants but also for the rest of the city" (Milbert, 2006, p. 302). This is reinforced by Kapstein Lopez (2009) who argues that social vulnerability is a direct consequence of a lack of material and symbolic resources that "create situations of deprivation, constraints on the degree of social participation and [a] psycho-social crisis of identity of the persons concerned" (p. 3) as well as social and economic segregation.

The new housing projects impact on dwellers' self-perception and play a role in creating new identities for these communities. A study by Rhabibadana (2007) focused on the alteration of squatters' worldview and attitude after upgrading their house under the Baan Mankong programme. He discovered that "having a permanent house significantly raises the community's status from the fringe, where it is called many contemptible names, to the developed community where its dignity is increased equally or more than other neighbourhoods" (pp. 2–3). Since the house can communicate its dweller's status (Senan, 1993), particularly in both physical and financial terms (Gram-Hanssen & Bech-Danielsen, 2007), the inhabitants take pleasure in their increased self-worth and feel that their status is significantly enhanced when their houses are better developed. The upgraded houses can also be "a way of rising up the social ladder [...] to acquire and conform to accepted tastes" (Dayaratne & Kellett, 2008, p. 60). This study found that new housing can replace the image of squalor of the squatter settlements, which not only leads to more respect in society but also to improvement in their social status, from being stigmatised and vulnerable to social exclusion to becoming included as full members of society.

New Housing and Increasing Financial Risk

The Baan Mankong programme insists on establishing strong financial systems for each project in which all participants must belong to a community savings group (Sopon, 2006). Combined with external subsidies, the finance "gives people a new power to

Figure 2. Participation meeting in a squatter's house. (Photograph: the authors)

change things" (CODI update, 2008, p. 3) and helps them to reduce their vulnerability to credit risks. However, this means that the participants have to take on large debts, which they must repay monthly within fifteen years. Some residents, whose savings were not sufficient, have been forced to borrow additional money from informal sources, at very high rates of interest, because the amount they are able to secure through a CODI loan is not enough for building a new dwelling (Onndum, Boonmatayat, Tulata, & Wongasa, 2007). Furthermore, since negative perceptions can lead to a sense of exclusion and cause informal dwellers to suffer low self-esteem, many Bangramard residents, like other low-income community dwellers, borrowed additional money from other sources to make their new houses as attractive as possible. They aim to definitively eradicate such pejorative images. These attempts to change their built environment not only signify that improvement of the housing helps to achieve social acceptance, but they also lead to increased status and self-identity (Holston, 1991).

Nevertheless, Chambers (2006) mentions that vulnerability differs from poverty because it is related to net assets. Poverty can be decreased by borrowing and investing, but "debt makes households more vulnerable" (p. 33). The burden of debt causes individuals more stress and means they have to work harder to gain extra income for their loan repayments. Their way of life tends to become more individualised, with diminishing engagement in collective actions and community participation which are key to many of the community's achievements. As one of the Bangramard leaders complained in a community meeting, "Now they've got their home it seems they are less interested in community works. They often disappear." The residents tend to be less united, spend longer hours working, and invest most of their energy addressing personal and family issues instead. Hence, taking a large loan not only increases family financial vulnerability, but also involves the risk of reducing community cohesion.

To decrease such vulnerabilities, some dwellers suggested that instead of the dramatic change from the informal dwellings to the new buildings, they could work on the new dwellings gradually while consolidating their funds. This would entail re-using materials from their informal dwellings and living in the dwelling during a longer construction period. Although such an approach would not address the stigma of a negative image within a short period, the process of consolidation could be more sustainable in the longer term.

It is also crucial that the local associations, including CODI and NGOs, should not emphasise only the financial system, but also address other issues that tend to create those vulnerabilities, in particular the trend of declining community cohesion. They could promote other collective activities that related directly to the residents' needs (such as income generation and community cooperatives), to maintain the strength of their collective actions.

Vulnerability in the Housing Design Process

A lack of understanding and of appropriate communication between the CODI architect and the residents in both Bangramard and Chonglom led to numerous difficulties in the participatory design process. Initially, Bangramard residents believed that if they let the CODI architect create the new housing design for them, it would be finished in a short period. They also believed that the application for building permission might be easier. The residents appreciated that such processes need to be done precisely and were complicated. Moreover, they had had many negative experiences with the local authority, because of threats of eviction in the past. Therefore, instead of using their collective power to confront this difficulty, they chose to depend on the CODI facilitator. They hoped that this could help them to avoid contact with officials and, because the local authorities and CODI agencies "are both government organisations it may be easier to negotiate" one resident commented. However, depending on the CODI designer was not as convenient as they thought.

The design meetings at Bangramard were arranged on four occasions, but the final design did not meet with the residents' requirements. The problems lay not in the design itself, but in poor communication. Due to the dwellers' inability to understand architectural drawings and technical jargon, what they envisaged regarding the house was not what the architect intended. The two sides failed to communicate effectively and the design revision process was problematic. Both the residents and the architect felt dissatisfied with the participation process and, consequently, the final design of the house did not match the design that they thought they had agreed upon. In the long run, every household ended up with a standard design and built their houses in the same shape and form, although diverse

Figure 3. The new housing in Bangramard: Each house is painted a different colour. (Photograph: the authors)

requirements were often presented during those meetings. However, now that most houses have been constructed, many residents are making great efforts to personalise their dwellings and make them distinctive (Figure 3).

In Chonglom, the residents also thought about the design of their future housing. As in Bangramard, they needed formal drawings to request residential planning permission. Nevertheless, without technical knowledge about how to produce the drawings and the accompanying information, they had to rely heavily on the CODI architect. Unfortunately, with over forty other communities to work with, the designer was unable to respond as the community had imagined. People had no choice but to wait, and construction has not yet started. Hence, the locus of participants' power in both communities was not under their control. In this sense, because it is hard for residents to cope with this situation, the vulnerability of the community has been increased through the housing design process.

While the number of housing units the CODI agencies have to deal with is large, the number of facilitators, particularly architects, is small.[1] Consequently a real participatory design process is difficult to achieve. Although the aim of CODI is to achieve "genuine participation," based on the ladder of participation introduced by Arnstein (1969) and developed by Choguill (1996), the evidence indicates that CODI agencies only simulate participation in tokenistic ways. We must conclude that unless the inhabitants have some

[1] In December 2008, there were only seventeen architects dealing with 1,826 communities (interview with Thongchai, head architect of CODI, December 2008).

control over decision-making, this cannot be regarded as "genuine-participation" (Luck, 2007; Toker & Toker, 2006).

To reduce such vulnerability, the first author volunteered to work with the residents, to find a solution to these problems. Based on discussions with NGOs and community leaders, we created alternative scale-models of the houses for Bangramard and arranged a design workshop for Chonglom, aiming to help the residents better understand the designs and to encourage them to realise their own capabilities and decisions. In turn, these collaborations led to a long and active period of dialogue and exchange of knowledge between the specific professional expertise of the author and the in-depth understandings and experiences of the local residents (Creswell, Hanson, Clark, & Morales, 2007). The outcome of this participation was that the Bangramard dwellers learnt to understand the architectural drawings and the Chonglom residents increased their confidence of engaging meaningfully in the design of their houses.

Conclusion: Knowledge is Power

Housing programmes have a significant capacity to reduce the vulnerabilities faced by low-income people, particularly the risks in their daily lives. The upgraded housing provides safer domestic environments and higher quality infrastructure, which supports the residents and helps to protect them from physical injury, illness and even accidental death. The permanent appearance of the new dwellings, which meets social norms of acceptability, is also an effective vehicle for declaring its occupant's new identity and hence facilitating the inclusion of such vulnerable people within wider society.

However, the burden of debt forces them to work harder and earn more money to repay their loans. This not only changes their way of life, from collective interdependence to increased individualism, but also weakens social cohesion and the collective dimension of communities. The complexity of bureaucratic procedures and the technical language used by the professional facilitators during the design process are other issues that make the residents more vulnerable to manipulation and leads to a loss of local autonomy. To cope with such problems, they learnt how to design the houses themselves. This leads to the conclusion that acquisition of knowledge is also an important strategy for reducing vulnerability. Therefore, it is worth reflecting on the ways these communities mitigated risks and reduced vulnerabilities; they are lessons for improving the situation of other vulnerable communities.

References

Arnstein, R. (1969). A ladder of citizen participation. *Journal of American Institute of Planner, 35*, 216–224.

Benchasup, C. (2005). *Baan Mankong Chumchon Sangsan Pattana 7–12* [Secured housing: Creative Development Community 7–12]. Bangkok: CODI.

Berner, E. (2000). Poverty alleviation and the eviction of the poorest: Towards urban land reform in the Philippines. *International Journal of Urban and Regional Research, 24*, 536–553.

Chambers, R. (2006). Vulnerability, coping and policy [Editorial Introduction]. *IDS Bulletin, 37*, 33–40.

Chantarapa, A. (2009). *เครือข่ายสลัมสี่ภาค ตัวตนและประสบการณ์การเคลื่อนไหว* [Four Region Slum Network: The self and experience of movement]. Bangkok: Lakthai.

Chaskin, R. (2002). The evaluation of "Community Building": Measuring the social effects of community-based practice. In A. Maluccio, C. Canali, & T. Vechiato (Eds.), *Assessing outcomes in child and family services: Comparative design and policy issues*. New York, NY: Walter de Gruyter.

Chiengtong, S., & Witayasomboon, I. (2009). *คนเล็กในเมืองใหญ่* [Small People in Big City]. Bangkok: Ruamtaweepon.

Choguill, M. B. G. (1996). A ladder of community participation for underdeveloped countries. *Habitat International, 20*, 431–444.

CODI. (2008). *คู่มือการจัดทำโครงการ บ้านมั่นคง* [Handbook for Project Arrangement of Baan Mankong]. Bangkok: Public Relations Section and Baan Mankong Office.

CODI update. (2008). *50 Community Upgrading Projects*. Bangkok: Community Organisations Development Institute.

Creswell, J., Hanson, W., Clark, V., & Morales, A. (2007). Qualitative Research Designs: Selection and Implementation. *The Counselling Psychologist, 35*, 236–264.

Daniere, A., & Takahashi, L. (1999). Poverty and access: Differences and commonalties across slum communities in Bangkok. *Habitat International, 23*, 271–288.

Dayaratne, R., & Kellett, P. (2008). Housing and home-making in low-income urban settlements: Sri Lanka and Colombia. *Journal of Housing and the Built Environment, 23*, 53–70.

Durand-Lasserve, A. (2006). Informal settlements and the millennium development goals: Global policy debates on property ownership and security of tenure. *Global Urban Development, 2*, 1–15.

Gram-Hanssen, K., & Bech-Danielsen, C. (2007). Housing and immigrants—Consumer aspects of the meaning of home. In P. Repo & E. Pylvänäinen (Eds.), *Proceedings of the Nordic Consumer Policy Research Conference*. Helsinki, Finland: Nordic Council of Ministers.

Hamza, M., & Zetter, R. (1998). Structural adjustment, urban systems, and disaster vulnerability in developing countries. *Cities, 15*, 291–299.

Holston, J. (1991). Autoconstruction in working-class Brazil. *Cultural Anthropology, 6*, 447–465.

Hyman, J. (2002). Exploring social capital and civic engagement to create a framework for community building. *Applied Developmental Science, 6*, 196–202.

Kapstein Lopez, P (2009, June). *Three Approaches to urban vulnerability: Social, physical and urban.* Paper presented at the international conference 'City Futures in a Globalizing World,' Madrid, Spain. Abstract retrieved from http://www.cityfutures2009.com/PDF/49_Kapstein_Paula.pdf

Luck, R. (2007). Learning to talk to users in participatory design situations. *Design Studies, 20*, 217–242.

McNeely, J. (1999). Community building. *Journal of Community Psychology, 27*, 741–750.

Milbert, I. (2006). Slums, slum dwellers and multilevel governance. *The European Journal of Development Research, 18*, 299–318.

Moser, C. (1998). The asset vulnerability framework: Reassessing urban poverty reduction strategies. *World Development, 26*, 1–19.

Onndum, B., Boonmatayat, C., Tulata, S., & Wongasa, V. (2007). *คนจนเมือง การเปลี่ยนแปลงโลกทัศน์และทัศนะที่มีต่อตนเองและสังคม กรณีศึกษาบ้านมั่นคง ขอนแก่น* [Urban poor: The change of worldview and attitude towards self and society, a case study of Baan Mankong Khonkaen] Bangkok: Unknown.

Rhabibhadana, A. (2007). *คนจนเมือง การเปลี่ยนแปลงโลกทัศน์และทัศนะที่มีต่อตนเองและสังคม* [Urban poor: The change of worldview and attitude towards self and society, the overview]. Bangkok: Unknown.

Rhabibhadana, A. (2009). *ชาวบ้าน ชาวเมือง ความยุติธรรมในสังคมไทย* [Local people, city people: Justice in Thai society]. Khonkaen: Wanida.

Saegert, S. (2004). *Community building and civic capacity* [Paper for the Aspen Institute Roundtable and Community Change]. Retrieved from http://www.aspeninstitute.org/sites/default/files/content/docs/roundtable%20on%20community%20change/CommunityBuildingCivicCapacity.pdf

Senan, Z. (1993). The house as an expression of identity: The case of the Palestinian house. *Forum, 2*(1), 3–10.

Sophon, P. (2006). *Social Capital and the Government Programme: A case study of Baan Mankong programme in Bangkok*. Bangkok: Thammasat University.

Stringer, E. T. (1999). *Action Research* (2nd ed.). Thousand Oaks, CA: Sage.

Thansettakit. (2008). *บอร์ดเปิดเค้กที่ดินรถไฟกว่า3 หมื่นไร่ทำเงิน* [The Board plan to make money from over 30,000 rai of the SRT's land]. Retrieved from http://www.thaicontractors.com/content/news/1721.html

Tipple, G. (2005). Housing and urban vulnerability in rapidly-developing cities. *Journal of Contingencies and Crisis Management, 13*, 66–74.

Toker, Z., & Toker, U. (2006). Community design in its pragmatist age: Increasing popularity and changing outcomes. *Journal of the Faculty of Architecture, Middle East Technical University, 23*, 155–166.

UN-Habitat. (2003). *The challenge of slums: Global report on human settlements*. London and Sterling: UN-Habitat.

Unger, A., & Riley, L. W. (2007). Slum health: From understanding to action. *PLoS Medicine, 4*, 295.

Viratkapan, V., & Perera, R. (2006). Slum relocation projects in Bangkok: What has contributed to their success or failure? *Habitat International, 30*, 157–174.

Wamsler, C. (2006). Tackling urban vulnerability: An operational framework for aid organisations. *Humanitarian Exchange, 35*, 24–26. Retrieved from http://www.odihpn.org/report.asp?id=2847

Shifting Local Vulnerabilities due to Housing Expansion

Case Study in the Colonial City of Seremban Municipality Area, Malaysia

Shaharudin Idrus,[1,2] Abdul Samad Hadi,[1] Abdul Hadi Harman Shah,[1,2] Ruslan Rainis,[2] and Ahmad Fariz Mohamed[1]

[1]Institute for Environment and Development (LESTARI), Universiti Kebangsaan Malaysia (UKM), Malaysia
[2]Geography Section, School of Humanities, Universiti Sains Malaysia (USM), Malaysia

Abstract

The process of urbanization in Malaysia has created conditions that have exposed urban residents and migrants to a myriad of new problems. These problems range from lack of affordable shelters and public amenities to insufficient economic livelihoods and a lack of a sense of belonging. These problems have made residents vulnerable to various shelter, health, safety, and social problems that undermine the city's efforts to provide better living conditions through housing. Yet, simultaneously with housing projects that are being designed and developed to solve existing problems, new problems arise that require a revision of earlier efforts, a symptom showing that housing programmes are not immune to the "wicked problem" discussed by Rittle and Weber (1973), Conklin (2001), as well as Brown, Harris, and Russell (2010), among others, where earlier urban problems metamorphose into new ones over time and space. This chapter presents such a situation in Seremban, Malaysia, where local vulnerabilities are shifting in form, rather than simply being remedied, despite planned efforts. We present a scenario of shifting vulnerabilities of the local residents over time, using analyses of housing expansion, derived from water consumption data, together with data on local problems of dengue and crime. Residents are considered to be vulnerable to various problems, due primarily to an inadequate supply of basic facilities, which subsequently leads to a shift towards poor health and crime. There has also been a shift in those who are vulnerable, from migrants and people on lower incomes who were among the initial vulnerable groups, to middle and upper income groups.

Key words: colonial city, housing expansion, local vulnerabilities, Malaysia, urbanization

Introduction

Housing is expected to contribute much to the liveability of the city. It is with this expectation that the planning of housing areas has taken centre stage in the urbanization process of many developing countries. Thus, development efforts focus on increased housing supply and improved housing conditions, assuming that this will lead to better communities. More than just a locational change, living in the city is an indication of progress and becoming modern in many developing post-colonial countries. Urbanization in these countries is therefore more than just the conversion of the rural to the urban; it is also an assumed path of progress for the locals, where the migration from the rural to the urban has always been a phenomenon to be expected and facilitated by the state. This has resulted in problems of housing the migrants, similar to problems of urban housing in developing countries that have been discussed by Turner (1967) and Tipple (2005), among others.

Vulnerability

Housing areas and their residents are vulnerable to a variety of problems. The plight of overcrowded housing areas, vulnerable to social and health problems, has been discussed early in the literature on housing (Myers, Baer, & Choi, 1996). There are several perspectives for looking at vulnerability and housing. The most common, perhaps, is housing for vulnerable people, including the poor (see for example Drakakis Smith, 1977), the elderly, and those with special needs. Current discussions of housing vulnerabilities also include public health perspectives (Lawrence, 2010). Settlements in Malaysia have become vulnerable due to various factors. Violence, diseases, natural and manmade physical hazards, and social unrest interrupt people's lives. The planning of residential areas often assumes that the factors which people are unable to deal with and that make them vulnerable will be mastered over time and with good planning. This paper argues that such linear thinking does not take new stressors and shocks that will replace the older stressors and shocks into account. The inabilities of the people themselves change over time, as they adjust to older stressors and are exposed to newer ones. In a sense, this is the "wicked problem" introduced by Rittel and Webber (1973). Just before independence, many residential areas were exposed to communist insurgency groups that attacked these fringe area settlements.

When this problem was solved, other problems that made individuals and entire communities vulnerable emerged. The classic definition of vulnerability itself refers to the capacity of a person to be "wounded," that is, the degree to which a system (or in our case an individual or a community) is likely to experience harm due to exposure to a hazard (Turner et al., 2003), but this term is now applied to a variety of other research contexts such as ecology, public health, poverty and development, livelihoods and famine, land use change, and climate impacts and adaptation (Fussel, 2007; Gow, 2005; Metzger, Rounsevell, Acosta-Michlik, Leemans, & Schröter, 2006; O'Brien et al., 2004). It has been shown that the ability of a system to cope with stresses and their consequences through various strategies or mechanisms constitutes a key determinant of system response, and ultimately, of impact on the system (Turner et al., 2003).

The primary research question inquires into whether housing efforts have reduced place and social vulnerabilities, or whether the efforts have replaced one type of vulnerability with another. The idea of housing is to reduce vulnerability, or at least exposure to the elements, if nothing else. The role of housing has since evolved and expectations have shifted from mere protection to the building of liveable communities.

Research Methods

Secondary data from various agencies (government and private) were the main source of information for the study. Large-scale temporal housing expansion data are difficult to access in Malaysia. Thus the study employed proxy data, using water supply and use data to trace housing growth. Since each individual household requires the installation of a water supply as a basic resource, besides electricity, housing growth through the physical expansion of residential areas as well as the actual occupancy of individual houses can be traced with this kind of material. Data were gathered from the planning section of the Department of Water Supply Division of the Seremban municipal area. Related household information was also gathered from water use reports. These include information such as ownership, the date when people enter their houses, and the ethnicity of the individual homeowner. From these data, the dynamics of housing development and expansion in the Seremban Municipal area were captured and later visualized through GIS-based housing maps of the area. Further data on crime and dengue were collected from the local police and health authorities, respectively. These were also mapped to show the spread of the problems and to identify locations of concern.

Housing Expansion in the Seremban Municipal Area

The Seremban municipal area (Figure 1) is located in the Linggi basin at the southernmost point of the extended mega urban region of the Bernam-Klang-Langat-Linggi basins.

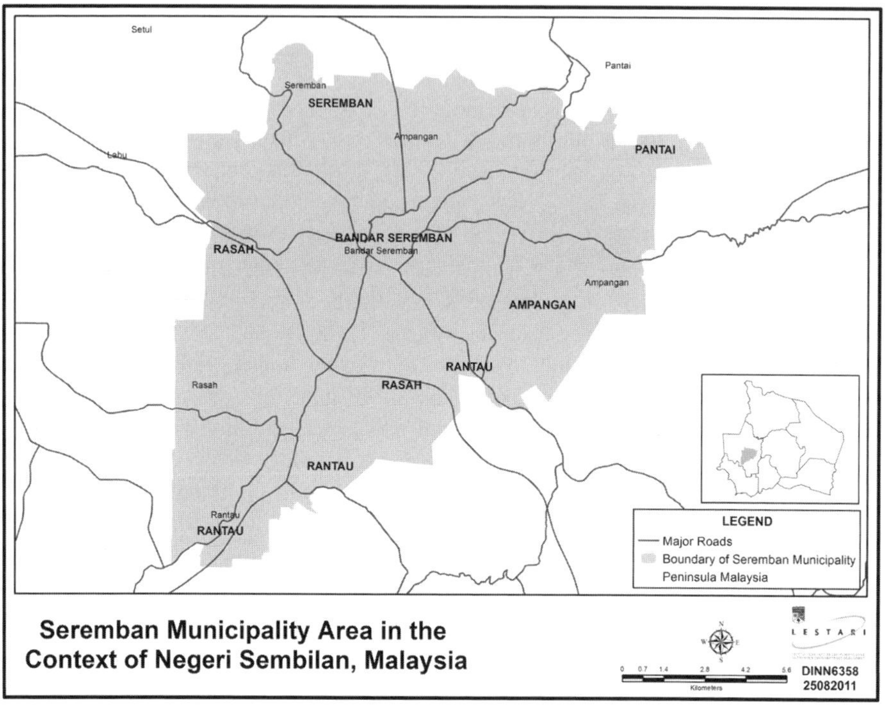

Figure 1. Seremban Municipality Area in the context of Negeri Sembilan, Malaysia. (Source: the authors)

The Seremban municipal area is an intermediate city with an estimated current population (2008) of around 400,000 (Seremban Municipal Council website, 2009). The population of Seremban has increased steadily over the last one hundred and ten years. In 1891, the total population was less than 3,000 people but had increased to 383,530 by the year 2000. The Malays and other local indigenous groups (Bumiputeras) were in the majority, with 46% of the total population, followed by the Chinese, with about 31%, and the Indians, with 19%. From these figures, the multi-ethnic nature of the Seremban District and Municipality is obvious. The total population is expected to increase to about 594,000 by the year 2020 (Malaysia, 2000).

The Seremban town was founded by the British colonial administration in the closing years of the 19th century and it served as the capital of Negeri Sembilan State within the Federated Malay States from 1895 to 1947. For over a century, Seremban, like many other towns with bustling economic activities, grew in area, following adjustments to its administrative boundary that took place in the decades after independence in the year 1957, to accommodate the demand of jurisdiction over the expanding urbanized areas forming on the urban fringes. Seremban is located about 70 kilometers south of the Malaysian conurbation hub, Kuala Lumpur. For decades, the municipality has been a major commuting hub of workers from the southern part of Malaysia to the centrally located primary development region of Klang Valley.

Housing growth in Seremban is strongly tied to Malaysia's development plan, The New Economic Policy (NEP), introduced in 1970, that aimed basically at eradicating poverty among all ethnic groups and restructuring the Malaysian society such that no one group would, at the end of the 20 year New Economic Policy period, be identified by any economic label (Malaysia, 1971). Numerous strategies, programmes, and activities were undertaken to meet the broad socio-economic goals and targets of the NEP. The impacts of the policy on urban areas were far reaching. The original time frame for the implementation of the NEP was between 1970 and 1990. Parallel to that, housing development in the Seremban municipal area was seen as one of the implementation instruments of the policy. In the early 1970s, not many housing schemes had been developed but, by the late 1970s and early 1980s, housing development had grown significantly, making these periods the inflection points in the housing market growth that boosted the economy of Seremban municipal area.

Housing Development and Shifting Vulnerabilities in the Seremban Municipal Area

Housing expansion patterns have to be understood within the context of the Malaysian urbanization experience. The urbanization of Malaysia can be divided into three phases, namely the nascent urbanization phase covering a period from about the 18th century to about 1948, after the Second World War. The bulk of the present urban centres, towns and cities were founded by the British colonial administration. Thus, in this phase most reports on the Malaysian urban centres were about their situations in the founding years. The second phase was that of pseudo-urbanization, covering the period of the years after the Second World War, the devolution years towards independence in 1957, and the initial years after independence, up to 1970. Overall, the trend in the number of houses built increased continuously from 1970 to 2000, to meet the demand from the increasing population, especially in urban areas (Malaysia 1971, 1976, 1981, 1986, 1991, 1996, 2001, 2006).

The concept of pseudo-urbanization, as introduced by McGee (1971), highlights the excessive rural to urban flows of unskilled workers searching for urban jobs in a fragmented labour market that is structurally incapable of absorbing the swelling urban labour force and, consequently, producing widespread poverty in cities. Some of the consequences were increasing numbers of spontaneous shelters in expanding squatter settlements, with widespread distribution of the poor and under-nourished. The impact of the rural to urban migration of workers from overcrowded villages would have been more severe without the development of land for commercial cropping and the opening of new settlements to accommodate poor villagers from over-crowded villages lacking economic opportunities; these settlements became known as the FELDA settlements (Malaya, 1956). These land schemes resulted in thousands of hectares of forested areas cleared for rubber and oil palm trees, together with housing schemes located within the plantations. In Malaysian history, these settlements were a repetition of the deforestation for commercial ventures undertaken by foreign companies under the British administration in the early decades of the 20th century (Teck Ghee, 1977).

The third urbanization phase covers the period from about the 1970s until today, during which the government has led the way for widespread development, working hand in hand with the private sector to bring the country out of the backwaters, and to create a developed country with high income by 2020 (Prime Minister's Department, 2010). The pursuit of industrialization for export has so far had far-reaching impacts on the entire country, but most notably in towns and cities. A spatial outcome of the process is the growth of the extended mega urban region in which the Seremban municipal area is located.

There is a simple linear housing expansion pattern, moving outward from the more established housing areas in the city centre (Figure 2). Housing development moved from

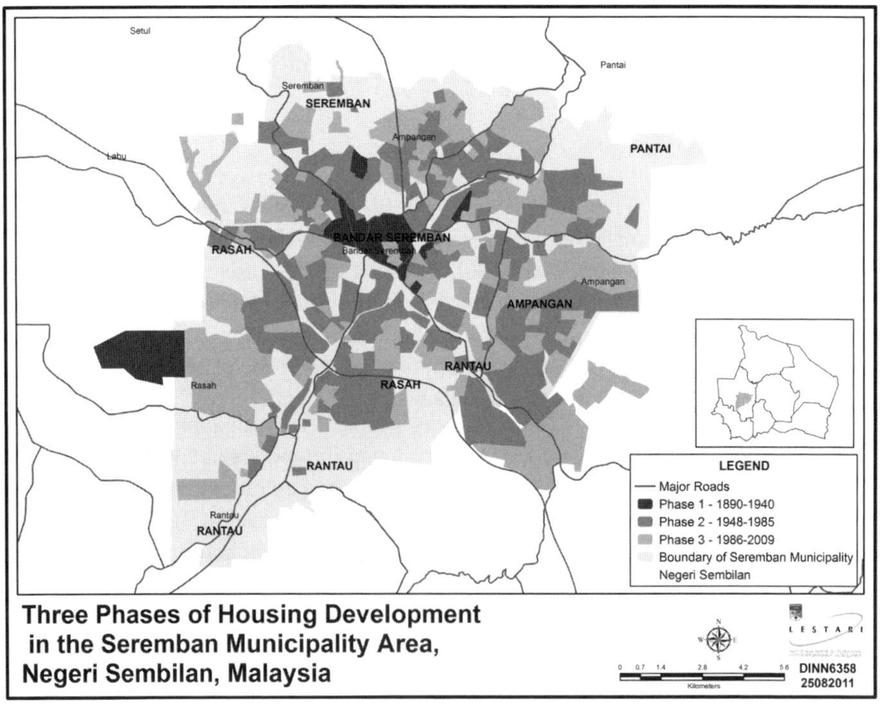

Figure 2. Three phases of housing expansion in the Seremban Municipality area. (Adapted from Idrus, Hadi, Harman Shan, & Rainis, 2010)

the central municipal area to the outskirts in the first urbanization phase, then further out in the subsequent urbanization phases. An examination of housing expansion at the scale of individual housing areas reveals more complex growth patterns.

Over the years between 1895 and 1948, housing development was confined to areas close to the town centre, because of land availability and the need to be close to the city centre; transportation was limited and basic household goods were more readily available in the town centre. Moreover, compact land use in the urban centre made it easier to police. The colonial administration policy at the time was to serve the interests of the administration. Housing was primarily for colonial administrators and officials, and their employees. The retail commercial sectors were in the hand of the immigrant Chinese and their housing needs were taken care of in the commercial centre. The rest of the population had little access to new housing and had to retreat to nearby urban villages.

Housing development during the independence era was still mostly confined to the town centre that housed economic activities such as retailing and wholesaling, as well as legal services, and health clinics. Shop houses were mainly two storey buildings. Commercial activity usually took place on the ground floor, while upstairs was where the family lived. Malaysian urbanists have reported that these early urban settlements offered goods and services in sundry shops with little specialization, in addition to their functions of providing security and law and order for the city dwellers as well as for people in the rural hinterland. The drivers of urban development and expansion at the time were resource exploitation and rubber growing in the hinterland. Town centres acted as collecting points. Such factors made compact urban housing areas close to the town centre sustainable.

Housing development and expansion in subsequent decades were restricted by land ownership. Thus, after independence in 1957, housing projects were largely on government lands in locations spread all over the municipal area. From the 1980s onwards, housing expansions seemed to be firmly sustained but new vulnerabilities arose, due to economic downturns and the Asian financial meltdown in 1998–1999.

Housing areas were sparse and far apart before independence, forming a clustered pattern. It was the period after the introduction of the NEP in 1971 that provided several fluctuations in national economic growth correlating with several tipping points of accelerated expansion in the pattern of urban housing construction. It took about ten years for Seremban to show a change from clusters of a few housing areas before NEP to an initial wider spread of housing projects, which led to an almost random pattern of growth. Given a large area in which to develop housing projects, government-led housing schemes spread all over the Seremban municipal area. The pattern continued to the end of the NEP in 1990. The New Development Policy era that followed NEP saw an increase in the clustering of housing projects, led by the private sector. This new housing growth pattern was an in-filling process that linked together previously separated housing estates. This in-filling increased the efficiency of residential services provided in the areas.

During the third urbanization phase, the Seremban municipal area has been growing outwards from the commercial centre to the periphery in a complex manner. Until the 1970s, the outward growth had been slow, since many of the economic activities in the municipality were medium-level retailing and support services, despite the availability of good physical and social infrastructures. Some residents commuted to Kuala Lumpur, about 80 kilometers away, for high-end shopping, and public transport helped commuters. However, since the 1970s, the municipal area attracted people to come and stay, following the opening of the country to foreign direct investments, in parallel with the opening of Malaysia to international investments. With the establishment of footloose branch companies by international industries from the West, and later from Taiwan and South Korea, the municipal area began to expand faster and outwards. With the establishment of industrial areas in the 1970s, hypermarkets, more support services, and recreational facilities moved in, along with new housing estates, to cater for new, expanding demands. New residential neighbourhoods sprang up and spread outwards, in all directions, to the rural periphery. These expansions were serviced by new state of the art infrastructures.

The eastern expansion of housing development outward from the city centre during this phase was towards Ampangan and Pantai, limited by the hills towards Kuala Pilah. The housing areas to the north were also extended, here limited by the Gala Forest Reserve. Further expansion is expected to be to the south, around the Rantau area, due to its flat, lowland areas. Accessibility, especially via roads, is a major force behind housing expansion.

The conditions that contributed to the vulnerabilities are caused by the role of the state, the markets, and society (Table 1). In the nascent phase of housing development, the focus of vulnerability issues was more on the incomplete physical infrastructures such as water supply, electricity, transportation, and commercial areas still under construction by British colonial administration, especially in the Seremban city centre.

Over time, housing areas grew steadily and houses that were built during the pseudo-urbanization phase were more on the periphery of urban areas, such as in new townships, as well as close to housing estates near the municipal centre. Residents' vulnerability during this phase was related more to insufficient physical and social infrastructures such as schools, transportation, and health care.

Community vulnerabilities shifted from exposure to lack of basic facilities and amenities, as well as threats of the communist insurgency, to housing abandonments and new problems of security. The levels of communities' vulnerability during the three phases of urbanisation are summarized in Table 2.

This table summarizes housing expansion vulnerabilities during Malaysia's three urbanization phases. The vulnerability components include housing characteristics drivers, and vulnerable groups.

Socio-Economic Vulnerability in the Phase of Extended Mega Urban Region: The Focus on Dengue and Crime Cases

Social vulnerability can be illustrated by the rising incidence of dengue fever and crime in the study area, impacting the area for more than 10 years. The dengue and crime cases

Table 1. Conditions giving rise to vulnerability of place and local people during the three phases of urban development

Phase 1: Nascent urbanization	Phase 2: Pseudo-urbanization	Phase 3: Extended mega urbanization
State approach • Colonial policy • City as administrative law and order centre • Urban centre founded by colonial British administration • Houses for British officers • Very limited private housing development	**State approach** • Independence • No specific housing policy • Limited housing development • Limited basic infrastructure	**State approach** • Housing policy • Demand for housing stock • Accessibility of basic infrastructures
Market characteristics • Very limited accessibility of financial facilities	**Market characteristics** • Limited accessibility of financial facilities	**Market characteristics** • Accessibility of financial facilities much easier • Private investors play a major role in housing development
Societal developments • Small population size • Limited accessibility to basic infrastructure • Widespread absolute poverty	**Societal developments** • Relatively low individual income • Mass migration from rural to urban areas • Increasing population density • Widespread absolute poverty	**Societal developments** • Rising middle class income • Migration from rural to urban areas continues • Urban areas as home • High population density • Relative poverty (5%)

Table 2. Shifting vulnerabilities during three phases of urban housing expansion in Seremban Municipality areas

	Seed stage (1950s–1970s) Nascent	In-filling stage (1980s–1990s) Pseudo	In-filling + Compact (1990s – today) Extended mega urban region
Characteristics	• Inefficient housing growth • Sprawl and disperse – seed • Squatters/slum • New villages • Vulnerable for rural migrants • National security concern – after Emergency in 1948 until 1960	• Housing "boom" period • Abandoned houses • Lack of infrastructures • Growing middle class • Economic melt-down in 1986/87 • Flat and low cost houses • More focus on terrace houses	• High density and dispersed • More waste produced • More complex regulations • 1990–1996: House buying "craze" • 1996/1997: Economic melt-down • Rise of gated communities • Rise of guarded communities • More greenery • Combination of urban and rural in "gentlemen orchards" • High demand • High price
Drivers	• Local • Limited to British capital • Not enough facilities • Low supply of houses • Demand from government officers • Supply by the government • Basic needs: buying flats • Rural to urban migration – adaptation process	• Local/international • Economic melt-down • Government – private partnership	• Global • Local • Private • Government link company (GLC)- driven
Resulting vulnerabilities: **(a) characteristics of the place**	**Place:** • Squatters • "Black Area" – support communist activities • Ghetto area	**Place:** • Over-crowdedness • Exposure to disease • Exposure to pollution	**Place:** • Dengue • Crime
(b) Social groups with particular vulnerabilities	**People:** • Villagers	**People:** • Investors, buyers, housing speculators, housing developers • Financial institutions	**People:** • Young people – difficult to buy houses • Elderly people • Disabled people
Levels of communities' vulnerability	Low-Income Group	Low-Income Group + Middle-Income Group	Low-Income Group + Middle-Income Group + High-Income Group

arose after the housing areas were built and established. The social dimension of vulnerability acknowledges the vulnerability of people, organizations, and even societies. Blaikie, Cannon, Davis, and Wisner (1994) argue that the most vulnerable groups are those who find it hardest to reconstruct their livelihood after a disaster. They find that, as a rule, the poor suffer more from hazards than the rich (see also Yohe & Tol, 2002). However, it is also possible for the rich to become more vulnerable over time, as is the case in this study, albeit for creeping crises rather than disasters. The time dimension is relevant, since reconstruction in poor areas can take a long time, which drastically affects the economy and livelihood of the area. Furthermore, the poorer population groups do not always have a choice of where to locate, thus they might have to live in risky areas, for example in dengue and crime-ridden areas.

The spatial analysis of dengue-related health cases shows the distribution of dengue cases for the years between 1996 and 2007, focusing on the Ampangan sub-district area. The number of dengue cases increased yearly, with the highest numbers recorded in the Taman Paroi Jaya, with 226 cases, followed by Taman Seremban Jaya, with 218 cases. Both areas are located in the middle class area. Dengue cases were easily linked to the

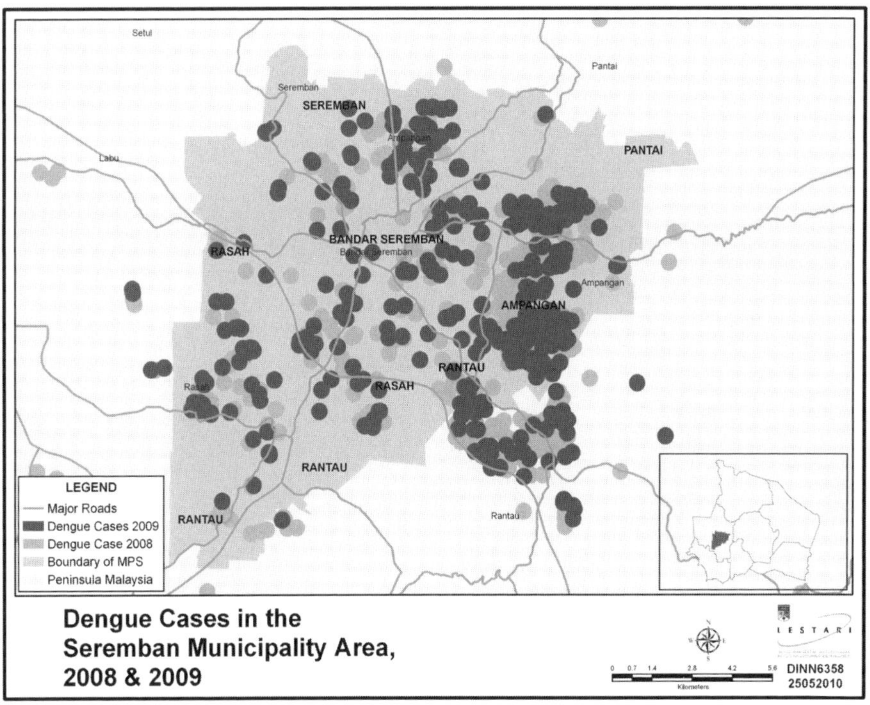

Figure 3. Dengue cases in the Seremban Municipality area in 2008 and 2009. (Adapted from Idrus et al., 2010)

planting of ornamental plants in gardens, where fresh water that collected at the base of the pots became mosquito breeding ponds. The dengue cases were spatially clustered into six areas, namely Taman Ampangan, Taman Paroi Jaya, Taman Seremban Jaya, Taman Senawang Jaya, Taman Rasah Jaya, and Taman Satria. These places recorded more than 100 dengue cases over the past 10 years (Figure 3).

The dengue problem has brought illness and even death to these neighbourhoods in recent years. The residents have, so far, been able to deal with the problems by adhering to guidelines set by the municipality's health officials to clean up breeding places of the *Aedes* larvae, opening up their compounds to regular fogging, and to bring people with dengue symptoms quickly to the hospital. To date, there has been no incidence of house owners in these areas wanting to move out to healthier neighbourhoods, nor have there been incidences where the value of houses in dengue affected areas has decreased. But the fear of dengue remains a threat to all households.

Turning now to the incidence of crime in the area, the identified crime cases were mainly related to robbery, armed robbery, theft, and murder. The distribution pattern of crime occurrences was dispersed but was located mainly in the city centre, and there

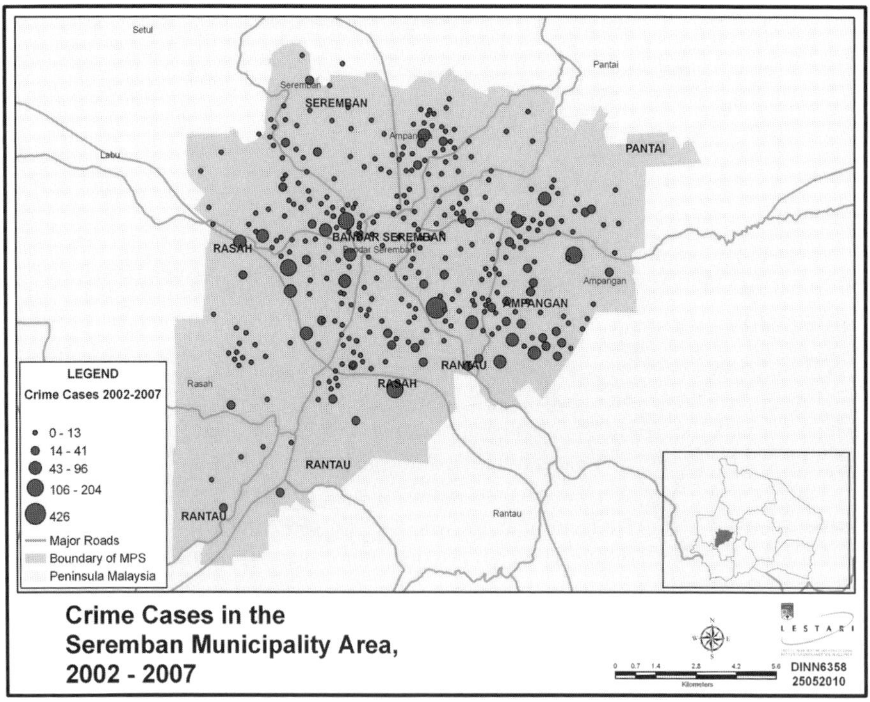

Figure 4. Crime cases in the Seremban Municipality Area in 2002–2007. (Adapted from Idrus et al., 2010)

mostly in high density shopping areas and housing areas (Figure 4). Taman Seremban Jaya recorded 426 cases, Taman Rasah Jaya 204 cases, Terminal One 145 cases, and others – Jalan Dato' Bandar Tunggal, Taman Tasik Jaya, and Jalan Tunku Munawir – recorded 142, 137, and 134 cases, respectively. In other areas, there were less than 120 cases of crime for the past 10 years. The residents of those areas with prevalent crime rates exhibited the "siege mentality" syndrome (so commented by the planner in the Seremban municipality). Areas most vulnerable to crime are the middle and upper middle class sections of the Seremban municipal area and also those sections in other lower middle income neighbourhoods. However, the move to make such housing areas gated, and to install security guards, reduced the threat to almost zero.

Petty thefts, however, have been rampant in the municipal area. Cases had been reported in terrace houses whose residents were from the low middle income groups. Residents in affected areas responded to the problem by establishing community watch groups whose members voluntarily took turns to look after their housing areas, especially from the middle of the night to dawn, during which time most of the problems occurred. The watch groups were linked to the police. Thus, crimes such as burglary, petty theft, and drug pushing, were taken care of. Once again, all housing areas with these problems have so far not experienced any decline in property values, nor have house owners wanted to sell off. Thus, vulnerability arising from crime has not yet produced mass sellouts or relocation.

The crime cases recorded show that people living in urban areas are even more vulnerable than before, and that the security infrastructure, such as police officers and private security guards, have to work closely with the residents of the areas.

Conclusions: Managing Shifting Vulnerability in the Seremban Municipality Area

This chapter presented the case of shifting vulnerabilities that result from housing growth in Malaysia, with the Seremban Municipality area as the case in point for Rittel and Webbers' "wicked problem" (1973). Individuals and entire communities are vulnerable to different stressors that change, rather than being eliminated over time. Such a situation provides food for thought for any planner who sees problems related to housing growth as being easy to solve. Shifting vulnerability during housing expansion in the study area must be understood before the entire housing expansion can be managed. One way of dealing with the issue is to understand the changing collaboration between the state with its agencies and institutions, market-banking and finance, and developers, including building contractors, and the society – the Seremban people.

The relationship between the state, market, and society needs to be seen in the context of time. During the British colonial administration, administrative and economic development policies mainly served the interests of the colonial administration. The local people received only marginal benefits; even the labour needed for development, especially in the construction industry, was imported from outside the country. In such circumstances, housing development that was built close to the town centre was for colonial needs. These beginnings of housing development at the time influenced the direction of future housing development during the independence years and especially in the four decades after 1970. This was the watershed for the developments in Malaysia, and the impacts on Malaysian society, even to this day, are varied and, among these impacts, is housing growth.

Managing housing provision and expansion in the Seremban municipal area and the country requires the close co-operation of the state, markets and the population. Today, there is some co-operation but what is needed most is to ensure the people are able to own houses that they can afford. In this sense, housing developers should be sensitive to people's needs and housing affordability. The state remains as the facilitator for housing provision. Even now, the government has stressed, in the 10th Malaysia Plan that more affordable houses are to be built for the poor. While special attention has been given to the poor, the market is still free to develop upper-end housing stock for the middle class and upper income groups. This might result in enclaves that segregate society even more. If a stronger co-operation is not developed between the state, markets, and people, new vulnerabilities will emerge in the future.

Acknowledgments

This study was supported by funding from the Ministry of Science, Technology and Innovation (MOSTI), Malaysia (Project Code: SF-06-01-02-SF0228) and Universiti Kebangsaan Malaysia Fundamental Grant (Code: UKM-XX-05-FRGS0001-2006), the Research University Grant (UKM-OUP-ASPL-6/2007), Ministry of Higher Education, Malaysia. We would like to thank the Seremban Municipal Council, and the Institute for Environment and Development (LESTARI), Universiti Kebangsaan Malaysia (UKM). Our gratitude also goes to Mr. Lokman Omar and Mr. Mohd Rafi Othman from Seremban Municipal Council for their help.

References

Blaikie, P., Cannon, T., Davis, I., & Wisner, B. (1994). *At risk. Natural hazards, people's vulnerability and disasters*. London, UK: Routledge.

Brown, V. A., Harris, J. A., & Russell, J. Y. (Eds.). (2010). *Tackling wicked problems through the transdisciplinary imagination*. London, UK: Earthscan.

Conklin, J. (2001). *Wicked problems and social complexity*. Retrieved from http://www.cognexus.org/wpf/wickedproblems.pdf

Drakakis-Smith, D. (1977). Housing the urban poor in West Malaysia: The role of the private sector. *Habitat International, 2*, 571–584.

Fussel, H. M. (2007). Vulnerability: A generally applicable conceptual framework for climate change research. *Global Environmental Change, 17*, 155–167.

Gow, G. A. (2005). *Policymaking for critical infrastructure*. Aldershot, UK: Ashgate.

Idrus, S., Hadi, A. S., Harman Shan, A. H., & Rainis, R. (2010). Neighbourhood Expansion and Urban Livability in Seremban Municipality Area, Malaysia. *Malaysian Journal of Environmental Management, 11*(1), 37–47.

Lawrence, R. J. (2010). Housing and health promotion: Moving forward. *International Journal of Public Health, 55,* 145–146.

Malaya. (1956). *The first Malayan plan*. Kuala Lumpur: Government Printers.

Malaysia. (1971). *The second Malaysia plan*. Kuala Lumpur: Government Printers.

Malaysia. (1976). *The third Malaysia plan*. Kuala Lumpur: Government Printers.

Malaysia. (1981). *The fourth Malaysia plan*. Kuala Lumpur: Government Printers.

Malaysia. (1986). *The fifth Malaysia plan*. Kuala Lumpur: Government Printers.

Malaysia. (1991). *The sixth Malaysia plan*. Kuala Lumpur: Government Printers.

Malaysia. (1996). *The seventh Malaysia plan*. Kuala Lumpur: Government Printers.

Malaysia. (2000). *Population and housing census 2000*. Kuala Lumpur: Government Printers.

Malaysia. (2001). *The eight Malaysia plan*. Kuala Lumpur: Government Printers.

Malaysia. (2006). *The ninth Malaysia plan*. Kuala Lumpur: Government Printers.

McGee, T. G. (1971). *The urbanization process in the third world*. London, UK: Bell.

Metzger, M. J., Rounsevell, M. D. A., Acosta-Michlik, L., Leemans, R., & Schröter, D. (2006). The vulnerability of ecosystem services to Landuse change. *Agriculture, Ecosystems and Environment, 114*, 69–85.

Myers, D., Baer, W. C., & Choi, S.-Y. (1996). The changing problem of overcrowded housing. *Journal of the American Planning Association, 62*, 66–84.

O'Brien, K., Leichenko, R., Kelkar, U., Venema, H., Aandahl, G., Tompkins, H., . . . Wes, J. (2004). Mapping vulnerability to multiple stressors: Climate change and globalization in India. *Global Enviornmental Change, 14*, 303–313.

Prime Minister's Department. (2010). *Government transformation programme: The road map*. Putrajaya: PEMANDU.

Rittel, H. W. J., & Webber, M. M. (1973). Dilemmas in a general theory of planning. *Policy Sciences, 4*, 155–169.

Seremban Municipal Council Website. (2009). Retrieved from http://www.mpnsns.gov.my/web/guest/home

Teck Ghee, L. (1977). *Peasants and their agricultural economy in Colonial Malaya, 1874–1941*. Kuala Lumpur: Oxford University Press.

Tipple, G. (2005). Housing and urban vulnerability in rapidly-developing cities. *Journal of Contingencies and Crisis Management, 13*, 66–75.

Turner, J. F. C. (1967). Barriers and channels for housing development in modernizing countries. *Journal, American Institute of Planners, 32*, 167–181.

Turner, I. I. B. L., Kasperson, R. E., Matson, P. A., McCarthy, J. J., Corell, R. W., Christensen, L., . . . Schiller, A. (2003). A framework for vulnerability analysis in sustainability science. *Proceedings of the National Academy of Sciences US, 100*, 8074–8079.

Yohe, G., & Tol, R. S. J. (2002). Indicators for social and economic coping capacity: Moving toward a working definition of adaptive capacity. *Global Environmental Change, 12*, 25–40.

On the Relationship Between Crime in Residential Areas and Street Segment Attributes

Antônio Tarcísio Reis and Celina Dittmar

School of Architecture/PROPUR – Federal University of Rio Grande do Sul, Porto Alegre, Brazil

Abstract

This paper examines the relationship between the occurrence of crime in residential areas, namely street robbery and residential burglary, perception of security, and street segment attributes such as connectivity, length, street lighting, appearance, escape routes, presence of security personnel, visual and physical barriers, visual and physical connections between residences and the public open space, land use and type of housing unit, in two residential boroughs in Porto Alegre, Brazil. Data related to occurrence of street robbery and residential burglary were collected with the Department of Public Security and through questionnaires administered to a sample of 123 residents in Menino Deus and 86 in Rio Branco. Data were analyzed in SPSS/PC according to street segments. Segments' connectivity and length were numerically and visually generated in Depthmap. Results reveal the impact made by segment characteristics such as connectivity, land use, and type of housing units on the occurrence of crime in residential areas, allowing a better understanding of the relationship between street robbery and residential burglary and some physical characteristics of the urban space.

Key words: crime, perception of security, residential burglary, segment analysis, street robbery

Introduction

Insecurity is not an exclusive characteristic of Brazilian cities, although, in some of these cities, the problem tends to be of a magnitude that seriously affects the quality of urban life of their inhabitants (Basso, 2002; Caldeira, 2000; Reis & Lay, 1996; Zaluar, 2002). Crime has been related not only to socio-economic or political variables but also to physical or spatial variables in various urban environments (Caldeira, 2000; Jacobs, 1961; Newman, 1972, 1976; Zaluar, 2002; Zanotto, 2002). Several investigators have addressed the issue of insecurity in Brazilian cities (Caldeira, 2000; Zaluar, 2002). However, their approaches tend to focus on the effects of insecurity on the urban space, as in the book "City of Walls: Crime, Segregation and Citizenship in São Paulo" (Caldeira, 2000), and not on the effects of urban form and configuration on the occurrence of crime.

On the other hand, some studies have shown that the urban configuration tends to play a role in increasing or decreasing opportunities for the occurrence of crimes in cities (Monteiro & Iannicelli, 2009; Reis, Portella, Bennett, & Lay, 2003; Sahbaz & Hillier, 2007; Shu & Huang, 2003). Nonetheless, it is necessary to better understand its effect on crime in the streets and in residences, especially in the context of residential neighbourhoods in a Brazilian city. In this sense, a detailed analysis on the level of a street segment may play an important role. This type of analysis was developed and initially used by Hillier and Sahbaz (2005) in their article entitled "High Resolution Analysis of Crime Patterns in Urban Street Networks: An initial statistical sketch from an ongoing study of a London borough," where they emphasize the importance of a detailed analysis of crime patterns in urban street networks to which many physical variables are related. A specific set of physical-spatial variables seems to have an impact on urban security. For example, the appearance of buildings and open spaces can affect the perception of security, with consequences for the use of open spaces and the actions of potential criminals. However, the effects of physical-spatial variables on the occurrence of crimes seem to be complex and not yet conclusively established, as already pointed out (Hillier & Sahbaz, 2005). It is therefore necessary to analyze these variables in greater detail. In this sense, an analysis at the street segment level (equivalent to a block or to a part of it) plays an important role, as it enables the analysis of relationships between the occurrence of crimes and the physical-spatial attributes of such street segments.

Moreover, the use of software such as Geographic Information Systems (GIS) helps in the description and analysis of crime at the segment level. Therefore, using this methodology, this chapter examines the relationship between the occurrence of crime in residential areas, including street robbery and thefts, and residential burglary, perception of security and attributes of street segments such as connectivity, length, street lighting, appearance, escape routes, presence of security personnel, residences with physical and visual barriers, visual and physical connections between residences and public open space, land use and type of residential unit, in the middle-class residential boroughs of Rio Branco (RB) and Menino Deus (MD) in Porto Alegre. Hence, a better understanding of the impact of urban form and configuration on the occurrence of crime can be generated, helping urban designers to improve security in residential areas.

Methodology

Data related to the occurrence of crime in the streets and residences were collected with the Department of Public Security (DPS) of the State of Rio Grande do Sul, for a period of ten months, and through questionnaires administered to a sample of 123 residents in the MD borough and 86 in the RB borough. This method also allowed the collection of residents' satisfaction levels regarding security. These boroughs are characterized by fenced or walled houses or blocks of flats in less crowded streets, and by non-residential buildings (i.e., offices, commerce, and services) or mixed-use buildings in the busiest streets, named in Figures 1 and 2.

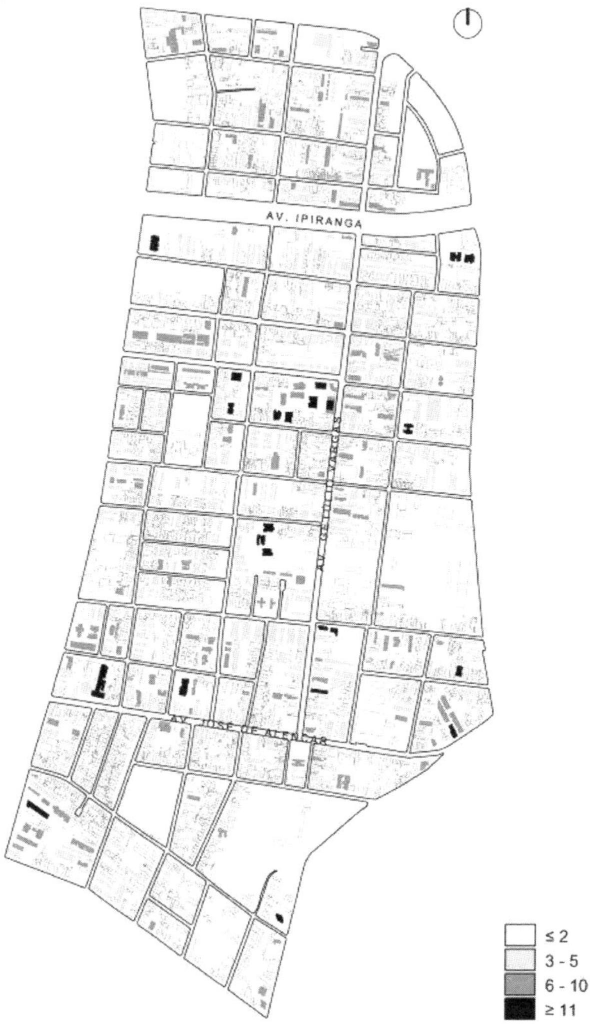

Figure 1. Height of buildings in Menino Deus borough. The legend indicates the number of floors in each building. (Source: the authors)

Figure 2. Height of buildings in Rio Branco borough. The legend indicates the number of floors in each building. (Source: the authors)

The number of floors (Figures 1 and 2) reveals the clear majority of buildings with no more than 2 floors in MD (1,120 residential and 563 non-residential) and in RB (820 residential and 591 non-residential), followed by: 3 to 5 storey buildings (MD: 573 residential and 257 non-residential; RB: 507 residential and 202 non-residential); 6 to 10 storey buildings (MD: 114 residential and 134 non-residential; RB: 147 residential and 123 non-residential); 11 to 15 storey buildings (MD: 11 residential and 6 non-residential; RB: 23 residential and 31 non-residential); and buildings with 15 or more floors in RB (5 residential and 16 non-residential).

The number of residential buildings is clearly greater than the number of non-residential buildings, considering up to five storey buildings. Residential buildings may include commerce and services on the first floor. In addition, the morphological description of the two residential boroughs studied include: The number of blocks in MD (89) and RB (58); the mean length of blocks in MD (120.5 meters) and in RB (119.5 meters); and the mean number of plots in a block (MD: 28.8; RB: 32.7).

The types of crimes considered in the survey were grouped into two categories: Crimes involving vehicles and pedestrians (pedestrian robbery, vehicle robbery, vehicle theft, and theft in vehicles) and crimes occurring in residences (residence burglary and theft in residence). Street robbery is measured using Hillier and Sahbaz's (2005) methodology of "time risk," which divides the total number of robberies in segments belonging to a certain band by the total number of segments in that band. The band is characterized by intervals of five meters for segments of 100 m or less, and by intervals of ten m for segments between 100 and 250 m. The "time risk" thus takes into account the length of the street segment where the crime occurs. Residential burglary, in turn, was measured by considering the "true risk rate for the bands;" here, the total number of burglaries is divided by the total number of residential units in each band, each one of these being composed of segments with an equal number of residential units, like a band made up of segments with one (1) residential unit or a band made up of segments with two (2) residential units (Hillier & Sahbaz, 2005). The crime rates in the streets and in residences were calculated and tabulated in a spreadsheet in ArcGIS with segment attributes for further analysis related to the physical attributes of the segments. Crimes obtained through the DPS of Rio Grande do Sul and through questionnaires to residents of the two boroughs were considered.

Information about land use, identified as residential or non-residential (i.e., offices, shops, bar/restaurants, and services, without regard to their periods of operation), and types of housing units, whether houses or blocks of flats, was obtained from the Department of City Planning. Features of the segment (equivalent to one block) such as street lighting (by

Figure 3. Crime on streets in Rio Branco borough (left) and Menino Deus borough (right). The dashed line indicates the exact boundaries of the borough and the number of crimes analyzed; the dots indicate the occurrence of crimes (according to DPS) registered through ArcGIS in the segments map. (Source: the authors)

the provision of streetlights), appearance (on the levels of maintenance, cleanliness, and personalization), presence of security personnel (through the existence of locations specifically provided for observation), physical and visual barriers, physical and visual connections between the housing units and public open space (number of openings in the buildings measured in meters), and the existence of escape routes for criminals (pedestrian crossings between buildings) were obtained from physical surveys. These were performed on 5% of the segments with the highest level and 5% of the segments with the lowest level of crimes involving pedestrians and vehicles, and crimes involving residences, representing 20% of the total number of segments in each borough. The information obtained through these physical surveys was tabulated in ArcGIS as attributes of the segments.

The data recorded in ArcGIS allowed the generation of maps and the visualization, for example, of the spatial distribution of Figure 3. In SPSS/PC, the criminal event data were statistically analyzed according to the attributes of street segments, using parametric tests of correlation (Pearson). The connectivity and length of the segments were generated visually and numerically with Depthmap, where the segments map was created from the axial map representing the street networks of the two boroughs (Figure 3).

Results

Comparison of Perceived Security in MD and RB

Residents' satisfaction levels with security in the MD and RB boroughs are similar, as evidenced by the lack of a statistically significant difference between levels. In both areas, the perception of insecurity is much greater at night, the period for which more than 50% of respondents in each of the two boroughs expressed dissatisfaction with the urban security (Table 1). In the 5% of the segments with the highest and the 5% of the segments with the lowest crime rates, 3.2 street lights per 100 m were found in MD and 2.7 street lights per 100 m in RB, suggesting that these mean street lights rates were not enough to provide adequate street lighting and security.

Occurrence of Crimes in the Streets and in Residences and Attributes of all Segments

Considering all segments (457) of the MD and RB boroughs, and the crime data obtained from questionnaires to residents of these boroughs and through the DPS, the existence of correlations between the "Time Risk Band" for street robbery and the following aspects related to the segments was investigated: Connectivity of segments; length of segments; integration value of axial lines (Rn – all the lines of the axial map are taken into consideration; and R3 – only considers the relationship between the reference line and the other lines to a maximum of three steps away, each step being characterized by a line); linear

Table 1. Residents' perceived security in the block during the day and night

Boroughs	Very secure	Secure	Neither secure nor insecure	Insecure	Very insecure	Mean rank values	Total
Residents' satisfaction levels with security in the block during the day							
Menino Deus	1 (0.8)	51 (1.5)	44 (35.8)	20 (16.3)	7 (5.7)	101.26	123
Rio Branco	0 (0)	40 (46.5)	35 (40.7)	10 (11.6)	1 (1.2)	110.34	86
Total sample	1 (0.5)	91 (43.5)	79 (37.8)	30 (14.4)	8 (3.8)		209
Residents' satisfaction levels with security in the block during the night							
Menino Deus	0 (0)	13 (10.6)	42 (34.1)	56 (45.5)	12 (9.8)	106.37	123
Rio Branco	0 (0)	6 (7)	27 (31.4)	50 (58.1)	3 (3.5)	103.03	86
Total sample	0 (0)	19 (9.1)	69 (33)	106 (50.7)	15 (7.2)		209

Notes. Mean rank values were obtained by the Mann-Whitney nonparametric statistical test and indicate the mean of ordinal values assigned by the respondents; values in parentheses represent percentage to total in each borough (123 and 86) and to the total sample of 209 respondents.

density (number of residential units over the length of the segment); ratio of residential to non-residential units; building density (number of buildings in relation to the length of the segment); and density of houses (number of houses in relation to the length of the segment).

The incidence of street crime (pedestrian robbery, vehicle robbery, vehicle theft, and theft in vehicles), according to data obtained from the DPS, is correlated to: Connectivity (Pearson Correlation, coef. = .144, sig. = .002); segment length (Pearson Correlation, coef. = .708, sig. = .000); density of buildings (Pearson Correlation, coef. = .094, sig. = .044), and density of houses (Pearson Correlation, coef. = .093, sig. = .048) (Table 2). Additionally, the incidence of street crime, according to data obtained through questionnaires, is correlated to: Segment length (Pearson Correlation, coef. = .572, sig. = .000); integration Rn (Pearson Correlation, coef. = .106, sig. = .023); density of buildings (Pearson Correlation, coef. = .139, sig. = .003); and density of houses (Pearson Correlation, coef. = .097, sig. = .038) (Table 2).

Correlation tests (Pearson) were also conducted between these variables (segment connectivity; segment length; integration value of axial lines, namely, Rn and R3; linear density; ratio of residential and non-residential units; density of buildings; and density of houses) and the rate of occurrence of crimes in residences (the True Rate for the Risk Bands). The occurrence of crimes in residences, according to data obtained from the DPS, is correlated with the density of buildings (Pearson Correlation, coef. = .108, sig. = .021) (Table 3). In addittion, the occurrence of crimes in residences, according to data obtained from the questionnaires, is negatively correlated to integration Rn (Pearson Correlation, coef. = –.252, sig. = .000) and positively correlated to density of houses (Pearson Correlation, coef. = .112, sig. = .017) (Table 3).

Table 2. Correlations between street crime and segments attributes in Menino Deus and Rio Branco

Segment (457) attributes in MENINO DEUS and RIO BRANCO		Time Risk Band for Street Robbery – quest	Time Risk Band for Street Robbery – DPS
Connectivity	Pearson Correlation	No correlation	.144
	Sig. (2-tailed)		.002
Segment length	Pearson Correlation	.572	.708
	Sig. (2-tailed)	.000	.000
Integration Rn	Pearson Correlation	.106	No correlation
	Sig. (2-tailed)	.023	
Density of buildings	Pearson Correlation	.139	.094
	Sig. (2-tailed)	.003	.044
Density of houses	Pearson Correlation	.097	.093
	Sig. (2-tailed)	.038	.048

Notes. Time Risk Band for Street Robbery – quest = street crime obtained through questionnaires; Time Risk Band for Street Robbery – DPS = street crime obtained through the Department of Public Security of the State of Rio Grande do Sul; all the segments (457) in Menino Deus and Rio Branco were included.

Table 3. Correlation between crimes in residences and segments in Menino Deus and Rio Branco

Segments (457) attributes in MENINO DEUS and RIO BRANCO		True Rate for the Risk Bands – ground floor units – DPS	True Rate for the Risk Bands – ground floor units – quest
Integration Rn	Pearson Correlation	No correlation	−.252
	Sig. (2-tailed)		.000
Density of buildings	Pearson Correlation	.108	No correlation
	Sig. (2-tailed)	.021	
Density of houses	Pearson Correlation	No correlation	.112
	Sig. (2-tailed)		.017

Notes. True Rate for the Risk Bands – ground floor units – quest = crimes in residences obtained through questionnaires; True Rate for the Risk Bands – ground floor units – DPS = crimes in residences obtained through the Department of Public Security of the State of Rio Grande do Sul; all 457 segments in Menino Deus and Rio Branco were included.

Occurrence of Crimes in the Streets and in Residences and Attributes of the Segments with Higher and Lower Crime Rates

The existence of a correlation between crime rates in the streets and in residences and the following attributes of the segments with highest and lowest crime rates in MD and RB was also investigated: Connectivity; integration (Rn and R3); linear density; ratio between residential and non-residential units; density of buildings; density of houses; street lighting rate (number of streetlights in relation to segment length); existence of vegetation that could provide a potential criminal hideout); segment length; appearance; escape routes; rate of surveillance locations for security personnel; rate of visual and physical barriers in residences (in proportion to the total length and to both sides of the street segment); visual and physical connections between the residences and public open space (in proportion to the total length and to both sides of the street segment); land use (residential or non-residential); and type of residential unit (house or block of flats).

With respect to street crime data obtained through the DPS, correlations between the rates of crimes in the segments and the following segment attributes were found: Length (Pearson Correlation, coef. = .680, sig. = .000); linear density (Pearson Correlation, coef. = .807, sig. = .000); lighting rate (Pearson Correlation, coef. = .771, sig. = .000); physical barriers (Pearson Correlation, coef. = .572, sig. = .003) and vegetation (Pearson Correlation, coef. = .675, sig. = .000) (Table 4).

For street crime data obtained through questionnaires, correlations between this data and the following segment attributes were found: Length (Pearson Correlation, coef. = .614, sig. = .001); physical barriers (Pearson Correlation, coef. = .563, sig. = .004) and vegetation (Pearson Correlation, coef. = .436, sig. = .033) (Table 4).

Table 4. Correlations between street crime rate and attributes of the segments with the highest and lowest rate of street crime in Menino Deus and Rio Branco

		Segment length	Linear density	Lighting rate	Physical barriers (%)	Vegetation
Time Risk Band for Street Robbery – DPS	Pearson coef.	.680	.807	.771	.572	.675
	Sig.	.000	.000	.000	.003	.000
	N	24	24	24	24	24
Time Risk Band for Street Robbery – quest	Pearson coef.	.614	No correlation	No correlation	.563	.436
	Sig.	.001			.004	.033
	N	24			24	24

Notes. Time Risk Band for Street Robbery – quest = street crime obtained through questionnaires; Time Risk Band for Street Robbery – DPS = street crime obtained through the Department of Public Security of the State of Rio Grande do Sul; only segments with the highest and lowest crime rate in the streets of Menino Deus and Rio Branco, a total of 24 segments, were included.

Table 5. Correlations between crime rates in residences and attributes of the segments with the highest and lowest crime rates in Menino Deus and Rio Branco

		Density of houses	Lighting rate	Physical barriers (%)
True Rate for the Risk Bands – ground floor units – DPS	Pearson coef.	−.555	No correlation	No correlation
	Sig.	.005		
	N	24		
True Rate for the Risk Bands – ground floor units – quest	Pearson coef.	No correlation	−.431	.433
	Sig.		.035	.035
	N		24	24

Notes. True Rate for the Risk Bands – ground floor units – quest = crimes in residences obtained through questionnaires; True Rate for the Risk Bands – ground floor units – DPS = crimes in residences obtained through the Department of Public Security of the State of Rio Grande do Sul; only segments with the highest and lowest crime rate in the residences in Menino Deus and Rio Branco, a total of 24 segments, were included.

With respect to crime rates in residences, obtained through the DPS, only a negative correlation with the density of houses (Pearson Correlation, coef. = −.555, sig. = .005) (Table 5) was found. Considering the data obtained through questionnaires, a negative correlation with lighting rate (Pearson Correlation, coef. = −.431, sig. = .035) and a positive correlation with physical barriers (Pearson Correlation, coef. = .433, sig. = .035) were found (Table 5).

Conclusions

The physical characteristics of the MD and RB boroughs, including the attributes of their street segments, caused no significant differences between residents' satisfaction levels with security there. As might be expected, the perception of insecurity is much greater at night, and similar in the two boroughs. Therefore, their physical characteristics, at least those that could affect the occurrence of crimes on the streets and in residences, are alike.

Taking into consideration all the street segments of MD and RB, the incidence of street crime, according to data obtained from the DPS, was affected by the connectivity and length of segments, density of buildings, and density of houses on the segments. In addition, the incidence of street crime, according to data obtained through the questionnaires, was affected by the length of the segment, the global integration value of the axial line where the segment is located (Rn – the higher the integration value, the greater the potential for movement), density of buildings, and density of houses. Thus, although there is some discrepancy between the results obtained from the two sources, it is noteworthy that the rate of occurrence of crimes involving pedestrians and vehicles (as calculated by taking into account the length of the segment) is also affected by the length of the street segment. Hence, one can deduce that the characteristics of longer segments seem to affect the rate of occurrence of crimes. In this sense, the greatest distance between connections with other segments could increase a potential criminal's visual control of the presence of people in the segment, reducing the surprise factor, that is, the possibility of people in other segments suddenly entering this longer segment.

The fact that the density of buildings, as well as the density of houses, seems to have affected the incidence of street crime (indicating that the greater the number of buildings and houses in a street segment, the greater the incidence of street crime) could be partly explained by the fact that there are more cars on the streets. Moreover, the possible effects of segment attributes on the occurrence of street crimes probably become clearer, if the categories that constitute such crimes, specifically, pedestrian robbery, vehicle robbery, vehicle theft, and theft in vehicles, are considered separately.

Examination of the correlations between street crime rates and attributes of the segments with the highest and lowest crime rates in MD and RB revealed that street crime appears to be affected by: The segment length (data obtained through DPS and questionnaires); the linear density (DPS); the lighting rate (DPS); the physical barriers (DPS and questionnaires) and by the vegetation (DPS and questionnaires).

The results for the segment length and the existence of vegetation (that could serve as visual protection for a potential criminal) are consistent with what might be expected, since longer segments with more vegetation would increase the opportunity for crime occurrence. However, the increase in linear density (number of housing units per length of the segment), with the consequent increase in the number of windows and doors facing the street, and especially, the increase of lighting rate, which increases visibility on the streets, could potentially reduce the incidence of street crime. In this sense, further investigations are necessary that, for example, consider the various types of street crime in addition to the time of crime occurrence.

The occurrence of crimes in residences is positively correlated with the density of buildings (according to data obtained from the DPS), the density of houses (data obtained through questionnaires) and negatively correlated to global integration (Rn – potential for movement; data obtained through questionnaires). This divergence between the results obtained using both sources of data indicates the need for further analysis. Moreover, the expectation is confirmed that increasing the potential for movement (translated into higher Rn integration levels) reduces the occurrence of crimes in residences, because, in principle, it would be harder to break into a house located on a busier street. This corroborates, for example, the conclusion drawn by Shu (2009, p. 102) that "houses located in those globally integrated areas, which are highly accessible, tend to be less vulnerable than those situated in globally segregated areas, which is also true in most case studies."

In turn, crime rates in residences were affected by the following attributes of the segments with highest and lowest crime rates in MD and RB: Negatively, by the density of houses (DPS); negatively, by the street lighting rate (questionnaires); and, positively, by physical barriers (questionnaires). In this case, it makes sense that increases in the number of residences and street lights in a street segment decrease the occurrence of crimes in residences, as residents tend to exercise some vigilance over neighbouring residences, and streets with higher visibility at night hinder the actions of a potential burglar. This is in line with findings attesting the beneficial effect of front door-to-front door inter-visibility on security (e.g., Shu, 2009). On the other hand, one would expect that the increased presence of physical barriers would decrease rather than increase the occurrence of crimes in residences. However, physical barriers that are also visual barriers may have the opposite effect on security, due to the lack of inter-visibility between buildings and between buildings and streets, thus negatively affecting visual surveillance either by people in residences or by people on the street. Hence, the identification of physical barriers that are also visual barriers would be necessary in a further investigation.

In conclusion, the examination of the relationships between the occurrence of crime in residential areas, including robberies and thefts on the streets, and residential burglary and thefts in residences, and attributes of street segments in MD and RB revealed the impact of some characteristics of the segments (e.g., length) that could be expected, according to our urban experiences and previous studies, and by other segment features (e.g., density of buildings and houses) that do not correspond to what could be expected. Therefore, for future investigations that either incorporate the data obtained for the MD and RB boroughs or use new data, it seems relevant to test the correlation between segment attributes and the distinct time periods of crime occurrence, as well as the categories of street crime used in this study, namely, pedestrian robbery, vehicle robbery, vehicle theft, and theft in vehicles. However, the results reported here help deepen the knowledge about how best to deal with urban design from the perspective of urban security in residential areas. This decreases the risk and vulnerability to crime and fosters urban identity and development in a changing world by favouring the use of open space, social interaction, and wellbeing in human habitats.

Acknowledgments

This investigation was undertaken with financial support from CNPq (National Council of Scientific and Technological Development), a Brazilian governmental institution sponsoring scientific and technological development.

References

Basso, L. (2002). *Influencia do Espaço Construído na Ocorrência de Crimes em Conjuntos Habitacionais* [Influence of built space on the occurrence of crimes in housing projects] (Unpublished master's thesis). Federal University of Rio Grande do Sul, Porto Alegre.

Caldeira, T. (2000). *Cidade de Muros: Crime, Segregação e Cidadania em São Paulo* [City of walls: Crime, segregation and citizenship in São Paulo]. São Paulo: Editora 34/Edusp.

Hillier, B., & Sahbaz, Ö. (2005). High resolution analysis of crime patterns in urban street networks: An initial statistical sketch from an ongoing study of a London borough. In A. Van Nes (Ed.), *Proceedings of the 5th International Space Syntax Symposium* (pp. 451–478). Delft: TU Delft.

Jacobs, J. (1961). *The death and life of great American cities*. New York, NY: Random House.

Monteiro, C., & Iannicelli, C. P. (2009). Spatial profiles of urban crimes: The role of morphology in a context of social inequality. In D. Koch, L. Marcus, & J. Steen (Eds.), *Proceedings of the 7th International Space Syntax Symposium* (pp. 080:1–080:11). Stockholm: KTH, Faculty of Architecture.

Newman, O. (1972). *Defensible space – Crime prevention through urban design*. New York, NY: Macmillan.

Newman, O. (1976). *Design guidelines for creating defensible space*. Washington, DC: US GPO, National Institute of Law Enforcement and Criminal Justice.

Reis, A. T., & Lay, M. C. (1996). Security in private and semi-private housing spaces in Brazil. In J. Nasar & B. Brown (Eds.), *Public and private places. Proceedings of the 27th annual conference of the Environmental Design Research Association* (pp. 117–123). Salt Lake City: EDRA.

Reis, A., Portella, A., Bennett, J., & Lay, M. C. (2003). Accessibility and security: Syntactic and perceptual analysis in two low-income housing estates. In J. Hanson (Ed.), *Proceedings of the 4th International Space Syntax Symposium* (pp. 44.1–44.12). London, UK: UCL, Space Syntax Laboratory.

Sahbaz, Ö., & Hillier, B. (2007). The story of the crime: Functional, temporal and spatial tendencies in street robbery. In A. Kubat, Ö. Ertekin, Y. Güney, & E. Eyüboglu (Eds.), *Proceedings of the 6th International Space Syntax Symposium* (pp. 022-01–022-14). Istanbul: ITÜ, Faculty of Architecture.

Shu, C. (2009). Spatial configuration of residential area and vulnerability of burglary: Case studies from UK and Taiwan. In D. Koch, L. Marcus, & J. Steen (Eds.), *Proceedings of the 7th International Space Syntax Symposium* (pp. 102:1–102:15). Stockholm, Sweden: KTH, Faculty of Architecture.

Shu, S., & Huang, J. (2003). Spatial configuration and vulnerability of burglary: A case study of a city in Taiwan. In J. Hanson (Ed.). *Proceedings of the 4th International Space Syntax Symposium* (pp. 46.1–46.14). London, UK: UCL, Space Syntax Laboratory.

Zaluar, A. (2002) (Ed.), *Da revolta ao Crime S.A.* [From revolt to crime corporation]. São Paulo: Editora Moderna.

Zanotto, K. (2002). *Segurança em Área Urbana Central: Configuração, Forma Urbana e Usuários* [Central urban area security: Configuration, urban form and users] (Unpublished master's thesis). Federal University of Rio Grande do Sul, Porto Alegre.

Social Vulnerability and Resilience of Young People

A Pilot Study in Fortaleza, Brazil

Zulmira Áurea Cruz Bomfim, Veriana de Fátima Rodrigues Colaço, Idilva Maria Pires Germano, Luciana Lobo Miranda, Andréa Filgueiras, Helenira Fonseca de Alencar, Letícia Leite Bessa, and Ticiana Santiago

Programa de Pós-Graduacao Em Psicologia, Universidade Federal do Ceará – UFC, Fortaleza-Ceará, Brazil

---

Abstract

This work discusses a pilot study of a research project[1] investigating the bio-socio-demographic profile of young people (age 14–24) attending public schools in Fortaleza (Ceará-Brazil). The aim is to investigate indicators of social risk and protective factors associated with conditions of social vulnerability, including economic, educational, work, and housing conditions, among others. Here we present a descriptive analysis of a minor sample (*n* = 64). The main risk factors are high school failure and need to stop school attendance to work and contribute to the family income. The main protective factors are related to self-esteem, self-efficacy, and perception of life chances. The results may reflect that positive resources are available and could help these young people mitigate obstacles in a resilient way.

Key words: pilot study, social risk, social vulnerability, youth

---

[1] "Adolescence and youth: Risk situation and protection networks in Fortaleza" – Exchange project (2009–2011) among graduate programs in Psychology of UFC (Federal University of Ceará) and UFRGS (Federal University of Rio Grande do Sul) financed by CNPq (National Council of Scientific and Technological Development).

Context of the Research

Knowledge about the demographic, social, cultural, economic, and health factors associated with young people's lives is the main basis for establishing public and private policies directed to their needs. In spite of this, a shortage of specific measures to evaluate social vulnerability of young people and protective factors that may lead to the promotion of their well-being is still being observed, especially in Latin America.

To fill this gap, some initiatives have gathered information about the conditions of young people's development in Brazilian cities. Several limitations imposed by economic and social contexts make Brazilian youths particularly vulnerable to risk factors such as unsafe sexual activity, risk of use of alcohol and illegal drugs, domestic and urban violence, and unemployment. Other factors also represent a risk to the well-being of young people are, having to repeat a year or years at school or dropping out of school and early entry into the labor market (Dell'Aglio, Cunningham, Koller, Borges, & Leon, 2007). Important research coordinated by the World Bank has included specific studies on youth at risk conducted in São Paulo, Recife, and Porto Alegre (World Bank, 2007). This model of research was later expanded to other cities like Manaus, Brasília, Belo Horizonte, and Presidente Prudente (Libório & Koller, 2009).

The research project, which includes this pilot study, has recently completed the data collection about risk and social protection from a sample of 1,140 young people from Fortaleza. This work expands on preexisting information about the current situation of young people in the northeast region of Brazil. Based on previous indicators of health, risk-taking behavior, school performance, institutional connectedness, and socioeconomic conditions, our evaluation shows that the region has not been successful in enhancing the well-being of young people (Dell'Aglio et al., 2007). Fortaleza is among the 200 Brazilian municipal districts with a high number of juvenile victimization, with a total of 378 deaths of young people from a total of 847 people murdered in the district in 2006 (Waiselfisz, 2008). This high percentage of victimization (41%) suggests that significant risk factors of juvenile segments of the society have been and are occurring to produce this undesirable scenario.

Therefore, precise information on social risks and forms of protection for young people in Fortaleza can help us understand which problems these people face and to design strategies to aid their development.

This paper presents a descriptive analysis of the pilot-study sample of 64 students selected from public schools and with low household income. In Brazil, students from public schools generally show lower economic and living conditions comparing to middle-class youth who tend to go to private schools. Therefore, public school students tend to be considered more at social risk and have more difficulties coping as a result of their unfavorable situations. This statement begs the following questions: Which risk and protection factors are present in such context? How do these young people positively adapt to the challenging conditions they face? To what extent are they vulnerable or resilient? The concept of psychological resilience addresses the maintenance of positive adjustment under significant threat or severe adversity. Resilience is not a personality trait but a dynamic process encompassing attributes of young people themselves, aspects of their families, and characteristics of their wider social environments (Luthar, Cicchetti, & Becker, 2000).

This paper aims to discuss a set of findings relating to risk and protection factors, social vulnerability, and dimensions of resilience suggested by the data were collected by means of a questionnaire. These results helped refine hypotheses which are to be explored and evaluate the questionnaire in preparation for its use for the major investigation.

Theoretical Basis

The concept of vulnerability has been discussed within different disciplines and approaches according to the focus on divergent dimensions, such as *risk* (risky events), *risk responses* (options for managing risk), and *outcomes* (welfare gain/loss) for individuals and populations (Alwang, Siegel, & Jørgensen, 2001). Part of the literature refers to "a state of high exposure to certain risks, combined with a reduced ability to protect or defend oneself against those risks and cope with their negative consequences" (United Nations, 2001). Social vulnerability is also defined as the difficulties that a person or a group may face and the extent to which they are able to resist or confront a particular threat or problem (Corraliza, 2000). These definitions imply three different aspects: (1) the existence of a potentially adverse event (exogenous and endogenous), (2) the capability of coping with certain situations (such as school dropout, precocious entry into the labor market, drug use, family disintegration, living on streets or child, and juvenile prostitution) due to lack of external and personal supporting resources, and (3) inability to adapt to the new situation caused by the risk (Busetta & Milito, 2010).

Although economic insecurity has been considered the main basis of social vulnerability, the answer to "what makes an individual or group vulnerable or at a greater risk of becoming vulnerable to a variety of social ills?" (United Nations, 2003, p. 1) demands a better understanding of individual and social dynamics interacting at micro, meso, and macro levels. Moreover, internal and external processes are at play in determining vulnerable or resilient pathways.

In addition, the exposure to risks and the ability to face them do not remain stable during one's life span but vary from one stage of life to another (United Nations, 2001). In fact, risk and protection of young people in transition to adulthood have been a central concern in academic and social policy, and impose special challenges for society in general. The discussion on youths' social vulnerability and risk addresses the ingredients for successful development, the spheres of influence resulting in positive and negative trajectories, and criteria for designing and implementing programs for young people (Roth & Brooks-Gunn, 2000). Research in the field of developmental psychology has accumulated increasing knowledge on the multiple interacting domains of the risk behavior of young people (Jessor, 1993; Jessor, Van den Bos, Vanderryn, Costa, & Turbin, 1995) and protective factors enhancing their healthy development (Ungar, 2006; Ungar et al., 2007;

Wigfield & Cambria, 2010; Zimmerman, Copeland, Shope, & Dielman, 1997). It is generally recognized that although a risk environment may lead to risk behavior, such as drug use, unsafe sex practices, participation in violent acts and illegal activities, this is not a linear equation since personal, historical, and cultural conditions interact to create different risk situations, evaluations and possibilities of coping with risky circumstances.

Studies of young people that investigate resilience from an ecological perspective tend to define resilience as health in spite of adversity, highlighting the ability of the individuals to minimize the adverse conditions of cultural deprivation, and improve their social and economic chances (Castro & Abramovay, 2002).

Youth vulnerability and resilience discussed here draw mainly from the ecology of human development (EHD) approach (Bronfenbrenner, 2000), but also from the historical-cultural theory (Vygostsky, 1995), which emphasize the role of social interaction in human development. The bio-ecological model proposes that human development occurs through processes of reciprocal interaction, among people, objects and symbols within and between one's changing immediate settings, as well as larger social contexts (Bronfenbrenner, 2000). The EHD approach asserts that human development can only be understood if properly contextualized by taking into account the dynamic interaction of four dimensions: Person, process, context, and time. The historical-cultural theory also explains human development on the basis of the growing person's immersion in his/her culture in daily social interactions.

These theoretical models help to reflect on the social vulnerability of young people, risk, and resilience in terms of a complex transactional process which encompasses multiple variables through interaction, and eschewing from a predetermined view of such phenomena.

Methodology

The pilot study was conducted in two schools which assist young people aged 14–24 years old, male and female. One of these schools was connected to a NGO (non-governmental organization) which teaches dance to young people in suburban areas in Fortaleza. The second school was a Catholic social project for young adult education. These schools were chosen because they attend a student population similar to the universe of respondents aimed by the study. Thirty-three participants (age 14–17) from the NGO dance program and 31 participants at the age of 18–24 from the Catholic school took part in the pilot study.

Regarding the sample's characteristics, the largest concentration of subjects are female (84%), single (94%), and in the age group 14–17 years old (52%). These numbers reflect the peculiarities of the schools chosen in this study whose target populations are mostly female (a dance school and a program aiming the education of young adults). Most of them (44%) identify themselves as of brown skin color, when added to the percentage of people who identified themselves as black (13%) the result is 56% of the total participants. Therefore the majority of the sample describes themselves as afro-descendents.

As the objective of the pilot study was to evaluate the questionnaire used to gather data from the main sample (composed of multiple choice questions and items of *Likert* scales), the central procedure was the collective application of the questionnaire. The questionnaire involves bio-socio-demographic information from adolescents and young people and their families, as well as themes related to education and work, health and life quality, risk behaviors (drugs, suicide, risky sexual habits, violence, etc.), and risk factors (violence at home and in the community, exposure to diseases/drugs, physical disabilities, discrimination, living in shelters or institutions for delinquent youth, life on the streets, conflict with the law, impoverishment/poverty, separation/loss in the family). It also explores social protecting factors (leisure, support network, family cohesion/satisfaction, friendships) and personal protecting factors (spirituality, values/morality, self-esteem/creativity, fulfillment/welfare, optimism, sense for life, humor, altruism/sociability, self-efficacy, and future aspirations).

For the group who took part in the NGO dance program, in addition to the application of the questionnaire, a "warm-up activity" was used employing the Circle of Culture methodology proposed by Paulo Freire (1983). This activity involved generating words which were related to the themes of the questionnaire and some related to youth's daily life, for instance: Young person, family, school, work, sexuality, drugs, violence, future, aspirations, media, religion, fear, and hope. The warm-up procedure was not possible in the school because the application of the questionnaire was conducted at night with students attending Education for Young Adults program (designed for students working during the day). Therefore, the available time was not enough for both activities. The results of the circle of culture were not used as data, serving only as a warm-up for completing the questionnaire. The final questionnaire was analyzed by means of descriptive statistics using the SPSS statistical program.

Main Findings and Discussion on Juvenile Vulnerability and Resilience

Among the risk and protection factors found in our pilot sample some results are highlighted here: School success/failure, parents' educational level, family income, self-esteem, self-efficacy, and perceived life chances. The literature on resilience has discussed several dimensions involved in positive outcomes and the well-being of young people in terms of the personal characteristics (e.g., IQ, self-efficacy), family characteristics, and social affective support network (school, community formal services, friends, peers; Koller & Lisboa, 2007).

In regard to the results concerning school life, 31% of participants answered that they had already failed school and 67% had not failed school yet. In addition, the results showed that the participants who had been reproved (n = 20), 35% had been once and 60% had been more than once. Furthermore, 82% of participants mentioned having a high identification with the school and 6% mentioned having a low identification. Moreover, 86% of participants said that they had a high level of trust in the school personnel, while

8% of the participants reported a low level of trust. Finally, when asked about the level of trust in school friends, 44% judged it high and 52% judged it low.

The high indication of school failure observed in the results is partly related to the fact that almost half of the students (48%) of the pilot study attended the Education for Young Adults program, which aims to help this population with previous educational problems. However, it was highlighted that the great majority of these participants value the school (83%) and trust the teachers (86%). In regard to the protective effect of institutional connectedness (Dell'Aglio et al., 2007), these numbers may indicate that the goal of repairing earlier learning damage is partly achieved by educational policies targeting mature age students and by encouraging trusting interpersonal relationships in school environments such as friends, teachers, supervisors among others.

More than half of the sample (53%) admitted to an interruption in school attendance at least once ("Have you ever had to stop attending school in order to work?"). This group's family average income (US$ 144.5) was higher than the income of the group that had never stopped going to school (US$ 99.4). This suggests that youth's earnings help to raise family overall income. Among the participants who answered who were working, it was found that a working as a domestic servant was the most common job (31%) and the income received by most of them only reaches the minimum wage, now around US$ 300 per month.

Relations were found between the degree of education of the parents and family income. If the father or mother has at least a secondary undergraduate degree, family income tends to increase gradually (Figure 1).

Economic difficulties were the most mentioned stressing factors: Economic level decrease (66%) as well as unemployment in the family (54%). Another factor was the death of a significant person (62%). References to hunger situations (18%) and prison in the family environment (15%) were also mentioned.

The items regarding self-esteem were obtained through a summarized version of Rosenberg Self-Esteem Scale (1989/1956) included in the questionnaire which revealed particularly high levels of self-esteem indicators: "I feel that I have a number of good qualities" (95%), "I have reasons to be proud in life" (95%); "On the whole, I am satisfied with myself" (92%). High self-esteem has been related to reporting more educational and higher occupational goals (Massey, Gebhardt, & Garnefski, 2008). Moreover, other studies have shown that higher self-esteem is related to more internal control beliefs regarding the future, greater goal progress, and greater probability estimations of attaining family and education goals, as reviewed by Massey et al. (2008).

A similar positive scenario was found in answers related to self-efficacy: "When I face a problem, I usually get to find several solutions" (88%); "I find it easy to persist in my intentions and reach my objectives" (88%); "I can solve most of the problems, if I do the necessary effort" (88%). Self-efficacy has been found to be strongly related to probability estimates of educational, occupational, and family goals (Trempała & Malmberg, 1998).

As Table 1 shows, the indicators of perceived life chances obtained through an adapted scale from Jessor, Donovan, and Costa (1990) suggest high levels of expectations in the future: The perspective of concluding high school (87%), the foresight of being healthy

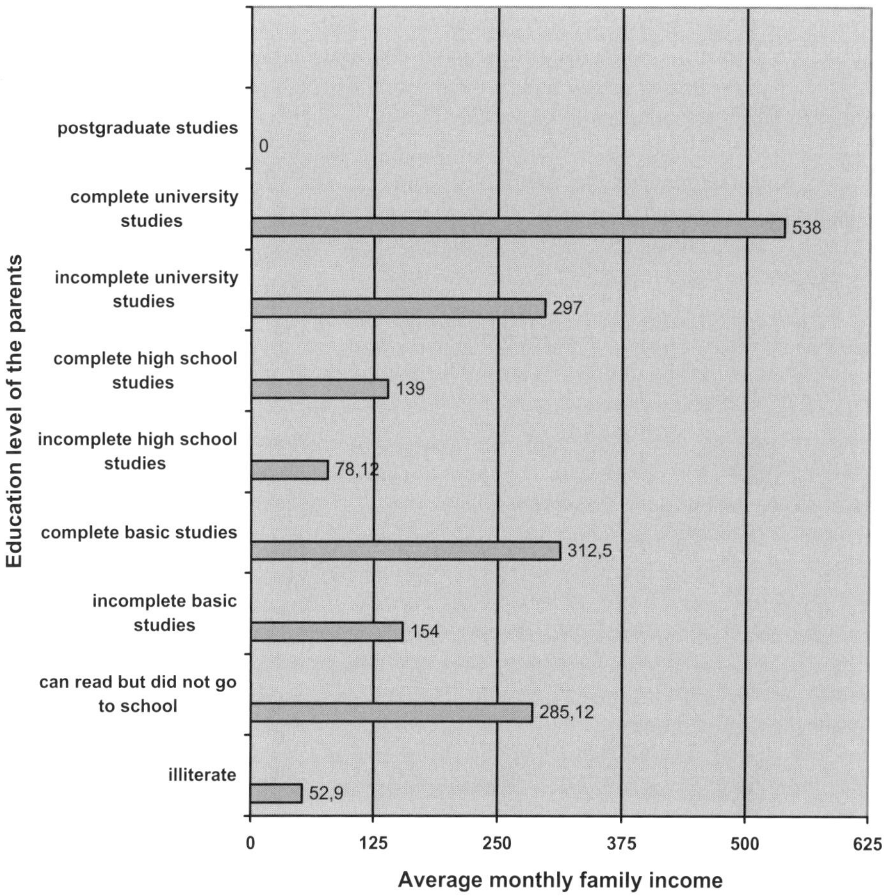

Figure 1. Family income (US$) and parents' education.

(84%), and of having supporting friends (90%). The high perception of the possibility of being admitted into university in the future (65%), in spite of previous and repetitive school failure is considered particularly suggestive of the influence of personal protective factors. As literature shows, young people's aspirations and the pursuit of personal goals are linked to both behaviors and overall well-being, not only for Western samples, but also for adolescents and youth from developing countries (Massey et al., 2008).

Data concerning self-esteem (Rosenberg, 1989/1956), self-efficacy and future expectations (Jessor et al., 1990) suggest a possible positive intercorrelation to be explored in the main study of the research project. The aspirations and goals that young people strive to achieve and the feeling of being able to work through the obstacles on the way have a

Table 1. Questions related to perceived life chances (n = 64)

Item		*f*	%*
Finish high school	Low	1	1.7
	Average	7	11.7
	High	52	86.7
Be admitted to university	Low	7	11.7
	Average	14	23.3
	High	39	65
Have a job that guarantees me a good quality of life	Low	1	1.7
	Average	17	29.3
Have my own house	High	40	69
	Low	4	6.9
	Average	11	19.0
Have a job that will give me satisfaction	High	43	74.1
	Low	4	6.9
	Average	9	15.5
Have a family	High	45	77.6
	Low	3	5.1
	Average	13	22.0
Be healthy most of the time	High	43	72.9
	Low	1	1.8
	Average	8	14
Be respected in my community	High	48	84.2
	Low	4	7
	Average	9	15.8
Have friends who will support me	High	44	77.2
	Low	1	1.7
	Average	5	8.6
	High	5	89.7

Notes. *Valid percentage (discarded students who did not answer). (Scale adapted from Jessor et al., 1990)

significant impact on behavior, self-perception, and well-being, therefore on academic and other sorts of resilience (Massey et al., 2008).

In short, in regard to school life, the high level of participants who have failed school is, in part, due to the fact that a large number of the participants were enrolled in Education for Young Adults program, which aims to help this population with previous educational problems. On the one hand, a high rate of school failure is undoubtedly a negative outcome for young people's development. On the other hand, most of them showed high identification with the school and a high level of trust in school staff and teachers. These results are considered protective measures employed by these young people, which in turn increases their resilience. The study also showed a conflict between school attendance and need to work which has implications for family income. Furthermore, on the one hand, Students who have never dropped out of school have lower family income, raising a hypothesis that those students have not been working and therefore have not been able to help their family financially. On the other hand, the data suggest that those who have dropped out of school have done so in order to work and contribute to their family income. The economic difficulties faced by their families appear as stress factors with implications for school performance.

These data show that this sample presents indicators of social vulnerability similar to observations pointed out by other studies on social vulnerability and exposure to social risk among Brazilian youth (Dell'Aglio et al., 2007; Koller & Lisboa, 2007; Libório & Koller, 2009). However, the group also showed favorable rates concerning protective factors especially when it came to issues of positive personal characteristics, such as self-esteem, self-efficacy, and perceived life chances. This web of protective and risk factors helps to appreciate that social vulnerability must be understood by taking into account the variety of individual attributes, contexts, and processes experienced by youths in the course of their lives (Bronfenbrenner, 2000/1979).

During the analysis some problems were noticed with the use of the questionnaire: The need of reduction of items, given that the extension of the questionnaire took a long time, making it tiresome and slow in its execution. As a result of this, it was reduced from 109 items to 77 items. The other modifications were related to the revision of some items to make them clearer, to the organization of some Likert scales in order to separate positive and negative sub-items and facilitate their analyses; and to alterations of repetitive sub-items.

In spite of those modifications, the validity of the instrument was confirmed for the objectives of the research. However, a significant effect of the "culture circle" activity on the participants' understanding or motivation was not verified. Nonetheless, it was considered important to maintain the activity whenever possible during data collection considering that it involved a lively and beneficial discussion among adolescents concerning interests and problems present in their daily lives.

Final Considerations

The pilot study was relevant for the quality of the final questionnaire and continuity of the research. The results helped to raise hypotheses to be explored during analysis of the data gathered from the final main sample. The participants belong to a vulnerable segment of the population that faces a complex set of obstacles to their development. The vulnerability verified in this pilot study reflects the EHD and historical-cultural theories which helped us to discuss the observed problems in a contextualized way, avoiding a determinist vision of development.

The young people studied here are part of contexts and situations characterized by lack of opportunities related to education, family environment and socioeconomic factors, such as high frequency of school failure, unemployment in the family, imperatives to get a job, and help raising family income, among others.

However, this does not mean that they are *determined* by those factors and environments; they have an active role in directing their lives. We support Vygotsky's (1995) notion that one is active and an agent of transformation of his/her own development. In daily social interactions it is possible for an individual to build his/her singularity. The meaning of risk and vulnerability and also of coping and resilience is dynamically constructed in ongoing social and cultural relations within which the individual is embedded.

We have highlighted the private experiences and the affective-volitional living of these young people in two educational environments, which resulted in answers that seem to indicate the participants' positive self-esteem and feelings of self-efficacy. Such results lead us to the network of interpersonal and institutional interactions related to the valorization of the self and to factors and situations of personal protection.

These results also help to critically consider the concept of social vulnerability, avoiding identifying it in a simplistic way (e.g., as poverty or low income). Social vulnerability of youth should be understood in a complex way, as product and producer of realities, that is, as a phenomenon that does not just restrict people's opportunities and lead to the reproduction of the *status quo*, but one that can produce new resistance forms and alternative manners of coping and leading personal and community life. The positive findings in self-esteem, self-efficacy, and future expectations indicators may reflect that protective resources are available and helping these young people to face and mitigate obstacles in a resilient way.

Promoting resilience among young people requires a variety of interconnected actions ranging from wider public policies to protection actions targeting micro and meso systems, including relations in family, school, community, and other settings. Although the definition of healthy development must be culturally oriented, researchers tend to encompass attributes like competence in academic, social, and vocational areas; confidence or a positive self-identity, connection or healthy relations to community, family, and peers, character or positive values, integrity, moral commitment, and caring and compassion (Roth & Brooks-Gunn, 2000). The main strategy is to focus on the individual's potentials and capacities, rather than just on problems. Efforts should be directed toward implementing those capabilities in the domains of leisure, organized activity participation, community place attachment, connectedness to school and other institutions.

There are no obvious ways to guarantee present and future well-being for youth; life is basically navigating through uncertain horizons. Nevertheless, social vulnerability stemming from socioeconomic disadvantages may impose additional damages and challenges for youth and society. Research and intervention should reflect on cultural contingencies as well as wider ethical and political dimensions of risk.

References

Alwang, J., Siegel, P. B., & Jørgensen, S. L. (2001). *Vulnerability: A view from different disciplines* (Social Protection Discussion Paper, 115). Washington, DC: The World Bank. Retrieved from http://siteresources.worldbank.org/SOCIALPROTECTION/Resources/SP-Discussion-papers/Social-Risk-Management-DP/0115.pdf

Bronfenbrenner, U. (2000/1979). *A ecologia do desenvolvimento humano: experimentos naturais e planejados* [The ecology of human development: Experiments by nature and design]. Porto Alegre, Brazil: Artmed.

Busetta, A., & Milito, A. M. (2010). Socio-demographic vulnerability: The condition of Italian young people. *Social Indicators Research, 97*, 375–396.

Castro, M. G., & Abramovay, M. (2002). Jovens em situação de pobreza, vulnerabilidades sociais e violências [Youth in situation of poverty, social vulnerability and violence]. *Cadernos de Pesquisa, 116*, 143–176.

Corraliza, J. A. R. (2000). *Vida urbana y experiencia social: variedad, cohesión y medio ambiente* [Urban life and social experience: Variety, cohesion and environment]. Madrid, España: Psicothema.

Dell'Aglio, D. D., Cunningham, W., Koller, S. H., Borges, V. C., & Leon, J. S. (2007). Youth well-being in Brazil: An index for cross-regional comparisons (World Bank Policy Research Working Paper No. 4189). Retrieved from http://ssrn.com/abstract=979661

Freire, P. (1983). *Educação como prática da liberdade* [Education as practice of freedom]. Rio de Janeiro, Brazil: Paz e Terra.

Jessor, R. (1993). Successful adolescent development among youth in high-risk settings. *American Psychologist, 48*, 117–126.

Jessor, R., Donovan, J. E., & Costa, F. (1990). Personality, perceived life chances and adolescent health behavior. In K. Hurrelmann & F. Lösel (Eds.), *Health hazards in adolescence* (pp. 25–410). New York, NY: de Gruyter.

Jessor, R., Van den Bos, J., Vanderryn, J., Costa, F. M., & Turbin, M. S. (1995). Protective factors in adolescent problem behaviour: Moderator effects and development change. *Development Psychology, 31*(6), 923–933.

Koller, S. H., & Lisboa, C. (2007). Brazilian approaches to understanding and building resilience in at-risk populations. *Child and Adolescent Psychiatric Clinics of North America, 16*, 341–356.

Libório, R. M. C., & Koller, S. H. (Eds.). (2009). *Adolescência e juventude: risco e proteção na realidade brasileira* [Adolescence and youth: Risk and protection in Brazilian reality]. São Paulo, Brazil: Casa do Psicólogo.

Luthar, S. S., Cicchetti, D., & Becker, B. (2000). The construct of resilience: A critical evaluation and guidelines for future work. *Child Development, 71*, 243–262.

Massey, E. K., Gebhardt, W. A., & Garnefski, N. (2008). Adolescent goal content and pursuit: A review of the literature from the past 16 years. *Developmental Review, 28*, 421–460.

Rosenberg, M. (1989/1956). *Society and the adolescent self-image*. Princeton, NJ: Princeton University Press.

Roth, J., & Brooks-Gunn, J. (2000). What do adolescents need for healthy development? Implications for youth policy. *Social Policy Report, 15*, 1–20.

Trempala, J., & Malmberg, L.-E. (1998). The anticipated transition to adulthood: Effects of culture and individual experience on Polish and Finnish adolescents' future orientations. *The Journal of Psychology, 132*, 255–266.

Ungar, M. (2006). Nurturing hidden resilience in at-risk youth in different cultures. *Journal of the Canadian Academy of Child and Adolescent Psychiatry, 15*, 52–58.

Ungar, M., Brown, M., Liebenberg, L., Othman, R., Kwong, W. M., Armstrong, M., & Gilgun, J. (2007). Unique pathways to resilience across cultures. *Adolescence, 42*, 287–310.

United Nations Economic and Social Council. (2001). *Extract from the report on the world social situation* (E/2001/104). Retrieved from http://daccess-dds-ny.un.org/doc/UNDOC/GEN/N01/516/73/PDF/N0151673.pdf?OpenElement

United Nations, Division of Social Policy and Affairs. (2003). *Report on the world social situation, Social vulnerability: Sources and challenges.* Retrieved from http://www.un.org/esa/socdev/rwss/docs/2003/fullreport.pdf

Vygotsky, L. S. (1995). *Obras Escogidas, Problemas del desarrollo de la psique* [Selected works, problems of psychological development]. Madrid, Spain: Visor.

Waiselfisz, J. J. (2008). *Mapa da violência dos municípios brasileiros 2008* [Map of violence in Brazilian municipal districts 2008]. Brasília, Brazil: RITLA, Instituto Sangari/Ministério da Saúde/Ministério da Justiça.

Wigfield, A., & Cambria, J. (2010). Students' achievement values, goal orientations, and interest: Definitions, development, and relations to achievement outcomes. *Developmental Review, 30*, 1–35.

World Bank. (2007). *Jovens em situação de risco no Brasil* (Relatório Técnico do Banco Mundial/32310-BR) [Risk situation of youth in Brazil. Technical Report of the World Bank], Brasilia, Brazil: Banco Mundial.

Zimmerman, M. A., Copeland, L. A., Shope, J. T., & Dielman, T. E. (1997). A longitudinal study of self-esteem: Implications for adolescent development. *Journal of Youth and Adolescence, 26*, 117–141.

Participation and Strategies of Risk Reduction

Participation

The Silver Bullet for the Acceptance of Renewable Energies?

Irina Rau,[1] Petra Schweizer-Ries,[1,2] and Jan Hildebrand[2]

[1]Research Group Environmental Psychology, Saarland University, branch office at the Otto von Guericke University, Magdeburg, Germany
[2]Research Group Environmental Psychology, Saarland University, Saarbrücken, Germany

Abstract

Using a multi-method approach, we look at the aspects and influencing factors of renewable energy technology (RET) acceptance and their relationships to various forms and opportunities for participation. We describe four levels of involvement: Information, consultation, cooperation, and citizens' control. Participation may be realised through a variety of methods, including information distribution, round tables, and opinion polls, all of which should be of high quality and available in all phases of RET introduction and use. We demonstrate a discrepancy between actual and desired participation options. The importance of participation for active acceptance is mediated by several aspects that are relevant in the context of environmental change, including justice and trust. A high potential for engagement among the survey respondents is demonstrated. In order to motivate people to become active supporters of a specific RET in their own community, the planning and decision-making process should be designed in a way that meets procedural and distributive justice criteria. Nevertheless, participation is not a guarantee for (active) acceptance, as other relevant factors also play a role.

Key words: acceptance, conservation ecological behaviour, justice, participation, renewable energies

Participation in the Use of Renewable Energy Technologies

Due to global development and growth, societies are facing an increasing need for energy and the associated risks of running out of energy and into distribution conflicts. Fossil and nuclear energy sources are finite and are climate-damaging or involve unpredictable dangers. Therefore, changes in these complex energy production and supply systems are essential, in order to enable development towards a sustainable energy community with reduced supply vulnerabilities and reduced risks of environmental and health damage. Decentralised renewable energy technologies (RET) play a major role in these new energy systems, necessitating popular acceptance of renewable energies in general, and of concrete, renewable energy plants on local levels specifically, the understanding of the complexity of the energy systems, as well as the active part everybody is contributing for its functioning (e.g., energy-saving behaviour, demand side management and load shifting).

The research described here is concerned with the social aspect of the expansion of the use of RET uses, where various societal levels play a role and, ideally, interlock. On the political level, development goals concerning the implementation of renewable energies have been formulated, which affect the composition of our future energy supply. The realisation of these goals takes place on the regional and local level through the construction of renewable energy plants, thus necessitating the acceptance of renewable energy use by those involved, or affected. In one of our research projects, "Acceptance of renewable energies and questions of social sciences,"[1] the majority of those consulted (94.9%) agreed that renewable energies should play a major role in Germany's energy supply in the future (Zoellner, Schweizer-Ries, & Wemheuer, 2008). On the local level, the degree of acceptance of a renewable energy plant depends on several parameters.

Planning and decision-making processes, and in particular their opportunities for participation, play a crucial role in the acceptance of the technological changes involved in supplying the national grid with increasing amounts of renewable energy. In this chapter, we will consider the role of participation, as well as the importance of several relevant aspects of planning and decision-making processes, including the wishes for and appraisals of participation at various levels, choices of concrete forms of participation, and the importance of perceived justice.

Theoretical Background of RET Acceptance and Participation

The Research Group Environmental Psychology developed, in cooperation with practitioners and other scientists, a basic definition and operationalisation of various dimensions of

[1] Funded by the German Federal Ministry for the Environment, Nature Conservation and Nuclear Safety; conducted by the Research Group Environmental Psychology from 2005 to 2008, FKZ 0327505.

acceptance, involving various forms and areas of acceptance (Zoellner, Schweizer-Ries, & Rau, in press). Building upon our model of action (Schweizer-Ries, Zoellner, & Rau, in press; Schweizer-Ries, 2008; Zoellner, Schweizer-Ries, & Wemheuer, 2008), the spectrum of acceptance is defined by two facets: Appraisal and action. It is a simplified model of human action (according to Kaufmann-Hayoz & Gutscher, 2001) that takes into account the complexity of the entire energy system. Because it is a systemic model, the interventions are also designed in a systemic way, based on cultural theories of transactions. Applied to the acceptance of RET, the possibilities of the appraisal of a certain object range from positive to negative, the (re)action manifests itself from passivity to activity. The combination of a positive appraisal of, in our context, renewable energies and little activity can be described as approval. A negative appraisal with a high level of activity can be described as resistance. In our understanding, a positive appraisal of renewable energies is the necessary precondition for them to be deemed acceptable. This is a crucial difference to the planning context, in which acceptance of a concrete undertaking is very often postulated, as long as active resistance is not apparent. Adding to a positive appraisal the activity leads to active acceptance in the form of support or commitment for renewable energies (Figure 1).

Furthermore, we differentiate three forms of acceptance: General acceptance of renewable energies, acceptance of various renewable energies (wind, biomass, solar) and active acceptance.

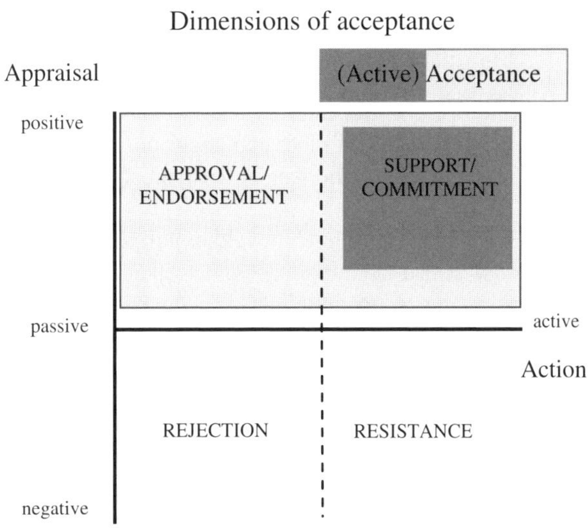

Figure 1. Definition of active acceptance as positive appraisal and active contribution (see also Zoellner, Rau, & Schweizer-Ries, 2009; Zoellner et al., in press).

- General acceptance of renewable energies is implied by a positive appraisal of renewable energies, in general, including the construction of a renewable energy plant in one's own surrounding, but not providing active support for the construction of a specific plant.
- The acceptance of the various technologies applies the same principle (positive appraisal in general, as well as accepting the construction of such an energy plant in one's own vicinity) for the three renewable technologies: Wind power stations, biomass plants, and photovoltaic fields.
- Finally, active acceptance is defined through the willingness to actively support a concrete RET or the construction of a renewable energy plant.

In order to illustrate and structure the complexity of the acceptance of renewable energies the aspects influencing these forms of acceptance can be classified in three areas:

1. TECHNOLOGICAL: Some aspects are linked to the technology itself. People fear, for example, accidents in biomass plants or birds being affected by wind energy plants.
2. LOCATIONAL: Other aspects depend on the location chosen for the plant and are associated with changes in the landscape, influences on local and regional identification, and concernment. A wind energy plant that interferes with the view of a church, an emblem or a landmark of the city, or other emotionally relevant landmarks, is an example for the impact of a poorly chosen location.
3. PROCEDURAL: Aspects concerning the planning and decision-making processes are of great relevance, including questions of justice and transparency, trust and opportunities for participation.

These aspects interrelate and also can intensify their influence on each other. For example, certain aspects can only take effect if other conditions are met – if biomass plants are constructed far from a residential area, the fear of accidents or odour nuisance becomes very unlikely.

Even though many key actors on different levels in political, administrative and economic spheres have acknowledged the importance of acceptance by the public, as well as by persons affected by any changes, opportunities for real participation are still rather rare. The direct link between participation and acceptance is rarely thoroughly analysed and explained, although it is sometimes referred to (e.g., Matthies, Homberger, Matthäus, Engelke, & Moczek, 2004; ÖGUT Strategic Group on Participation, 2004; Schweizer-Ries, 2011; Zoellner et al., 2008). In order to address this issue, we build on an understanding of participation that is adapted from Arnstein (1969), and simplified by Paul (1987), Lüttringhaus (2003), and Fischer, Schophaus, Trénel, and Wallentin (2003). Even though it is a slightly artificial distinction, we decided to differentiate two groups of persons involved in these processes: The persons who involve others through participation and those who participate (Figure 2). Some of the actors, representatives of an environmental organisation,

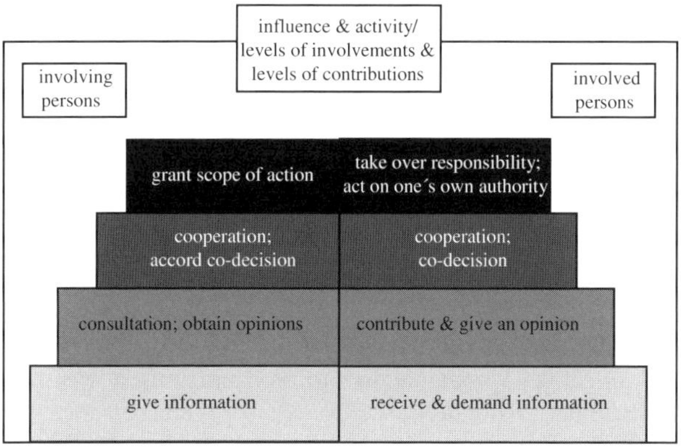

Figure 2. Participation pyramid. (Adapted from Rau et al., 2011)

for example, are, on the one hand, participants and on the other, they involve or activate other persons, for example, their co-members, during the process.

Participation can be realised to varying extents, referred to as increasing levels of participation. The higher the level, the more responsibility is asked of participants, along with increased active contribution to the process.

Unlike Arnstein (1969), who developed the famous ladder of participation, we include "information" on the lowest level of participation, because we see information giving and information processing as a very important and demanding process that is especially important for supporting transparency (an important aspect of ascribing fairness).

On the level of "consultation," citizens are asked for their opinions. Arnstein (1969) names both levels – information and consultation – "tokenism," which means that there is no real sharing of power. This is an important aspect in the whole discussion on participation – to what extent are decision makers willing and able to share power?

At the level of "cooperation," the public receives the opportunity to co-decide, at least in some parts of the planning process. On this level, according to Arnstein (1969), the real participation begins, building upon a real partnership.

At the highest level, those involving the citizens delegate the decisions and share their power. The persons involved accept the partnership and shared responsibility. When it becomes a citizens' project, they act under full self-authority, ideally supported by those who have political, administrative and financial power. This level can also be initiated by the citizens themselves, which can then turn into self-reliance and citizen control (as Arnstein calls it) for the planned RET, or taking the initiative to construct for example, a biomass plant to supply a village. We subsumed citizens' own projects in the highest level (see Figure 2), unlike Lüttringhaus (2003), who added a separate level on the participants' side for these kinds of projects in the hands of citizens.

With regard to the purpose of participation, Lane (1995) distinguishes between participation as a means and participation as an end:

> Participation may be a means to improve project effectiveness through the use of local information to correctly specify problems and needs, improve solutions, avoid misunderstandings. [...] Participation may however be seen as an end itself. It may have an intrinsic merit, if it increases self-esteem, confidence, and the individual's sense of power. (p. 183)

It also can help people to identify with and be proud of community developments. According to Minkler (2000), participation can be characterised as follows:

1. Recognition of limitations of expert knowledge and as a consequence, valorisation of local knowledge and recognition of community members as experts for their situation.
2. Tendency to be driven by community priorities rather than external expert or internal project priorities.
3. Emphasis on community strengths and its problem-solving capacity, use of local resources, and supporting local development.

Lane (1995) points out that in cases where participation is offered, it is very common in the implementation phase of projects and the beneficiaries are involved by contributing resources (e.g., time, active personal involvement). She concludes: "However, this is only one stage in the development process, and we must also consider participation in decision making, in implementation and maintenance, in benefits, and in evaluation of both successes and failures" (p. 182).

But participation as an end means "co-determination and power sharing throughout the [...] programme cycle" (Nelson & Wright, 1995, p. 4), where users are treated as equal partners from the initial idea of the project or programme, for example, energy supply development until the outcome evaluation and they receive, additionally, long term access to resources and decision making. "Effective participation implies involvement not only in information collection, but in analysis, decision-making and implementation – implying devolution of the power to decide" to those that would be affected most by possible risks (Pretty & Scoones, 1995, p. 160, see also Villalobos Montoya & Schweizer-Ries, 2005). This can also become a matter of empowerment, allowing especially vulnerable and powerless citizens to participate and to experience having influence and control over their lives.

One goal of the research referred to here[2] was to gain a deeper understanding of the different levels and forms of participation in Germany, related to the introduction of

[2] The main part of the data stem from the project „Aktivität & Teilhabe – Akzeptanz Erneuerbarer Energien durch Beteiligung steigern" [Activation and Participation – Increasing Acceptance of Renewable Energy through Involvement] (FKZ 0325052), in which we worked with the "Institut für Zukunftsstudien und Technologiebewertung" (IZT, Berlin) and the "Zentrum Technik und Gesellschaft" (ZTG, TU Berlin) as project partners in a very constructive interdisciplinary exchange.

RET; how they are perceived, evaluated, and used. Which wishes concerning participation on different levels exist, how are the existing offers evaluated, and which forms of participation should be offered? Furthermore, the relationship between participation and the acceptance, especially in its active form, of changes in the complex energy system was analysed. Does participation influence active acceptance and by which other psychological constructs is this relationship mediated? Here, for example, perceived justice, trust, local identity, and competence were operationalised and included, and their contribution to this relationship was specified.

In this chapter, we concentrate on the clarification of the participation levels and their concrete forms, and also present a brief analysis of the relationship between the different forms of perceived justice and active acceptance.

Multi-Method Approach

A multi-method approach for the data collection was applied to address the research questions described here. Questionnaires were distributed in three regions in Germany: Altmark, Harz, and Steinfurt county; the Altmark served as a pre-test, and in the other two regions, several communities were chosen for the surveys, with a total sample of $N = 859$. Furthermore, in order to intensify the research concerning aspect of concrete forms of participation, additional surveys were conducted in the city of Magdeburg ($N = 323$). In our approach, the research team visited the communities, distributed the questionnaires to the households (randomly and balanced for specific criteria), and used field diaries to note the reactions of interviewees and their impressions of the local situation, as additional information on the processes. A week later, the questionnaires were collected, or participants used prepaid envelopes to return the completed documents. With this method, a return rate of 29% was achieved for the first set of regions and 51% for Magdeburg.

Fourteen interviews were conducted with central actors (in operation, construction management, municipal representatives, regional and local experts, and citizens). Finally, the results and practical field recommendations were discussed and verified by practitioners and other researchers during several workshops. Biomass plants, wind power stations and large-scale photovoltaic power plants were the concrete technologies addressed in this study.

For the survey, in order to help clarify the relationship between participation and acceptance and to describe the planning and decision-making processes, several other constructs, including justice, trust and competences were investigated. In this chapter, we will concentrate on perceived justice (for the others see e.g., Rau, Zoellner, Nolting, Rupp, & Keppler, 2011). In the major section of the questionnaire, the survey participants were asked to give their estimations of various constructs (the appraisal and estimation of different aspects as well as wishes concerning the processes) on a five-point Likert scale with 1 representing "completely inapplicable," 2 "partly inapplicable," 3 "I am undecided," 4 "partly applicable," and 5 "completely applicable."[3] To distinguish the category "I am undecided" from having divergent information and actually not knowing, the category "I don't know" was added.

In the questionnaire's next section, the respondents were asked to evaluate concrete forms of participation and their relevance for future planning and decision-making processes.[4]

The use of these standardised, open format questions provided the opportunity to make additional comments for example, concerning future opportunities for participation. The questionnaire closed with some personal information (socio-demographic variables like age, gender, and educational background), always respecting the participants' anonymity.

The samples from the various communities and regions did not differ systematically. Most of the respondents had a General Certificate of Secondary Education and had completed an apprenticeship. Their ages ranged from 36 to 75 years. There was a slight preponderance of male respondents.

Results

Result I: Discrepancy Between Actual and Desired Participation Options

A clear desire for participation at all four levels of participation (see above, Figure 2) is evident. Wishes for information and consultation show the highest level ($M = 4.55$ and $M = 4.51$), followed by the wish for cooperation ($M = 4.17$). An obvious decline occurs for the level of acting autonomously ($M = 3.51$).

A comparison reveals that the wishes for participation exceed the appraisal of the actual participation options on all four levels of participation (see Figure 3).

Result II: Distinguishing Between Options for Participation

In principle, all forms of participation appear to be of importance for the realisation of planning and decision-making processes. The categories "rather important" and "important"

---

3 Despite the frequently-stated criticisms of the "convenient middle," we consciously decided to offer a "middle" response category. This was based on the assumption that, particularly in the context of process evaluation and in the field of renewable energies, possible uncertainties concerning the information and facts could make a clear statement difficult. Furthermore, this offered a reasonable possibility for those who could not, or did not want to, decide pro or contra, even though they considered it very carefully.

4 For this, five-point scales were also used, ranging from "good" to "bad" and from "very important" to "unimportant." In the additional survey in Magdeburg, the respondents were asked about the probability of use of different forms of participation (like round tables, excursions, flyers, etc.). The possible answers were "very unlikely," "rather unlikely," "undecided," "rather likely," and "very likely."

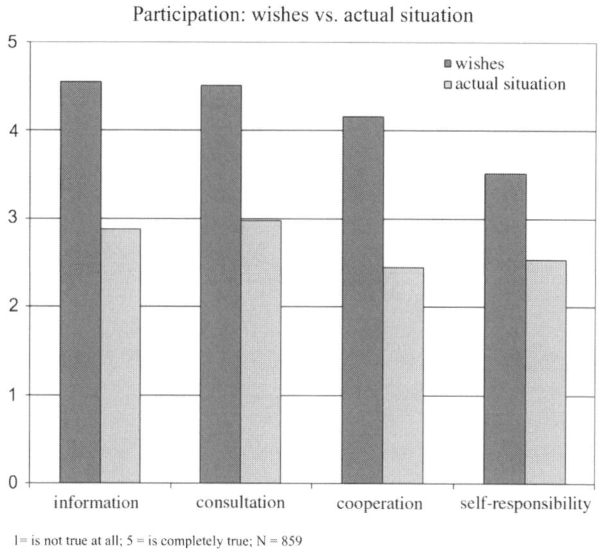

Figure 3. Differences between the actual and desired participation options for Renewable Energy Technology.

were chosen most often, with the percentage of answers in the other categories clearly declining towards the negative pole ("unimportant"). The option of financial participation shows a different distribution profile, with 25% "undecided" representing the category with the most answers.

Visiting the plant (46.9%), information through TV, radio, and other media (45.8%), an information event (43.1%) and an opinion poll (41.2%) are the forms of participation estimated as "very important," with more than 40% of the responses. Nevertheless, other options, each with a result of about 35%, were deemed "very important" including a round table (37%), website (34.2%), and informative flyers (32.9%). Less than 30% of the respondents deemed expert discussions (29.1%) to be important in future planning and decision-making processes (Figure 4 illustrates the distributions on the basis of some exemplary results). Additional information from the interviews revealed that expert discussions are important, in principle, but that it is crucial whether the source is considered to be reliable, and who consults the experts.

Result III: Probability of Using Participation Options

For the question of the probability of using one of the participation options under real-life conditions, the options related to information are also the most probable ones (Figure 5).

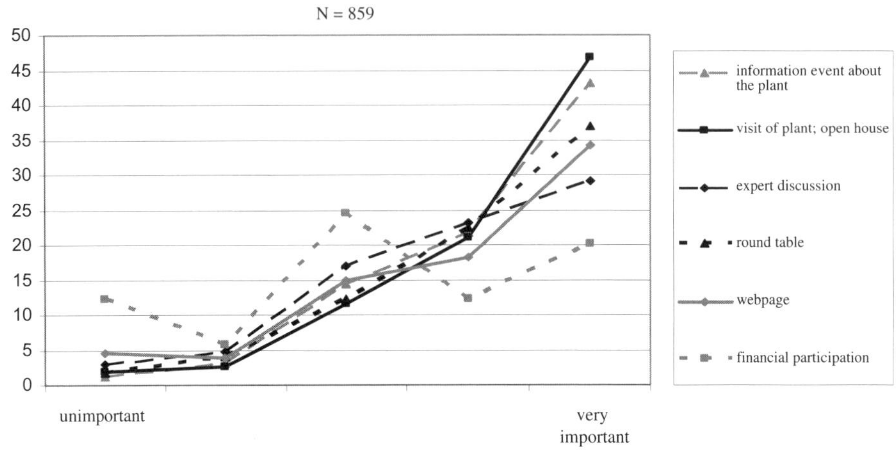

Figure 4. Assessment of the importance of participation option.

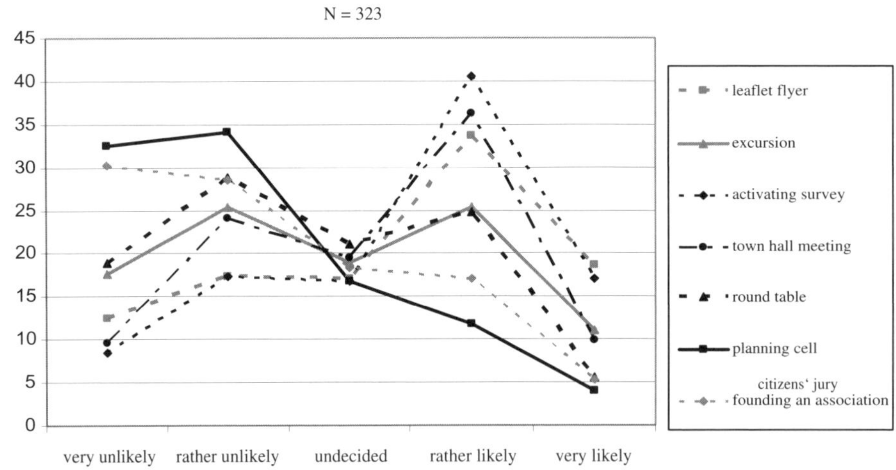

Figure 5. Probability of using various participation options under real-life conditions.

Responses do not occur at the extremes of the spectrum nearly as often; the most common category at the positive end of the spectrum is "relatively likely." The highest probability of use is for "activating survey" with 40.6% of those questioned estimating it to be "rather likely," town hall meeting (36.2%), and flyers (33.7%). Excursions and round tables follow, both with about 25%. The more elaborate forms of participation, such as the founding of an association (17.1%) and citizens' survey (11.8%) are the least probable.

Result IV: Assignment of Procedural and Distributive Justice Are Related to Active Acceptance of RET

We conducted a regression analysis, using the two forms of perceived justice, procedural and distributive, and with technology-related acceptance as the assumed influencing factor on the dependent variable active acceptance.

The results show a positive relation of the technology-related acceptance on the active acceptance ($\beta = .33$, $p < .001$). Distributive justice has a relation comparable with active acceptance ($\beta = .32$, $p < .001$) and procedural justice also has a highly significant, although smaller, relation with active acceptance ($\beta = .11$, $p < .01$). The regression model can explain 36% of the variance in the active acceptance (corr. $R^2 = .357$, $F = 94{,}246$, $p = .000$, see Figure 6).

Discussion: More Transparency and Options for Professionalised Participation in the Introduction and Use of RET Are Needed

Various persons, with their personal perspectives and interests, are concerned about changes in the energy system, such as the construction of renewable energy plants; companies or individuals constructing the plant, planners, politicians on the local, regional, and national level, administrative personnel, citizens' initiatives, and local residents all have their own experiences, motives, views, and competencies. The occurrence of conflicts between the various actors on various levels is thus not surprising.

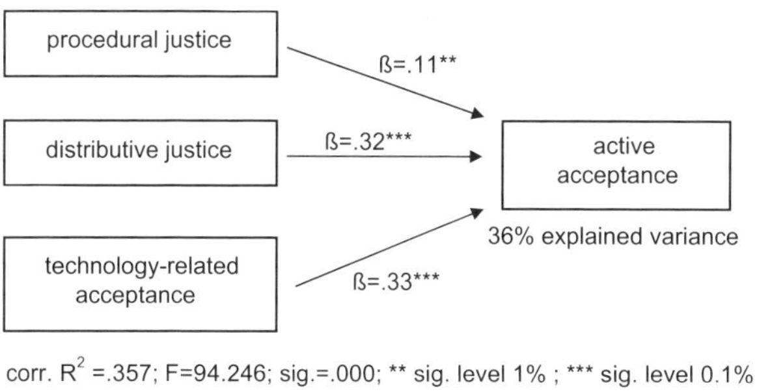

Figure 6. Influence of procedural and distributive justice on active acceptance of Renewable Energy Technology.

Against this background, and in the context of the expansion of renewable energies, planning and decision-making processes play a central role in achieving a high level of acceptance by concerned people. Wishes relating to participation among the respondents are high – people want to be informed and asked about their opinion regarding the planned changes in the complex energy system, including energy production, consumption, and management. These two levels of participation constitute the basis of a good planning process and are thereby essential (Figure 2: The two lowest steps of the participation pyramid). The survey respondents additionally expressed the wish to be involved in the decisions (cooperation: step 3). The desire to take responsibility and actively design the changes in one's own surrounding for example, by participating in planers' board meetings, seems to be limited by factors such as time, limited personal resources and competition with other private issues. In this context, it is important to consider that "dealing with" or "thinking about" renewable energies only represents a small part of daily life. Nevertheless, there is an increasing number of communities in Germany that have started citizen-initiated projects to implement RET (e.g., the bioenergy village of Juehnde).

Depending on the concrete local situation, a variety of aspects becomes effective in the judgement of the RET implementation. Also, whether somebody considers actively supporting the construction of an RET correlates more with the acceptance of the specific RET ($r = .46$; $p < .01$) than with the acceptance of renewable energies in general. Considering the divergence of the wishes concerning information and the estimation of the level and quality of information offered, the imparting of more, and reliable, information about the planning processes constitutes a means to create knowledge and transparency. This can already contribute to resolving some of the concerns about the RET and possible associated risks for the community.

In addition to the relevance of participation on the various levels, the importance of a variety of participation methods is evident. Methods that mainly provide information ("information about the planned changes"), such as contributions by different media, for example, have the highest priority. They constitute the basis for further steps of participation, such as round tables or visits to existing plants, through which proposed changes become more concrete and questions can be addressed ("discussion of the planned changes") as they arise. Finally the question of how to involve citizens in decisions about a changing energy supply remains. Recent developments provide a solid basis for visions of energy-sustainable communities (Schweizer-Ries, 2008), which will not be accepted and supported if appropriate participation is lacking (Rau et al., 2011).

The estimation of financial participation shows a different pattern; it is evident that this option is determined by certain underlying circumstances and individual prerequisites. As it is a relatively new instrument, experience with this form of participation is still lacking. Financial participation is currently not very common and not everyone can afford to make a financial contribution, in order to take part. Consequently, it must be examined and further adapted.

Even though most of the forms of participation are estimated to be of great importance for the design of planning and decision-making processes, this does not imply that people will take part in them; in social life, other activities compete and resources such as time and capacities are limited. In addition, some people who did not feel concerned and some of these questioned expressed resignation, because they felt of a lack of genuine influence. This is a very important point and corresponds to the question of sharing power: Is there a true desire for cooperation and citizens' participation, including the handing over of responsibilities and control?

This resignation is intensified through unfavourable motive assumptions, whereby opportunities for participation offered by planning boards are not seen as sincere efforts to involve, but rather as manipulative instruments (see discussion of "tokenism" above), based on prior negative experiences. These assumptions of dishonesty lead to non-participation.

External moderation would be expedient in extremely controversial cases, in order to resolve deadlocked positions and patterns of perception. In any case, it is important to design participation professionally; otherwise, a negative process may develop and result in the opposite of what was intended.

At this point, we would like to return to our opening question of whether participation is the silver bullet for the acceptance of renewable energies. The importance of participation seems to be mediated by a variety of aspects that are relevant in the context of environmental change, including justice. It may be realised with several methods, such as flyers, round tables, and opinion polls. Participatory measures should be of high quality, in order to be of value for the process. Moreover, systematic embedding of the planning process in the regional context, structures and processes is a supporting factor. If the activation of people is not only related to a concrete technical realisation of RET, but also initiates or is embedded in a broader movement, like Agenda 21 or Energy Sustainable Regions, which develop mission statements and communal or regional energy concepts, there is a greater chance that reasonable and durable changes will be initiated.

All in all, a high potential for engagement among the survey respondents is obvious. In order to motivate people to become active supporters of a concrete RET in their own community, the planning and decision-making process should be designed in a way that criteria for procedural and distributive justice are met (e.g., Leventhal, 1980). Providing the necessary information, creating transparency and a foundation of trust contribute to a just process and outcome.

Participation is not a guarantee for (active) acceptance, because other relevant factors also play a role. For instance, active acceptance also depends on the acceptance of the specific technology that will be introduced (Figure 6). As the aspects described cannot explain all the variance in active acceptance, the importance of competences, local identity, and trust will have to be examined in further analyses.

References

Arnstein, S. R. (1969, July). A ladder of participation. *Journal of the American Planning Association 1969, 35*, 216–224.

Fischer, C., Schophaus, M., Trénel, M., & Wallentin, A. (2003). Die Kunst, sich nicht über den Runden Tisch ziehen zu lassen. Ein Leitfaen für Bürgerinitiativen in Beteiligungsverfahren [The art of maintaining one's ground at the round table. A guide for citizens' initiatives in participation processes]. Bonn, Germany: Stiftung Mitarbeit.

Kaufmann-Hayoz, R., & Gutscher, H. (2001). Transformation toward sustainability: An actor-oriented perspective. In R. Kaufmann-Hayoz & H. Gutscher (Eds.), *Changing things – Moving people. Strategies for promoting sustainable development at the local level* (pp. 19–25). Basel, Switzerland: Birkhäuser.

Lane, J. (1995). Non-governmental organizations and participatory development: The concept in theory versus the concept in practice. In N. Nelson & S. Wright (Eds.), *Power and participatory development: Theory and practice* (pp. 181–191). London, UK: Intermediate Technology.

Leventhal, G. S. (1980). What should be done with equity theory. New approaches to the study of fairness in social relationships. In K. Gergen, M. Greenberg, & R. Willis (Eds.), *Social exchange: Advances in theory and research* (pp. 27–55). New York, NY: Plenum.

Lüttringhaus, M. (2003). Voraussetzungen für Aktivierung und Partizipation [Requirements for activation and participation]. In M. Lüttringhaus & Richers H. (Eds.), *Handbuch Aktivierende Befragung. Konzepte, Erfahrungen, Tipps für die Praxis* (pp. 66–72) [Handbook activating survey. Concepts, experiences, tips for the practice]. Bonn, Germany: Verlag Stiftung Mitarbeit.

Matthies, E., Homberger, I., Matthäus, S., Engelke, P., & Moczek, N. (2004). *Lokale Agenda-Prozesse psychologisch steuern.* [Psychological Management of Local Agenda 21 Processes]. Lengerich, Germany: Pabst Science.

Minkler, M. (2000). Using participatory action research to build healthy communities. *Public Health Reports, 11*, 191–197.

Nelson, N. & Wright, S. (Eds.). (1995). *Power and participatory development: Theory and practice*. London, UK: Intermediate Technology.

ÖGUT Strategic Group on Participation. (2004). *Worksheets on participation 2. Benefits of participation processes from the point of view of the various groups of agents*. Vienna, Austria: ÖGUT.

Paul, S. (1987). Community participation in development projects. *World Bank discussion paper*. Washington, DC: World Bank.

Pretty, J. N., & Scoones, I. (1995). Institutionalizing adaptive planning and local level concerns: Looking to the future. In N. Nelson & S. Wright (Eds.), *Power and participatory development: Theory and practice* (pp. 157–169). London, UK: Intermediate Technology.

Rau, I., Zoellner, J., Nolting, K., Rupp, J., & Keppler, D. (2011). *Projektabschlussbericht: Aktivität und Teilhabe – Akzeptanz Erneuerbarer Energien durch Beteiligung steigern* [Final report: Activation and participation – Increasing acceptance of renewable energy through involvement]. Berlin, Germany: BMU. FKZ: 0325052.

Schweizer-Ries, P. (2008). Energy sustainable communities: Environmental-psychological investigations. *Journal of Energy Policy, 36*, 4126–4135.

Schweizer-Ries, P. (2011). Socio-environmental research on energy sustainable communities: Participation experiences of two decades. In P. Devine-Wright (Ed.), *Public Engagement with Renewable Energy: From Nimby to Participation* (pp. 187–2002). London, UK: Earthscan.

Schweizer-Ries, P., Zoellner, J., & Rau, I. (in press). Soziale Aspekte einer nachhaltigen Energieerzeugung [Social Aspects of Sustainable Energy Production] In L. Kruse (Ed.), *Natur, Naturwahrnehmung und Naturschutz: Nachhaltige Entwicklung aus der Sicht der Psychologie.* [Nature, nature perception and nature conservation: Sustainable development from a psychological perspective]. Reihe Naturschutz und Freizeitgestaltung. St. Augustin: Academia Verlag.

Villalobos Montoya, C., & Schweizer-Ries, P. (2005). Getting people involved: The relationship between community participation and project sustainability. *Proceedings of the 20th European*

Photovoltaic Solar Energy Conference and Exhibition. Barcelona, Spain: WIP – Renewable Energies.

Zoellner, J., Rau, I., & Schweizer-Ries, P. (2009). *Akzeptanz Erneuerbarer Energien und sozialwissenschaftliche Fragen* [Acceptance of Renewable Energy and Questions of Social Science] (final project report). Magdeburg, Germany: Otto von Guericke University.

Zoellner J., Schweizer-Ries, P., & Rau, I. (in press). Akzeptanz Erneuerbarer Energien. [Acceptance of Renewable Energy] In T. Müller & M. Schütt (Eds.), *Fachpublikation zum Recht der Erneuerbaren Energien im Strombereich*. [Science publication on legislation for Renewable Energies in the area of electricity]. Würzburg, Germany: Universitäts-Verlag.

Zoellner, J., Schweizer-Ries, P., & Wemheuer, C. (2008). Public acceptance of renewable energies: Results from case studies in Germany. *Journal of Energy Policy, 36*, 4136–4141.

Confronting Ignorance

Coping With the Unknown and Surprising Events in the Remediation of Contaminated Sites

Alena Bleicher and Matthias Gross

Department of Urban and Environmental Sociology, Helmholtz Centre for Environmental Research – UFZ, Leipzig, Germany

Abstract

Many "risky" events elude the frameworks of actuarial prediction because there is not enough historical experience to fall back onto in evaluating them. This appears to be especially true for environmental decision-making, where growing knowledge on environmental processes opens up ever new knowledge about what is not known. Ignorance is therefore actively created alongside knowledge. Instead of following the traditional view, in which ignorance is seen as detrimental *per se*, we try to analyze how ignorance is specified and can even help to serve as a productive resource. By so doing, we address some strategies used to cope with situations involving ignorance, building on examples from the remediation of areas containing multiple contaminant sources and plumes related to past industrial activities in densely settled urban and semi-urban areas. Analysis of these processes that deal with the unknown indicates that planning and policy-making benefit when limits to knowledge are openly acknowledged and communicated. Such processes can be seen as a lynchpin for successful remediation processes of contaminated land.

Key words: contaminated land, decision-making, public policy, remediation, risk and ignorance

Introduction: Risk, Complexity, and Nonknowledge

Many definitions and concepts of risk exist that vary by application, situational context, and scientific discipline. However, risk generally refers to the (often quantified) possibility of suffering loss. The management of risk has become a standard for making decisions that transform the future into a calculable and known object based on known and observable facts. Ignorance, in turn, is something that lies outside of what risk refers to in most definitions. Cleaning-up and managing contaminated areas in a timely and effective manner in regions characterized by deindustrialization and economic restructuring, however, involves considerable pressure to act in spite of ignorance – or, alternatively, not to act at all. When new knowledge is acquired about contaminants, it often leads to greater knowledge about what is *not* known, since many of the contaminants are very hard to detect in soil and groundwater investigations conducted prior to site remediation, or are only discovered *in situ* during the actual clean-up process. Quite often, contaminants only come to light (again) by surprise, that is, by accident when, for instance, contaminants are mobilized by floods (see Frickel & Vincent, 2011) or by construction activities on site or close to the site, as we discuss in this chapter. In general, surprising events result from unknown or little understood interactions between elements of the piece of land or the technological installation under scrutiny (see Casti, 1994; Küppers, 2002; Mitchell, 2009). Thus understood, surprising events that make actors aware of their own ignorance are the outcome of very complex processes.

Many regions subject to these types of complex events around contaminations have to make the reuse of brownfields attractive to investors, even if remediation of a site is more expensive than locating new operations on greenfield land. Thus, dealing with ignorance almost becomes the norm. This aspect highlights the general limits of the field of risk studies and has even led to a general questioning of risk as a meaningful concept during the last 20 years (see Dowie, 1999; Green, 2009; Groves, 2009; Pender, 2001). In the following, we seek to contribute to the debate on risk and ignorance by looking more closely at how actors appear to deal successfully with the fact that one never knows what to expect in restoration and remediation processes on contaminated sites in densely populated regions. In a sense, then, dealing with ignorance becomes "normal." One is forced to deal with the unknown in such a way that a foundation is established for acting, despite the limitations imposed by risk assessments and available scientific knowledge.

Using an example from a major research program related to the remediation of regionally contaminated aquifers in eastern Germany, we focus on some of the stages in a clean-up process undertaken in the industrial area of the municipality of Weissandt-Gölzau in the German state of Saxony-Anhalt.[1] We illustrate some of the strategies used by the heterogeneous actors involved to deal with "useful" ignorance, with a special emphasis on how they handled surprises and re-negotiated previously agreed plans. Willem Vanderburg (2000, p. 91) has referred to ignorance as useful "if an awareness of its existence has positive consequences, including a sense of the limitations of a specialty and the motivation to engage in 'conversations' wherever possible." We agree on this point and introduce the notion of nonknowledge to emphasize its positive connotation. Thus understood, we

1 The following sections of the chapter draw heavily on Bleicher and Gross (2011) as well as on Gross and Bleicher (2012). For a more detailed description and analysis of other events and aspects from the cases highlighted in this chapter, we ask our readers to consult these sources. For general discussions on the topic of coping with ignorance and surprise in various fields of ecological restoration and design, see Gross (2010).

depart from the common view, in which ignorance is seen as detrimental and, instead, analyze how ignorance is used as a productive resource in decision-making processes.

Nonknowledge, defined as the "natural" opposite side of knowledge, can be identified in documents and interview sequences at those points where analysis reveals a decision to act after, or in light of, an event that was communicated as "surprising" and where the actors became aware of the fact that they did not know enough about something. Theoretically speaking, nonknowledge exists when actors know that, in reference to a specific research question or a problem posed, not enough knowledge is available and they then define this knowledge as crucial for decision-making. This is a notion of the unknown that goes back to the classical sociologists Georg Simmel (1906) and Max Weber (1978). Weber, for instance, commented on a statement by a physiologist who said that he knew nothing about the spleen, referring to lack of knowledge about one special aspect of the spleen. In Weber's view, however, the physiologist "knew a good deal about the spleen – its position, size, shape, etc.; but he could say nothing about its function, and it was his inability to do this that he called 'ignorance'" (Weber, 1978, p. 15). Ignorance here refers to a much more specific realm of the unknown. Since the original German word Weber used for ignorance is *Nichtwissen*, a term pointing to the symmetry between the known and the unknown (*Wissen* and *Nichtwissen*), in the following we will use the term nonknowledge as a more literal translation of the German original, and also to avoid the many negative connotations inherent in the term ignorance (see Gross, 2010).[2]

Data and Methods

The findings presented in this chapter are part of a larger research program entitled *Research on Remediation of Regionally Contaminated Aquifers* (acronym based on the German title: SAFIRA II). In this program, scientists develop technical solutions and management concepts for the remediation and revitalization of contaminated sites. Based on a step-by-step approach, innovative technologies and large-scale technical devices are being used to investigate extensive and complex contaminated areas. As members of staff of this major project (funded by the *German Federal Ministry of Education and Research*, BMBF), the authors were able to obtain easy access to one of the research sites, and to make contact with the stakeholders involved in the remediation process. The research site, an industrial area, is located in the federal state of Saxony-Anhalt within the community of Weissandt-Gölzau.

Research data on the remediation process come from various sources, including informal meetings, observing activities at the site, minutes from previous meetings, published material, evaluations of archive materials, contracts, field notes, and semi-structured

[2] During the last decade, many scholars have discussed different shades of ignorance as a social phenomenon warranting detailed empirical study. For various disciplinary perspectives, ranging from the neurosciences to religious studies, as well as geography, sociology and economics, see Abbott (2010), Chua (2009), de Krom and Mol (2010), Kuhlicke (2010), Pushkarskaya et al. (2010), Smithson (2010), Son and Kornell (2010), and Wehling (2011); and the literature cited therein.

interviews with 17 stakeholders involved in the revitalization process. We began our work by interviewing scientists previously involved in other revitalization programs, as well as policy makers and decision makers in several federal states in Germany, in order to gain an overall understanding of the topic. Most of the information used in the writing of this chapter stems from numerous field trips to the site, undertaken between April 2006 and December 2007. We often went to the site unannounced, in order to observe activities there at first hand. In general, our methodological approach was informed by grounded theory in the tradition of Glaser and Strauss (1980). This involved looking systematically at our interview transcripts and observation notes with the aim of formulating some general findings. The transcribed interviews and other material for this research were codified using the qualitative text analysis program MaxQDA (see Kuckartz, 2007). We identified a set of recurring themes that were compared and discussed until a coding scheme was agreed on.[3] In addition, we used the standard methodology of content analysis to identify text passages relating to pre-defined themes and categories, based on the theoretical concepts of nonknowledge and ignorance introduced above. Nonknowledge exists when an actor is nonetheless able to identify the point of reference of what is not known – as in Weber's example above. Our overall coding scheme contained categories developed from the interviews and categories developed from the theoretical concept. As part of the analysis, the coded segments were linked to each other, in order to locate possible causal or temporal relationships between the categories. In the course of this work, we found that an expectation of surprising events (as a trigger for becoming aware of one's ignorance) was a part of the actors' own strategies.

Complexities in the Remediation of an Industrial Site

The municipality of Weissandt-Gölzau is located in Mitteldeutschland (Central Germany), a region in which, by the end of the 19th century, coal mining had become the main industry (see Figure 1). In the 1920s, chemical industries based on brown coal refining were established there in order to produce distillate fuel and tar, along with a carbonization plant. As was common practice at that time in all industrialized countries, the residual products from these industrial processes were dumped close to the production facility, without any special treatment. In addition, all flammable liquids and chemicals were literally dumped into the ground during the period of Allied bombing at the end of World War II, in order to prevent the factories from burning down. One long-lasting effect of these activities is that the region still has to deal with the – often forgotten – consequences of these activities.

During the 1950s and 1960s, processing activities in Weissandt-Gölzau switched from coal to oil. The production of polyethylene film began, accompanied by a mechanical engineering facility. Since 1990, the commercial enterprises located in Weissandt-Gölzau

[3] This work was done with the generous help of our former student assistants Sebastian Bellstedt, Johanna Hilsberg, and Franziska Werner.

Figure 1. Location of Weissandt-Gölzau in the German state of Saxony-Anhalt. (Source: the authors)

have been successfully restructured within the framework of the wider socio-economic changes occurring in the region after the fall of the Berlin wall. Today, Weissandt-Gölzau hosts two major companies that are active on the global market, as well as a few dozen smaller ones. The two main companies are a polyethylene film production company and a mechanical engineering plant. In the restructuring and consolidation phase, which lasted about 10 years, not all parts of the industrial area were used – some were left deserted as brownfield sites. It was only during the early 2000s that the larger companies on the site began to expand onto the contaminated brownfields within the industrial area. Due to existing knowledge about the site's former industrial use (and almost certainly also to mistrust of previous environmental practices), private companies located there were keen to establish the degree of contamination. Since much of the contamination originated from the chemical industries of the 1920s, records of accidents and waste disposal causing contamination were very rare. In the course of economic restructuring after World War II, during the 1950s and 1960s and especially after 1989, many parts of the industrial facilities were rebuilt, buildings were demolished, sinks filled and new buildings constructed in the same place. Very little documentation (archive materials or maps) about these changes exists today; most of the information that otherwise probably would have been available disappeared when new administrative structures were established after 1990. Many of the events can only be reconstructed from interviews with people who have lived in the area long enough to remember some of the activities on these sites. Between 1990 and 1996, local enterprises contracted specialist engineering companies to investigate water and soil contamination in some parts of the area. No significant contamination was found on these tracts of land. However, experts in local and regional agencies were aware that further investigation of the remaining tracts would be needed. Because there was little investment in public infrastructure in the industrial area, the political representatives of the municipality decided, in 2002, to improve the existing roads and walkways as well as the sewage and rainwater drainage systems, in order to make the site more attractive for investors. On the basis of existing knowledge about the area's industrial history and a number of unexpected encounters with contaminants during site developments since 1990, both the municipality and the relevant regional and district authorities were aware that construction activities in the area would have to deal with contamination of unknown dimensions.

One of the most important activities at the site was a project carried out by the municipal authorities to improve the infrastructure. It was the largest investment project undertaken to date and was accomplished by a heterogeneous group of actors. Although investigations into contamination had been carried out in advance, diverse types of hitherto unknown contaminants were discovered not long after construction work began (Gemeinde Weissandt-Gölzau, 2007). Given the site's history and the actors' anticipation of surprises in relation to finding contaminants, one could say that the actors involved were aware of their own ignorance (contamination of unknown dimensions) – they expected the unexpected. As the principal manager of the contracted local construction company explained: "Sometimes different activities took place simultaneously. Sometimes, though, we didn't know what to plan for, because there were no studies we could build on." This awareness, of course, influenced each decision made by the stakeholders involved.

Nonknowledge and the Limits of Risk Assessments

The remediation project was divided into several sub-projects, each of which proceeded according to the following scheme: (1) systematic investigation by the consulting company, (2) planning by the engineering company, (3) application for building permits and authorization for construction from the administration, and (4) construction. The steps had to be undertaken sequentially, but due to the time frame and implementation pressures involved, they were realized almost simultaneously. The task of each company was defined precisely. All the companies carried out their work in the area at the same time and all the actors we interviewed stated that they knew before they began that they would have to cooperate with one another, should they unexpectedly encounter contaminants – surprises which all the actors involved anticipated. Given the predetermined time frame of the project and the limited budget available, the actors knew that they would have to make decisions based on

Figure 2. Metal container (12 t) filled with tar, encountered unexpectedly during the remediation process. (Photograph: the authors)

nonknowledge. They knew that it would be useless to try to know things that could not be known at the time a decision was required. To wait for reliable information would have meant extending operations beyond the project time frame and frittering away the chance to promote new investments and further economic development.

However, this means that the actors involved had to expect that surprising things could happen. Thus, trying things out in a carefully thought-out way appeared to be a feasible way to move forward. In this spirit, the reconstruction of the first road began in March 2006 and proceeded as planned. In November 2006, a large number of old electrical cables were found whose removal did not pose serious problems. In July 2006, the renovation of the second major access road was started. Here, difficulties arose some two months later, when bulldozers exposed a deposit of liquid tar. Because of the toxic fumes and the risks involved with coming into contact with this material, workers had to wear protective clothing during the following 3 months. In both cases, none of the actors had expected to find items such as cables or tar beneath existing roads. When the construction of the first of four new roads began, the workers were immediately confronted with the discovery of tar-filled metal containers which were hard to date (the largest one, found in April

2007, was 12-m long and weighed 12 t, see Figure 2). Soil contaminated with tar and solvents was found in crumbling basement foundations dating from the early 20th century. Although the removal of these materials slowed down some of the other activities for a while, quick disposal solutions were found, at least for the metal containers, while construction work in other places continued unhindered. Because it was known that such containers can easily catch fire and explode and that expert knowledge is needed to handle them, special companies capable of opening the container and handling them safely for transportation elsewhere had to be contracted at very short notice. All these incidents pushed up the project costs, but the overall process continued. Nevertheless, the last phase of the project was completed on schedule in October 2007.

In October 2006, a sub-project was started to construct a rainwater reservoir. During the initial phase, in November 2006, some common domestic waste was found and the first soils removed were found to contain small amounts of liquid tar. A few weeks later, after the construction company had removed the topsoil, workers reported that, in their opinion, the subsoil did not appear to be "normal." The actors involved knew, then, what they had not known (nonknowledge) earlier and were able to use this awareness as a basis for further planning and action. Subsequently, an engineering company was consulted. The company took samples and evaluated the soil, concluding that the entire area was heavily contaminated with liquid tar and various solvents. Specialist companies established that the tar came from a nearby landfill, where residues from the carbonization plant had been stored since the 1920s. Thus, the nonknowledge that emerged from a general state of ignorance could be transformed into knowledge. As a result, the construction work continued, but a layer of soil measuring more than one meter in depth had to be removed. The original plan thus changed and was later adapted to each new situation as it arose; this often went hand in hand with obtaining a new building permit. Commenting on this incident, the representative from the engineering company in charge of the project stated: "The discovery of tar meant that the whole philosophy of the project had to be rethought. A completely new plan had to be drafted *ad hoc*." In a similar vein, the representative from the project management explained:

> Construction and cleaning-up has to be understood as a continuous process. They [the workers on the ground] started to remove the material, and when they did so they found that the subsoil was totally different from what we expected. The problem is that tar, which turns to liquid at a certain temperature, makes it necessary to use another type of water barrier for the rainwater reservoir. [. . .] At times like these, we have to act first and get the permits changed later.

A representative of the major engineering office involved added, on the same theme: "Sometimes there is no time left. Then we have to decide ad hoc. A strategy needs to be changed as fast as lightning." These citations from our interviews show that actors were prepared for unexpected events, so they could make decisions quickly and flexibly. In other words, the actors involved agreed on what was not known and took this into account in their subsequent activities; that is, they decided to act in spite of their nonknowledge. Nonknowledge, however, can be developed into new knowledge, which is introduced into the next stage of the process, to subject it to further observation and assessment.

It became a challenge to finish the project within the prescribed time frame, between January 2006 and November 2007. By the end of the project, much of the contamination had been successfully removed. Nevertheless, the actors were aware that they could only be certain that all contaminants had been removed only in those places where they had actually worked with their machines and equipment. This awareness was also based on preliminary calculations regarding the movement of liquid tar – there is always the possibility that a cleaned-up area will be re-contaminated. In the next section, we will take a closer look at the strategies adopted by the various actors to deal with the unknown.

Nonknowledge and Strategies of Coverage

Whenever scientists and engineers claim to be able to deliver hard facts and genuinely reliable knowledge, they may often overestimate their own methodological capacity. Experts' willingness to disclose the limits of knowledge in their communication with other stakeholders can thus be understood as an important part of the process of implementing successful development projects. As the interviews with actors and the analysis of minutes from planning meetings reveal, the strategy that was adopted to deal with unexpected events was based on precise definitions of what was unknown (as in Weber's case of the physiologist's nonknowledge about the spleen), thus making it possible to devise flexible, cushioning measures around these unknown factors. These measures included a change of permits (administration), a change of plans (engineering companies) and, last but not least, a certain amount of flexibility in redeploying capital for remediation.

One major factor which helped to generate such a strategy was an explicit agreement on shared objectives made at an early stage of the project. In addition to personal acknowledgements of what was not known, nonknowledge was also expressed in the contracts made between the actors. Special clauses in these contracts and permits – so-called collateral clauses – specifically mention the unknown. Collateral clauses are agreements between joint contractors to pool their guarantees in handling a large project. The objectives are to provide others with recourse to the contractor's warranty, to continue the project in the face of unexpected changes (German: *Auflagenvorbehalt* and *Nachtragsangebot*). The strategies involved in making use of these warranties we call "strategies of coverage." These strategies were developed specifically for this site's remediation process.

A further important issue related to successfully coordinating such projects is the institutionalization of contacts and information exchange. Although this might seem obvious at first sight, regular consultation of all institutions and actors involved, for the purpose of exchanging information, discussing new developments, and agreeing on strategies for adapting to new situations is not the norm in many projects dealing with contaminants. Consultation also implies that all the actors involved must communicate their own nonknowledge, understood not in terms of a failure but rather as a normal way of dealing with contaminants.

Conclusion and Outlook

The approach taken to cope with contaminants originating from industrial activities that occurred up to 100 years ago, which we have described in this chapter, is one that is able to successfully lead project activities, despite the presence of unknown factors. Seen in this way, dealing with nonknowledge in the course of the remediation of contaminated sites appears as a normal occurrence rather than as an anomaly or an indication of failure. Central to strategies for dealing with nonknowledge used by actors involved in remediation projects is to take nonknowledge *actively* into account in the decision-making process. This approach has become a rule-of-thumb for actors dealing with contaminated land, since clean-up plans for a site include awareness of unanswered questions. It renders obsolete the idea that mistakes and failures are always based on decision makers' wrong choices or other human mishaps. Put another way, dealing with unknowns is not simply a process of trial-and-error or of learning from failure, because errors and failures suggest that mistakes have been made. Instead, the stakeholders and actors we observed took the impossibility of avoiding ignorance seriously, so that there was no target for blame or "finger pointing." Many of the activities we observed seem to explicitly include ignorance and nonknowledge as part of the planning process. This encouraged the development of innovative ways of dealing with ignorance by reducing complexity and making full use of the potential and resources of the actors involved, to achieve a common goal. The actors achieved a state of preparedness, found organizational solutions for cushioning the impacts of surprising events, and thus developed strategies for coping with the unknown.

Given the enormous number of unknown factors involved in the dynamics of contaminated landscapes, acting in spite of ignorance by acknowledging nonknowledge will perhaps become even more important in years to come. In order to act successfully, the relevant social networks need to incorporate what Weick and Sutcliffe (2007) have called "collective mindfulness." This mindfulness involves acknowledging ignorance as a potentially useful resource and not merely as the detrimental absence of knowledge. Thus understood, straightforward and honest acknowledgement of the limits of knowledge among the actors involved can be viewed as an important part of successful remediation projects, where ignorance is a daily reality to be faced and addressed. After all, clean-up processes instigated to deal with such contaminants demand fast and flexible governance structures rather than blame and finger pointing. Perhaps the case discussed here can even be seen as an indication that the paradigm change from a rationality of insurance and risk to what Collier and Lakoff (2008) call the rationality of "distributed preparedness" is indeed taking place. Official rhetorics may still call for more safety and time-consuming precautionary measures, but the real world of everyday decision-making in contaminated brownfield settings seems to be moving in a different direction, offering creative strategies to successfully cope with inevitable ignorance and surprise.

References

Abbott, A. (2010). Varieties of ignorance. *The American Sociologist, 41*, 174–189.

Bleicher, A., & Gross, M. (2011). Response and recovery in the remediation of contaminated land in eastern Germany. In R. A. Dowty & B. L. Allen (Eds.), *Dynamics of disaster: Lessons on risk, response and recovery* (pp. 187–202). London, UK: Earthscan.

Casti, J. (1994). *Complexification: Explaining a paradoxical world through the science of surprise*. New York, NY: Harper.

Chua, L. (2009). To know or not to know? Practices of knowledge and ignorance among Bidayuhs in an "impurely" Christian world. *Journal of the Royal Anthropological Institute, 15*, 332–348.

Collier, S. J., & Lakoff, A. (2008). Distributed preparedness: The spatial logic of domestic security in the United States. *Environment and Planning D: Society and Space, 26*, 7–28.

de Krom, M., & Mol, A. P. J. (2010). Food risks and consumer trust: Avian Influenza and the knowing and non-knowing on UK shopping floors. *Appetite, 55*, 671–678.

Dowie, J. (1999). Against risk. *Risk Decision and Policy, 4*, 57–73.

Frickel, S., & Vincent, M. B. (2011). Katrina's contamination: Regulatory knowledge gaps in the making and unmaking of environmental contention. In R. A. Dowty & B. L. Allen (Eds.), *Dynamics of disaster: Lessons on risk, response and recovery* (pp. 11–28). London, UK: Earthscan.

Gemeinde Weissandt-Gölzau. (2007). *Weissandt-Gölzau: Seit 130 Jahren Industriestandort* [Weissandt-Gölzau: Industrial site for 130 years]. Weissandt-Gölzau, Germany: Official Information Book for Potential Investors.

Glaser, B. G., & Strauss, A. L. (1980). *The discovery of grounded theory: Strategies for qualitative research*. Hawthorne, NY: Aldine.

Green, J. (2009). Is it time for the sociology of health to abandon "risk"? *Health, Risk & Society, 11*, 493–508.

Gross, M. (2010). *Ignorance and surprise: Science, society, and ecological design*. Cambridge, MA: MIT Press.

Gross, M., & Bleicher, A. (2012). Jenseits der Zurechnung auf Entscheidungen: Nichtwissenskommunikation am Beispiel Altlastensanierung [Beyond the attribution to decisions: Communicating nonknowledge using the example of the restoration of brownfields]. In N. Janich, A. Nordmann, & L. Schebek (Eds.), *Nichtwissenskommunikation in den Wissenschaften* (pp. 209–234). Frankfurt am Main, Germany: Peter Lang.

Groves, C. (2009). Future ethics: Risk, care and non-reciprocal responsibility. *Journal of Global Ethics, 5*, 17–31.

Kuckartz, U. (2007). *Einführung in die computergestützte Analyse qualitativer Daten* [Introduction to computer based analysis of qualitative data]. Wiesbaden, Germany: VS Verlag.

Kuhlicke, C. (2010). The dynamics of vulnerability: Some preliminary thoughts about the occurrence of "radical surprises" and a case study on the 2002 flood (Germany). *Natural Hazards, 55*, 671–688.

Küppers, G. (2002). Complexity, self-organisation and innovation networks: A new theoretical approach. In A. Pyka & G. Küppers (Eds.), *Innovation networks: Theory and practice* (pp. 22–52). Cheltenham, UK: Elgar.

Mitchell, M. (2009). *Complexity: A guided tour.* Oxford, UK: Oxford University Press.
Pender, S. (2001). Managing incomplete knowledge: Why risk management is not sufficient. *International Journal of Project Management, 19*, 79–87.
Pushkarskaya, H., Liu, X., Smithson, M., & Joseph, J. E. (2010). Beyond risk and ambiguity: Deciding under ignorance. *Cognitive, Affective, & Behavioral Neuroscience, 10*, 382–391.
Simmel, G. (1906). The secret and secret societies. *American Journal of Sociology, 11*, 441–498.
Smithson, M. (2010). Understanding uncertainty. In G. Bammer (Ed.), *Dealing with uncertainties in policing serious crime* (pp. 27–48). Canberra, Australia: Australian National University Press.
Son, L. K., & Kornell, N. (2010). The virtues of ignorance. *Behavioural Processes, 83*, 207–212.
Vanderburg, W. H. (2000). *The labyrinth of technology.* Toronto, ON: University of Toronto Press.
Weber, M. (1978 [1921]). *Economy and society: An outline of interpretive sociology.* Berkeley, CA: University of California Press.
Wehling, P. (2011). Vom Risikokalkül zur Governance des Nichtwissens [From calculating risk to the governance of nonknowledge]. In Gross, M. (Ed.), *Handbuch Umweltsoziologie* (pp. 529–548). Wiesbaden, Germany: VS Verlag.
Weick, K. E., & Sutcliffe, K. M. (2007). *Managing the unexpected: Resilient performance in an age of uncertainty.* New York, NY: Wiley.

Structural Change in Former Mining Regions

Identifying Potentials and Building Capacities

Jörn Harfst, Gerd Lintz, and Peter Wirth

Leibniz Institute of Ecological Urban and Regional Development, Dresden, Germany

Abstract

In times of rapid global economic change, mono-structured mining regions can be regarded as especially vulnerable entities. At the end of mineral exploitation, such regions and their inhabitants often face enormous challenges for their economic, social, and environmental future. In this situation, mining regions are relying increasingly on their own capacities and potentials to master structural adjustment. One way to tackle such far-reaching changes is the identification and utilisation of post-mining potentials. In particular, for regions characterised by small and medium-sized towns, this is a complex and challenging task. This paper discusses how actors in such regions can enhance their capacity to realise new development options through cooperation. The case study of three eastern German regions confirms the importance of using post-mining potentials. It also highlights differences in the forms of cooperation that correspond with different approaches to post-mining utilisation.

Key words: capacity building, cooperation, mining region, post-mining landscapes, structural change

Introduction

For mining regions, resource exploitation is often the main source of employment and income. This makes them particularly vulnerable to intensifying competition in a rapidly globalising economy. In recent decades, in particular traditional European centres of the mining industry have been rendered unprofitable, due to new competitors on the world market and cutbacks in national subsidies. The damage is considerable: Mining regions have frequently experienced the complete closure or substantial down-sizing of mining and related industries. This leads to de-industrialisation, high unemployment, and out-migration.

Creating a new development path for such regions is a complex and challenging task. The legacy of mining such as old buildings, facilities, and infrastructures, as well as compromised landscapes and environmental pollution, does not seem to be a good point of departure. However, the end of mining also provides an opportunity to turn to more sustainable regional development, fundamentally changing the relationship between people and economy, on the one hand, and the environment on the other, in terms of environmental repair, re-use of land and facilities, and more environmentally-friendly workplaces.

Today, new development has to be realised in a situation where academic and political debates on structural change in old industrial regions have lost their importance, as national and European funding for regional support programmes has generally decreased. In this context, mining regions are relying increasingly on their own capacities and potentials to master structural adjustment. One way to deal with such far-reaching changes can be to review the mining legacy, in order to identify and utilise hitherto unrecognized potentials.

There are good examples of how to deal with mining legacies in many countries, as a conference on “Post-Mining Landscape” recently demonstrated (IBA, 2010a). In Xuzhou, China, for instance, lakes created in subsidence areas are being used as fish ponds or recreational areas. In Ostrava, Czech Republic, a complete new city centre has been built on the site of an old mine and coking plant. Another well-known case is the Eden Project in Cornwall, UK, where an old china clay pit is being re-used as an eco visitor attraction. One of the most prominent comprehensive regional approaches is the International Building Exhibition (IBA) Emscher Park in Germany, which took place in the Ruhr District between 1989 and 1999 (e.g., Eckart, Ehrke, Krähe, Eckart-Müller, 2003). Here, the ecological and cultural restructuring of an old industrial region was seen as a necessary basis for comprehensive renewal. Overall, 89 projects were realised, for example, slag heaps were integrated into green landscape concepts and old mining facilities adapted to accommodate education and innovative services. While the IBA Emscher Park has received much international attention and is seen as a success story by many, it cannot be denied that the framework conditions were almost ideal: The project received significant financial and political backing and took place in a large metropolitan setting with high population density.

In contrast, this paper is concerned with mining regions where small and medium-sized towns predominate, with often only limited capacity to act, and which attract little political attention. Earlier research has shown that, in general, the ability to deal with structural change depends strongly on the mobilisation of local and regional actors (e.g., Kunze, 1997). Looking for the factors that contribute to successful structural change in old industrial regions characterised by small and medium-sized towns, Lintz and Wirth (2009a) have already found indications that cooperation between actors, led by key personalities, can decisively enhance regional capacities for action, and thus foster regional development.

This paper focuses on this factor and asks *how*, in practice, actors in the regions under investigation build capacities to identify and utilise their post-mining potentials through cooperation. The empirical basis for research is provided by three former eastern German mining regions, namely Mansfeld-Südharz, Zwickau-Lugau-Oelsnitz, and the Lusatian Lakeland (Lausitzer Seeland).

This paper has six sections: The following section describes post-mining potentials and their possible use-value. The third deals with capacity building through leadership and cooperation. Section four presents case studies of the utilisation of post-mining potentials and established modes of cooperation. A fifth part compares and discusses the three examples, highlighting differences and similarities. Finally, core findings are presented and reflected on.

Background: Post-Mining Potentials

When talking about the legacies of mining and its related industries, one often thinks of environmental damage in the form of persistent pollution of water, soil, and air. Combined with the difficult economic and social consequences of structural change, all these factors culminate not only in structural weaknesses but also in a difficult "black" image for such regions. Nevertheless, there are relicts from the industrial past that have the potential to attract new investment and, perhaps more importantly, to become part of a new vision for the sustainable future development of such areas. Generally, two groups of post-mining potentials have been identified (Černič-Mali, Marot, & Klančišar-Schneider, 2010):

- *Natural potentials* appear in the form of natural elements modified by mining. These potentials can be the transformed landscapes left behind after mining activities, such as slag-heaps and open-cast pits. While, at the first glance, such conditions do not seem a good basis for development, many former mining regions have found innovative solutions and approaches for using these potentials. Remediated landscapes are often spectacular forms that can be used for recreational purposes, with newly-created lakes and parks or forests, and which often include features such as hiking or cycling trails. Other utilisations of post-mining landscapes focus on energy production. After mining has ceased, large-scale land reserves can be used for renewable energy projects focusing on biomass production on former slag heaps, while rising mine water is a potential source of geothermal energy production.
- The second group, *cultural potentials*, has also been utilised. These include technical heritage, infrastructure, and buildings for processing, as well as mining settlements. In addition, non-materialized potentials such as mining events, mining traditions, and mining identity also belong to this group. Museums – whether a show mine or an exhibition – are one of the most common uses of heritage; other examples include mines as concert venues, swimming pools,

religious buildings, storage areas, venues for conferences and weddings, or eco-housing. Last but not least, old transport facilities from mining times – especially railways – offer a set of options for re-use.

Both groups of potentials can be found in nearly all former mining regions. Their utilisation depends partly on framework conditions such as national laws, policy aims, the financial situation of local governments, subsidies, and availability of other funding opportunities (i.e., national energy policies). Local demand conditions are also important, for example, in the fields of retailing, services, agriculture, education, recreation, and research. Identifying and realising such post-mining potentials is a complex challenge. The necessary changes have to be initiated and executed. Creative and attractive ideas are needed, complemented by specialised knowledge and detailed planning. In the end, the measures must be physically implemented and long-term use must follow. All these tasks require considerable financial resources.

Background: Capacity Building Through Leadership and Cooperation

The previous section showed that identifying and using potentials of old mining regions is a highly demanding undertaking. The affected regions struggle to cope with structural problems, due to the decline of the mining industry and the ensuing economic and fiscal weaknesses. The once helpful old networks and actor structures have often become obsolete. In the case of small and medium-sized towns, the situation is even more acute. While their size makes them easier to manage and comprehend, small administrations with few highly qualified personnel can be seen as a distinct disadvantage (see Lintz & Wirth, 2009b).

Following a policy- and planning-related research approach, the main question in this context is how the capacities of regional actors can be strengthened, especially local and regional authorities. In the context of local and regional development, the term capacity can be defined as the general ability to identify and solve development problems, with explicit reference to a wide range of factors. Based on the work of Jänicke (2002) on environmental policy, capacity depends firstly on the actors with their strengths, competencies, constellations, personal will, and skills. Secondly, the strategies actors choose are important. For this investigation, it is decisive whether they are willing to cooperate or not. Thirdly, structural framework conditions play a role that encompasses political and institutional aspects (e.g., the remit of municipalities in a multi-level politico-administrative system), economic and technological aspects (e.g., possibility of eliminating toxins), and cognitive informational aspects (e.g., general knowledge about good practices in the use of old mining potentials).

Given that there is no clear overriding responsibility for development on the regional scale ("no actor is in charge"), self-organisation is the predominant form of governance (see, e.g., Chhotray & Stoker, 2010). The identification and use of old mining potentials can serve as a good topic for regional actors to establish cooperative networks for building the required capacity by joining forces in tackling problems. In the case of shared problems, such as regional tourism development based on the past mining, joint concepts can be developed with coordinated measures to form a comprehensive regional tourism destination with a variety of attractions and joint marketing. Actors also can jointly finance projects or lobby for funding. In all events, they can share their experiences about solving common problems.

All in all, there is a strong case for cooperation among regional actors, to build up the capacity to act and to make the best of local and regional potentials. This is particularly true for old industrialised areas (Kunze, 1997, p. 9). Collaboration can take various and complex forms, for example, they can involve different kinds of actors (Blotevogel, 1997), from within or beyond the region, and they can vary in the degree of formalisation.

Much of the literature deals with the factors influencing cooperation, including the literature on networks, partnerships, and governance (e.g., Feiock, 2007; Healey, 1998). As potential participants are largely independent, the perceived costs and benefits of collaboration play a vital role. Actors will only participate if they perceive the benefits of cooperative projects to be greater than the costs of the project and the costs of the cooperation process itself (Lintz & Leibenath, 2005).

Trust also plays a vital role, because cooperation might involve the risk of cheating. Given these factors, leadership often seems to be particularly important. As Diamond and Little (2005, p. 85) note, regeneration requires leaders "who not only are themselves creative and have technical expertise, social and organisational skills, and the ability to 'sell' the message, but also provide inspirational motivation." Such a key actor can thus influence the perceived costs and benefits of collaboration and potentially bring together the required partners.

As mentioned, one factor for successful progress in tackling structural change in small and medium-sized towns is capacity building through cooperation led by key actors. To gain a better understanding of cooperation in identifying and using post-mining potentials, three questions are important: What role do post-mining potentials play in regional development? How is cooperation organised? Does cooperation indeed help improve the capacity to identify and use the potentials of previous mining activities?

Results of the Case Studies: Three Post-Mining Regions in Eastern Germany

A qualitative case study approach (Yin, 2002) was employed to answer these questions. Three former mining regions in eastern Germany, Mansfeld-Südharz, Zwickau-Lugau-Oelsnitz, and the Lusatian Lakeland were investigated (see Figure 1 and Table 1). All are characterised by predominantly small and medium-sized towns and have to cope with the consequences of mine closure and the accompanying structural changes in their economies.

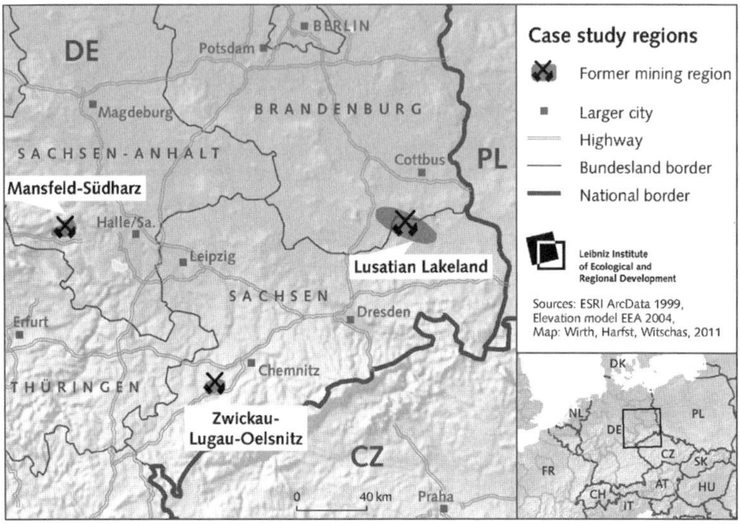

Figure 1. Location of regions under investigation. (Source: the authors)

Table 1. Facts on the regions under investigation

	Mansfeld-Südharz	Zwickau-Lugau-Oelsnitz (FLOEZ)	Lusatian Lakeland
Administrative status	Administrative District	Association of 9 municipalities	Association of 10 municipalities
Inhabitants (2007)	155,000	161,000	106,000
Population density (2007) (Germany: 230 people per km^2)	109 p./km^2	597 p./km^2	82 p./km^2
Population development (1990–2007) (Germany: +3%)	−20%	−18%	−29%
Type of mining	Underground	Underground	Open cast
Material mainly extracted	Copper shale	Hard coal	Lignite
Cessation of mining activities (year)	1990	1978	1999

The regions are the German participants in the EU CENTRAL EUROPE project "ReSource" (INTERREG IVB). The authors of this paper are scientific partners in the project. The investigation combined document analysis (i.e., strategy papers, plans), statistical analysis, and expert interviews with local and regional stakeholders. In the course of the investigation, 22 interviews were conducted. The evaluation of the interviews is based on a qualitative data analysis. The authors also attended eight regional workshops and events, gathering further information through observation.

Mansfeld-Südharz

The administrative district "Landkreis Mansfeld-Südharz" is located in the state of Saxony-Anhalt and its largest towns are Sangerhausen (31,000 inhabitants) and Lutherstadt Eisleben (25,500). The region contained one of the largest deposits of copper shale in Central Europe. In the 1960s, at the peak of production, the regional mining industry employed around 40,000 people. The closure of the mines after German unification in 1990 has left the region struggling, up to the present. Because it is, in many ways, at the periphery of economic development and it lacks infrastructures necessary for boosting regional economic development, Mansfeld-Südharz has been marked by constantly high unemployment rates and a very problematic demographic development ever since the end of mining.

The administrative district Mansfeld-Südharz has neither an overall strategic concept for regional development nor an organisational setting to coordinate individual initiatives. Nevertheless, the region has been relatively successful in utilising some of the region's post-mining potentials. A mining museum, a touristic mining railway and a show mine are already in place. Other projects have identified existing potentials of former mining industries (i.e., project "Copper Traces," a database for the documentation of mining legacies), and explored ways to utilise them as a potential for regional tourism.

The only overall planning instrument for the district is the "Integrated Rural Development Concept" of 2006, created in the context of the EU LEADER programme, which has targeted the region since 2002. The LEADER action group (LAG) has coordinated the realisation of 19 projects with a variety of foci, worth about €3 million up to 2006. In its use of EU funds, the region shows strong, informal, project-based networking, formed around thematic issues and existing informal network structures within the region. The LAG serves as an important nodal point for various ideas and actors in the district, especially because of its ability to convert ideas into projects and to involve other actor groups, such as the miners' association. Even though the LEADER initiative excludes the region's urban core municipalities, it has been important for regional renewal. Another key actor is a private "Training Centre of Industry and Commerce." The business manager of the centre is a former mining engineer with many contacts in the old, mining-related networks of the Mansfeld-Südharz region. Cooperation among these actors and their networks has significantly increased the region's capacity to utilise post-mining potentials, even without having a specific strategic development concept in place.

Zwickau-Lugau-Oelsnitz

The area of Zwickau-Lugau-Oelsnitz (Saxony) is the most extensive former hard-coal mining region in eastern Germany. Mining ended in 1978 and the area today has about 160,000 inhabitants. The nine municipalities in the region (FLOEZ) actively cooperate in regional development.[1] The area has relatively good development potentials: The regional centre, Zwickau (95,000 inhabitants), has a diversified economy today, because it is part of an automotive manufacturing cluster. It encompasses the research capacities of a University of Applied Science; thus, the region is relatively well-equipped for successful economic development, some 30 years after the end of mining. Nevertheless, the FLOEZ area definitely has to deal with the same challenges that are confronting many other eastern German mining regions, including high unemployment rates and a problematic demographic structure.

To act on these common problems, the municipalities founded the FLOEZ network in 2005, formalising a long-standing informal cooperation. The initiative focuses strongly on the utilisation of post-mining potentials. Fields of action include the re-use of mining land, utilising the agricultural and forestry potentials of the former mining areas (renewable energies), and the tourism potentials of the mining legacy. Projects include preservation of an old shaft tower ("Martin Hoop Shaft" in Zwickau), a joint tourism marketing label ("Traces of Black Gold") and other pre-investment activities.

The FLOEZ project provides a solid, formalised basis for regional development and cooperation on both the local and regional level. The main planning document is the "Integrated Development Strategy for the Zwickau-Lugau-Oelsnitz Area" of 2006. The FLOEZ steering group includes all important stakeholders from the region's administrative institutions, such as mayors and regional planning associations. FLOEZ was also supported by the Economic Region of Chemnitz-Zwickau,[2] the biggest regional development body in south-western Saxony. Cooperation influences capacity building in the region in a variety of ways: Firstly, a better perception of regional requirements by state authorities, secondly, purposeful access to state funding and, thirdly, the opportunity for larger common projects.

In the framework of cooperation, one mayor in particular, can be described as a key promoter of the process. He established a wide network of contacts for mining questions, involving universities, foundations, external experts, mining authorities, as well as other mining regions across central Europe (MINEC). Nevertheless, the FLOEZ scheme is not without its difficulties: The most problematic aspect remains the lack of funding for regional projects.

---

1 The name "FLOEZ" is short for "Future for Lugau-Oelsnitz-Zwickau." The name alone indicates the focus on mining traditions – "Floez" is the German term for "coal seam." FLOEZ is a network of public bodies without transference of rights to other bodies.

2 The Economic Region of Chemnitz-Zwickau was a public limited company of two cities and four administrative districts that managed the tasks of project coordination and regional marketing. Following the county reform in Saxony in 2008, the company was liquidated in 2010. An adapted cooperative structure is currently under construction.

Lusatian Lakeland

The Lusatian Lakeland (Lausitzer Seenland) is located on the border between the states of Brandenburg and Saxony; Hoyerswerda (40,000 inhabitants) and Senftenberg (27,000 inhabitants) are the largest towns. After World War II, the region became the "energy centre" of the German Democratic Republic, with many open-cast pits, power plants, coking plants, and briquette factories. After 1990, this industrial complex declined, and population figures in the Lusatian Lakeland sank from 151,000 in 1990 to 106,000 in 2007. Since mining finally ceased in 1999, groundwater has been filling the residual open-cast pits and changing the landscape into a "lakeland," giving the region its new name.

Despite environmental damage, deindustrialisation, high unemployment rates and population loss, the region is on its way to a modest comeback – as a new tourist destination. The whole area will include 21 lakes with a final total of 13,000 hectares of water. Nine of the lakes are to be connected by navigable canals to support tourism and the development of sporting activities. The first lakes are now completely filled and some canals are finished. Marinas, beaches, bicycle trails, and other tourist attractions, but also nature conservation areas for secondary habitats are being prepared or are already under construction. Some impressive projects like "The Lying Eiffel Tower" (a former lignite mining conveyor-belt bridge), "Bio-Towers" (former wastewater treatment facilities), and "The Landmark" (a new visitors' tower made of steel) are now well-known attractions in the region.

Because the Lusatian Lakeland is situated in two German federal states and consists of nine municipalities in three counties, finding modalities for regional cooperation was essential. By the mid 1990s, several communal associations had been founded to promote the touristic use of former mining land. These associations amalgamated step by step into two large associations, one in Brandenburg and the other one in Saxony. Cooperation between the two associations occurs in the form of a working group. Since 2003, a regional development concept (RDC) has been the conceptual basis for cooperation. All the main aspects of further development, such as the touristic function of each lake and the system of navigable canals, are laid down in the RDC.

It would not have been possible to implement the RDC without a rehabilitation programme for the mining landscape. Since 1994, this task has been carried out by the Lusatian and Central German Mining Management Company (LMBV),[3] a corporation of the Federal Republic of Germany that was launched as a national enterprise to rehabilitate the eastern German lignite areas. Another development factor in Lower Lusatia was the IBA "Fürst-Pückler-Land"[4] which took place between 2000 and 2010. The IBA acted as a kind of think tank and promoter to create new perspectives for the former mining

[3] In German: Lausitzer- und Mitteldeutsche Bergbauverwaltungsgesellschaft (LMBV). The limited liability company is financed by the national budget, with a total financial volume of about €9 billion to date.

[4] In German: Internationale Bauausstellung Fürst-Pückler-Land (IBA). Funding for the IBA was provided by the state of Brandenburg and five administrative districts in Brandenburg. Saxony cooperated with the IBA on the basis of common projects.

region (IBA, 2010b). The IBA, which defined itself as a workshop to shape new landscapes, saw the Lusatian Lakeland as one of the key projects.

For the continued development of the Lusatian Lakeland, the cooperative approach, involving actors from several spatial levels (local, regional, state) and several fields of action (politics, business, administration, civil society), is crucial, because no single actor possesses the capacity to handle the complex process as an individual. Without the support of the federal government and of the Brandenburg and Saxony state governments, the development of regional capacities would have been possible at only a very low level. Although the restructuring of the region can be seen as a promising approach, the path from mining to tourism is still long.

Discussion: Comparison of the Case Studies

The empirical results show that, in all cases, *post-mining potentials* play an important role in regional development, albeit in various forms and degrees of integration into overall development concepts. In Mansfeld, the main focus was on culture. The region is an example for the selective use of typical elements of the mining scenery (show mine, mining railway, etc.). Whilst post-mining affairs are primarily part of rural development concepts, their integration into more general regional development strategies is weak there.

The FLOEZ initiative has a broader approach to the legacies of mining, including both cultural (mining museum) as well as natural aspects (land use on slag heaps). It encompasses the search for renewable energy sources and the re-use of mining land. The RDC in FLOEZ must be seen as an ambitious but sectoral concept with limited impacts on the municipalities involved. But, as a part of the regional development strategy of the much stronger Economic Region of Chemnitz-Zwickau, it has an official status and is supported by the Saxon government.

The most complex and comprehensive case is the Lusatian Lakeland. In this region, the complete landscape left behind by mining is seen as a potential for further development. Regional development is connected with industrial heritage projects and the re-use of mining infrastructure. The renewal of the mining landscape is the crux of the regional development strategy. All the concepts at the local and regional level, as well as the activities of the state rehabilitation company, are dedicated to this overall strategy. Post-mining potentials form the key factor of the change.

A comparison of the cases reveals obvious differences in the use of post-mining potentials. They have several causes. The type of mining plays an important role. There are clear differences between open cast (which left large pits and giant excavators) and deep mining (which left tunnels, slag heaps, and shaft frames). But the processing industries and their architecture also make a difference to the potential offered. The date of closure – long ago or recently – influences the condition of legacies.

The cases also differ considerably in their *forms of cooperation*. This is true for the examples of Mansfeld-Südharz and FLOEZ, both of which exhibit restricted use of post-mining potentials. The former is characterised by low formal steering capacity, due to weak forms of cooperation between municipalities and the district administration. Nevertheless, some successful, relatively informal and project-based forms of cooperation between actors from different fields can be found in the region.

In contrast to Mansfeld, the FLOEZ region has developed formal, strongly institutionalised cooperative structures, embedded in a hierarchically structured system of several policy levels. The network was established with the specific aim of developing post-mining potentials for regional development, but funding is very difficult because of the limited resources of the region and its municipalities.

The Lakeland, where the most extensive restructuring efforts have been undertaken, has the most intensive formal and informal cooperation between municipalities, district authorities, the state governments of Saxony and Brandenburg and, last but not least, the IBA Fürst-Pückler-Land, as an independent think tank. The Lakeland accordingly involves many actors from various politico-administrative levels, creating a particularly complex cooperation setting.

The approaches to development in all the regions investigated depend strongly on the ability of key actors to create networks. In Mansfeld-Südharz, one actor plays a decisive role by keeping alive old networks from the times of active mining. A strong leader was also identified in the FLOEZ region, whose power derives from his political mandate. In the Lakeland, the number of actors involved and their resources are far greater. Here, the coordination of the groups and administrative procedures has been by far the biggest challenge. While some actors involved in this process stand out, a single leading actor in charge of the entire network cannot be identified.

Although every region has its own approach to cooperation, due to different substantive and actor-related regional situations, all three examples demonstrate the importance of building capacities through cooperation between regional actors. They can thus identify and use post-mining potentials. In Mansfeld-Südharz, competence developed to implement the European LEADER+ programme. This would not be considered extraordinary if it were the result of the actions of established actors from local policy and administration. But under the weak organisational and conceptual framework conditions described above, the acquisition of competence has been based on a network of actors coordinated by a civil society organisation.

In the FLOEZ case, cooperation played an important role in attracting attention to post-mining problems, long after mining had ceased. The challenge was to achieve a better appreciation of regional requirements by state authorities. The common initiative and integrated development concept provided the basis for a discussion of post-mining problems and potentials.

In the Lusatian Lakeland, where rehabilitation by the state company (LMBV) has been a key driver in regional development, the difficulty was to focus the endeavours of a multiplicity of actors on a common goal. Regional cooperation among actors from different administrative levels and fields of activity facilitated the vision and realisation of a new Lakeland with navigable canals.

Conclusion

Due to their economic mono-structure, mining regions are particularly vulnerable to rapid global economic change. Once regional mining can no longer compete, new pathways towards more sustainable development need to be found. The regions then often rely on identifying and using the potentials of the mining heritage, a complex and challenging task. However, the affected municipalities – in particular small and medium-sized towns – are often unable to cope with the challenge. They therefore need to build additional capacities to tackle the problems of structural change.

The investigation shows that in all of the three eastern German case study regions, post-mining potentials were identified and used to create and improve the basis for a new path of economic development. Thus, for instance, landscapes were rehabilitated, nature conservation areas created, infrastructures upgraded, and new opportunities for tourism established, albeit without guaranteeing automatic job creation. These considerable achievements are clearly based on cooperation between regional actors. Their joint efforts produced the required capacity for finding ideas, as well as for planning and financing projects. In addition, and depending on general framework conditions, all three regions adopted their own approach to cooperation, in keeping with their individual projects for exploiting post-mining potentials. The Lusatian Lakeland, for instance, has the strongest and most complex cooperative approach to managing a comprehensive transformation of the entire region. In contrast, the Mansfeld-Südharz region shows a more selective approach to the use of post-mining potentials, based on relatively informal networks of cooperation, in which key actors seem to play a particular role.

All in all, the investigation reveals various approaches to successful capacity building through regional cooperation in the three small and medium-sized post-mining regions. For comparable regions and for policy-makers at the national level, this suggests that cooperation tailored to individual framework conditions pays off and is worth supporting. Moreover, the investigation suggests that cooperation among the affected actors might help mitigate the general impacts of global change on local regions. A major challenge for regional actors, however, is cooperating to reduce regional vulnerability before real problems occur.

Acknowledgments

This chapter is based on investigations conducted in the framework of the project "ReSource," supported by the EU CENTRAL EUROPE programme (INTERREG IVB, No. 1CE084P4). The Leibniz Institute of Ecological and Regional Development was responsible for coordinating research activities within the project. We would like to thank the academic partners in the project for inspiration and scientific discussions: Wolfgang Fischer and David Osebik (Graz), Barbara Černič-Mali and Naja Marot (Ljubljana), Antonín Vaishar and Milada Štastná (Brno), Gergely Horváth (Budapest), and Sylwia Dołzbłasz (Wrocław). Furthermore, the authors would like to thank two anonymous reviewers for their valuable comments.

References

Blotevogel, H.-H. (1997). Strukturwandel und Regionalpolitik im Ruhrgebiet. [Structural change and regional policy in the Ruhr area]. In H.-J. Axt (Ed.), *Strukturwandel in Europa* [Structural change in Europe] (pp. 125–140). München, Germany: Südosteuropa-Gesellschaft.

Černič-Mali, B., Marot, M., & Klančišar-Schneider, K. (2010). European initiative analysis. Unpublished manuscript. Retrieved from http://www.resource-ce.eu/en/project-results/scientific-support/10-eia-thematic-report-introduction-to-european-initiative-analysis.html

Chhotray, V., & Stoker, G. (2010). *Governance theory and practice. A cross-disciplinary approach.* Basingstoke, UK: Palgrave Macmillan.

Diamond, J., & Little, J. (2005). *Management of regeneration.* London, UK: Routledge.

Eckart, K., Ehrke, S., Krähe, H., & Eckart-Müller, I. (2003). *Social economic and cultural aspects in the dynamic changing process of old industrial regions. Ruhr district (Germany), Upper Silesia (Poland), Ostrava region (Czech Republic).* Münster, Germany: LIT Verlag.

Feiock, R. C. (2007). Rational choice and regional governance. *Journal of Urban Affaires, 29*, 47–63.

Healey, P. (1998). Building institutional capacity through collaborative approaches to urban planning. *Environment and Planning A, 30*, 1531–1546.

IBA Fürst-Pückler-Land (Ed.). (2010a). *Bergbau-Folge-Landschaft/Post-Mining Landscape.* Berlin, Germany: Jovis.

IBA Fürst-Pückler-Land (Ed.). (2010b). *IBA Katalog 2010: Neue Landschaft Lausitz* [IBA catalogue 2010: New landscape Lusatia]. Berlin, Germany: Jovis.

Jänicke, M. (2002). The political system's capacity for environmental policy: The framework for comparison. In H. Weidner & M. Jänicke (Eds.), *Capacity building in national environmental policy. A comparative study of 17 countries* (pp. 1–18). Berlin, Germany: Springer.

Kunze C. (Ed.). (1997). *Strukturkrisen altindustrialisierter Regionen und ihre Bewältigung in West- und Osteuropa* [Structural crises of old industrial regions and their handling in Western and Eastern Europe]. Leipzig, Germany: Leipziger Universitätsverlag.

Lintz, G., & Leibenath, M. (2005). Cross-border decision-making processes regarding Natura 2000: Some theses on key factors. In M. Leibenath, S. Rientjes, G. Lintz, C. Kolbe-Weber, & U. Walz (Eds.), *Crossing borders: Natura 2000 in the light of EU-enlargement. Proceedings of an International Workshop Held in Dresden* (pp. 97–105). Tilburg, Netherlands: ECNC.

Lintz, G., & Wirth, P. (2009a). Erfolgsfaktoren der Sanierung und Entwicklung von Bergbauregionen [Factors of success of rehabilitation and development in mining regions]. *GeoScape, 4*(Suppl.), 222–238.

Lintz, G., & Wirth, P. (2009b). The importance of Leitbilder for structural change in small towns. In W. Strubelt (Ed.), *Guiding principles for spatial development in Germany* (pp. 75–96). Berlin, Germany: Springer.

Yin, R.-K. (2002). *Case study research. Design and methods* (3rd ed.). Thousand Oaks, CA: Sage.

Strengthening Pro-Environmental Behavior

Does Binding Communication Increase Consideration of Future Consequences and Decision-Making Concerning Pro-Environmental Behaviour?

Christophe Demarque, Thémis Apostolidis, and Robert-Vincent Joule

Laboratory of Social Psychology, University of Provence, Aix-en-Provence, France

Abstract

This research was intended to study the impact of binding communication (Joule, Py, & Bernard, 2004), a procedure that is at the crossroads between persuasion and commitment, on consideration of future consequences (CFC; Strathman, Gleicher, Boninger, & Edwards, 1994) and individuals' decision-making in the field of environmental protection. Since pro-environmental behaviours often involve an investment directed towards the future, we hypothesized that time perspective and especially the consideration of future consequences, traditionally regarded as a personality trait, could be reinforced in a situation of social influence. The goal was to make people more sensitive to long-term consequences of their behaviours, since this sensitivity leads to an increased setting up of pro-environmental behaviours. The results showed an increase of subjects' CFC score after a binding communication procedure. We discuss the practical implications of these results as they relate to communication about environmental issues.

Key words: binding communication, consideration of future consequences, future time perspective, pro-environmental behaviour

Future Time Perspective and Environment

Since the founding work of Lewin in 1942, time perspective has been the concept of choice for studying the relationships between individuals and time from a psychosocial perspective. In his earliest work in this area, Lewin considered it as an essential structuring element of the psychological field. Many studies conducted in recent years in social psychology (Karniol & Ross, 1996; Strathman & Joireman, 2005; Zimbardo & Boyd, 1999) have focused particularly on future time perspective and on the role of the experience of future time in the determination of behaviour.

Studying future time perspective seems particularly relevant in the field of pro-environmental behaviours. Indeed, the "psychological time" factor may enlighten us about these behaviours' dynamics, as they take place in the future and require anticipation. The development, in the twentieth century, of new technologies with beneficial effects in the short term, but more uncertain ones in the long term, quickly raised questions about the long-term environmental consequences of human actions. For instance, although nuclear plants enable the production of electricity with low greenhouse gas emissions, they generate waste with several thousand years of shelf life, hence raising the question of waste treatment. This increased awareness of future consequences at a collective level lead to the formulation of the precautionary principle, insisting on the dimension of irreversibility: Even though a degradation of the environment could happen very quickly, for instance through an accidental event like an oil slick, a return to the *status quo ante* could take a very long time (Moser, 2009). Therefore, the precautionary principle involves *de facto* responsibility toward future generations, as well as the notion of sustainable development. Initially evoked within the Brundtland report in 1987, this development is defined as "a development which meets the needs of the present without compromising the ability of future generations to meet their own needs." Interestingly, anticipation and future time register predominate in this definition: Sustainable development is situated by definition in the long run, since it aims at the adoption of anticipatory behaviours, establishing the principle of a consideration of our behaviours' consequences and of future generations' needs when we act today on the environment. Thus, because of their nature, environmental behaviours imply, for the individual, an investment oriented toward the future.

If the question of environmental protection has found endorsement in the majority of the population, individual pro-environmental behaviours are still difficult to initiate (Barr, 2004). This is explained partly by the specific temporality of environmental issues but also by the fact that environmental behaviours, because they concern limited resources, may be considered as social dilemma situations. This idea was expressed by Hardin (1968) in his article *The tragedy of the commons*. According to Hardin, the management of limited resources will always lead to conflicts between individual and collective interests, even if a suitable institutional intervention enables a regulation of these conflicts (Ostrom, 1990). Beyond the individual versus collective dichotomy, Platt's work (1973) on the idea of social trap emphasises that environmental behaviours are generally characterised by short-term benefits and long-term costs. Therefore, these behaviours may be considered as temporal dilemmas (Joireman, 2005). Since the perception of losses is more intense than those of the benefits of equivalent gains (Tversky & Kahneman, 1974), we may suppose that a loss in the present is hard to consider, even over a future benefit. This set of characteristics makes it difficult for an individual to create a link between a behaviour and its consequences.

The Importance of Considering the Future Consequences of Our Actions

This difficulty is problematic. Indeed, Stern (2005) reminds us that "People who do not see connections between their behaviour and such consequences or who believe that their actions are so insignificant in the scheme of things as not to matter will not be motivated to act by an internalized sense of obligation" (p. 10788). According to Stern, making people aware of their behaviours' consequences for valorised objects makes it is possible to influence individual behaviour, in the limits determined by the context, habits, and personal capacities. Therefore, we have chosen to focus more specifically on a particular aspect of future time perspective to study pro-environmental behaviours: The degree of consideration for future consequences, measured through the consideration of future consequences scale – CFC (Strathman, Gleicher, Boninger, & Scott Edwards, 1994). According to Strathman et al., individuals differ in the way in which they foresee the consequences of their acts, some focusing on the long-term consequences, while others do not see beyond the immediate consequences. Therefore CFC measures a particular dimension of future psychological time experience, namely time extension, since the scale measures the extent to which the subjects are able to perceive the consequences of their behaviours over time. According to the specific temporality of environmental issues, CFC could be a relevant construct to explain the setting up of pro-environmental behaviours.

This general hypothesis has been confirmed by research employing the CFC scale in the domain of environmental protection. Indeed, these studies showed that the individuals who obtained a high score on the CFC[1] scale declared they felt more concern for environmental problems, had pro-environmental attitudes, and stated that they either followed, or intended to follow to a greater extent, "eco-citizen" patterns of behaviour (Joireman, Lasane, Bennett, Richards, & Solaimani, 2001; Strathman et al., 1994), in particular in the areas of recycling (Ebreo &Vining, 2001; Lindsay & Strathman, 1997) and the use of public transportation (Joireman, Van Lange, & Van Vugt, 2004). If the studied behaviours were diverse (in terms of cost, visibility, automaticity, etc.), positive albeit weak correlations were observed, generally between 0.15 and 0.30. A few experimental results are available. Subjects were exposed to hypothetical dilemma situations where they had to manage some fictitious natural resources (Joireman, Posey, Truelove, & Parks, 2009; Kortenkamp & Moore, 2006). Subjects who scored high on the CFC scale were significantly more cooperative and managed resources by taking the collective interest more into account.

[1] A high score indicates a strong tendency to focus on the future consequences of their acts in making behavioural choices rather than focusing on immediate consequences.

Influencing CFC Through Binding Communication

Since pro-environmental behaviours often involve, because of their nature, an investment directed toward the future (Joireman, 2005), we hypothesized that time perspective, traditionally regarded as a personality trait, could be reinforced in a situation of social influence. Indeed, one cannot ignore the socially regulated character of this psychological construct, whose development among individuals is largely based on social learning (Trommsdorff, 1983). Therefore, we can assume that the relation to time is the result of dynamic interrelations between people and their environment, as well as their social characteristics, including: Social class, cultural capital, and family relationships (Agarwal, Tripathi, & Srivastava, 1983; Bourdieu, 1977; Fieulaine, 2006). In the same vein, Zimbardo and Boyd (1999) consider Time Perspective to be a construct determined by relatively stable inter-individual differences, acquired via cultural and social belonging factors, but it is also a construct determined by the characteristics of the situation. Furthermore, a contextualised approach is needed to better understand the role of CFC in initiating pro-environmental behaviours. Indeed, according to Lewin, the behaviour (B) should be considered as a function (*f*) of the person (P) and of her/his environment (E): B = *f* (P, E).

Taking into account the results described above, the aim of our study is to discover whether we can make people more sensitive to the long-term consequences of their behaviours, since this sensitivity leads to an increased setting up of pro-environmental behaviours (Joireman, Strathman, & Balliett, 2006). In other words, if it is well-established that CFC plays a contextualising role, by offering a cognitive frame for individuals' psychological activities, it seems important to us to insist on its contextualised character (or dependency on the context). Our research thus intends to study the impact of binding communication (Joule, 2000; Joule, Py, & Bernard, 2004) on CFC and individuals' decision making in the field of pro-environmental behaviours. Binding communication is a procedure that stands at the crossroads between persuasion and commitment: In a situation of classic persuasive communication (Hovland, Janis, & Kelley, 1953; Perloff, 2003), participants receive a persuasive message defending a given position that we want them to adopt; in a situation of binding communication (Girandola & Joule, 2008; Joule, Bernard, & Halimi-Falkowicz, 2008; Joule, Girandola, & Bernard, 2007), participants receive the same persuasive message, but only after they have fulfilled a simple request (preparatory action) matching the message. For instance, if the message promotes waste separation, participants are requested to wear a badge that promotes waste separation just before they read the message. This situation is based on the foot-in-the-door technique (Freedman & Fraser, 1966). A situation of binding communication thus differs from a situation of persuasive communication by the inexpensive act that participants are requested to perform. The act must be, according to consistency theories, consistent with the position defended in the message. Much research, notably in the area of eco-citizenry, has demonstrated that binding communication is more effective than persuasive communication in eliciting changes in behaviours (Joule et al., 2008). With regards to this set of theoretical elements, we hypothesised that we could influence the subjects' CFC scores and decision-making in the context of a binding communication procedure. More precisely, we have chosen to study a costly behaviour, governed by volition and characterised by great social visibility: Manning a stand for an environmental awareness organisation. In order to test these hypotheses, we carried out an experimental study.

Procedure

The subjects were asked to read a document about environmental issues (the so-called new awareness document about environmental issues developed by the French Agency for the Environment and Energy Control (*Agence de l'Environnement et de la Maîtrise de l'Energie – ADEME*). Once they had agreed to read it, they completed the 12 items of the consideration of future consequences scale (step *T1*). At this time, we explained to the subjects that this questionnaire had in fact nothing to do with our work and was distributed at the request of another PhD student. They were also informed that they would be asked to answer another questionnaire via email a week later (which was, in fact, the CFC scale again).

Then we presented a persuasive message to the experimental subjects ($N = 56$) about the consequences of pro-environmental behaviours and their benefits for future generations, according to two conditions:

- Direct presentation of the persuasive message, without preparatory action (condition 1 "persuasive communication," $N = 19$);
- Presentation of the message after a committing procedure that involved a preparatory action. The action consisted in producing arguments which stress the importance of the preservation of the environment, in order to protect future generations (condition 2 "binding communication," $N = 37$).

When they had finished their reading, we asked them for their opinion about the text. We finally asked them if they would agree to man a stand for the ADEME. Then, they had to fill in the CFC scale again a week later, via email (step *T2*).

A group of control subjects ($N = 20$) answered the questionnaire without the persuasive message, then we made the target query. They also filled in the questionnaire a week later. Subjects were female students at the Université de Provence.

Our dependent variables were the mean differences on the CFC scale between T1 and T2 and the agreement to man the ADEME stand. We anticipated that the subjects' average score on the CFC scale would be higher in T2 than in T1, in the two experimental conditions vs. Control (Hypothesis 1). Specifically, we anticipated that the difference between T1 and T2 would be higher in the "binding communication" situation than in the "persuasive communication" situation (Hypothesis 2). Concerning the willingness to man the stand, we hypothesized that it would be higher in the experimental conditions than in the control condition (Hypothesis 3) and higher in the "binding communication" situation than in the "persuasive communication" condition (Hypothesis 4).

Results

With respect to our first dependent variable, we observed the expected results. Globally, in our experimental conditions, average scores in T2 were higher than scores in T1 (3.48 vs. 3.60, $t(55) = -3.14, p < .01$), a difference we did not find in the control condition. These results confirm our first hypothesis. More precisely, this difference was significant and more pronounced only for subjects who obtained low scores on the CFC scale[2] (3.03 vs. 3.24, $t(27) = -3.61, p < .001$). According to our second hypothesis, in condition 1 ("persuasive communication"), there was no significant difference between T1 and T2, whereas in condition 2 ("binding communication"), the average scores in T2 were higher than in T1 (3.48 vs. 3.68, $t(36) = -4.07, p < .001$).

For our second dependent variable, only 1 subject out of 20 (5%) agreed to man the ADEME stand in the control condition. There were 3 subjects out of 19 (15.8%) in the condition 1 ("persuasive communication") and 10 subjects out of 37 (27%) in the condition 2 ("binding communication"). We did not observe a significant difference between the control and the experimental conditions. However, a tendential difference was observed between the control and the binding communication condition ($\chi^2 = 2.75$, $p < .10$). This result confirms only partially our third hypothesis and refutes the fourth one. Nevertheless, we observed that the reinforcement of the CFC score between T1 and T2 was stronger among the 13 experimental subjects who agreed to man the ADEME stand (3.41 vs. 3.73, $t(12) = -3.72, p < .01$). On the other hand, the difference was not significant among the other experimental subjects (3.49 vs. 3.56).

Discussion

These results, although exploratory, have a dual interest, both theoretical and practical. First, they give us information about the future time perspective status. According to these previously unseen effects, a so-called personality trait could be reinforced, at least temporarily, in an experimental socially influential situation. This reinforcement of the CFC score seems to indicate that the subjects were more sensitive to the future consequences of their behaviours after a binding communication procedure (at least subjects with a low score in T1). Thus, we consider that future time perspective, and especially CFC, should not be considered strictly as a stable and cross-situational disposition. Adopting a similar point of view, Lewin (1942) asserted that time perspective is largely dependent on the social context in which the

[2] We dichotomized "CFC+" and "CFC−" subjects with the aid of a median split.

individual evolves, and he observed a narrowing of future time perspective in a crisis situation, such as unemployment. Whereas a dominant personality-based approach often considers time perspective as a trans-situational personal trait, we tried to adopt a more Lewinian perspective of ecological psychology: CFC should be considered as a dynamic psychological construct, since individuals have relations of circular interdependence with their environment. It also should be noted that a classical persuasive situation was not sufficient to detect changes. On the other hand, as binding communication required an active participation of the subjects, it allowed us to observe a reinforcement of the subjects' scores. Thus, the active participation resulting from the preparatory action was necessary. This placement in a socially influential situation also seemed to be necessary for the CFC to intervene as a facilitator for decision-making. Furthermore, when the subjects accepted the final request, we observed the most significant reinforcement of the CFC score. These results again underscore the necessity of taking into account social context and behavioural commitment when we study the relation to time.

However, this work obviously involved important limitations and should be considered as an exploratory study. There were relatively few subjects in each condition, and they were exclusively female students. For this first attempt, we used a convenience sample, which restricts the scope of our results. Nevertheless, previous studies in the field of binding communication, using diverse populations (Joule et al., 2008) allow us to conjecture that similar results would be observed in other samples. We can also point another limitation: We did not observe strong effects, although significant, of binding communication on the CFC score and decision-making. Furthermore, it would be necessary to set up longitudinal studies in order to study the persistence of the observed effects over time.

Despite the fact that they are preliminary, these results open up interesting new perspectives in CFC research, around the idea of double contextualisation (contextualising role/contextualised effect, Fieulaine, 2006). On the other hand, if we manage to influence the sensitivity to future consequences, it could have some practical implications in the field of awareness of environmental issues. Indeed, we saw that people who were sensitive to long-term consequences were more concerned with environmental issues and acted accordingly. Thus, the recognition of consideration for the future consequences of behaviour could prove to be relevant in composing persuasive arguments and putting in place strategies for modifying behaviour. Moreover, future research will have to develop a more contextual approach to the study of links between CFC and pro-environmental behaviour, taking into consideration the dynamics and the issues of the social communicative situation. In this context, the binding communication situation presents an interesting heuristic value.

Acknowledgments

Financial support for this research was provided by the French Agency for the Environment and Energy Management (Agence de l'Environnement et de la Maîtrise de l'Energie – ADEME) and the Provence Alpes Côte d'Azur Region (PACA).

References

Agarwal, A., Tripathi, K. K., & Srivastava, M. (1983). Social roots and psychological implications of time perspective. *International Journal of Psychology, 18*, 367–380.

Barr, S. (2004). Are we all environmentalists now? Rhetoric and reality in environmental action. *Geoforum, 35*, 231–249.

Bourdieu, P. (1977). *Algérie 60. Structures économiques et structures temporelles* [Algeria 60. Economic structures and temporal structures]. Paris, France: Editions de Minuit.

Ebreo, A., & Vining, J. (2001). How similar are recycling and waste reduction? Future orientation and reasons for reducing waste as predictors of self-reported behavior. *Environment and Behavior, 33*, 424–448.

Fieulaine, N. (2006). *Perspective temporelle, situations de précarité et santé: Une approche psychosociale du temps* [Time perspective, situations of precariousness and health: A psychosocial approach of time] (Unpublished doctoral thesis). Université de Provence, Aix-en-Provence, France.

Freedman, J. L., & Fraser, S. C. (1966). Compliance without pressure: The foot-in-the-door technique. *Journal of Personality and Social Psychology, 4*, 195–202.

Girandola, F., & Joule, R. V. (2008). La communication engageante [The binding communication]. *Revue Electronique de Psychologie Sociale, 2*, 41–51.

Hardin, G. (1968). The tragedy of the commons. *Science, 162*, 1243–1248.

Hovland, C. I., Janis, I. L., & Kelley, H. H. (1953). *Communication and persuasion. Psychological studies of opinion change*. New Haven, CT: Yale University Press.

Joireman, J. A. (2005). Environmental problems as social dilemmas: The temporal dimension. In A. Strathman & J. Joireman (Eds.), *Understanding behavior in the context of time* (pp. 289–304). Mahwah, NJ: Erlbaum.

Joireman, J. A., Lasane, T. P., Bennett, J., Richards, D., & Solaimani, S. (2001). Integrating social value orientation and the consideration of future consequences within the extended norm activation model of proenvironmental behavior. *British Journal of Social Psychology, 40*, 133–155.

Joireman, J. A., Posey, D. C., Truelove, H. B., & Parks, C. D. (2009). The environmentalist who cried drought: Reactions to repeated warnings about depleting resources under conditions of uncertainty. *Journal of Environmental Psychology, 29*, 181–192.

Joireman, J. A., Strathman, A., & Balliet, D. (2006). Considering Future Consequences. In L. J. Sanna & E. C. Chang (Eds.), *Judgments over time* (pp. 82–99). New York, NY: Oxford University Press.

Joireman, J. A., Van Lange, P. A. M., & Van Vugt, M. (2004). Who cares about the environmental impact of cars? Those with an eye toward the future. *Environment and Behavior, 36*, 187–206.

Joule, R.V. (2000). Pour une communication organisationnelle engageante : vers un nouveau paradigm [For a new binding organisational communication: Toward a new paradigm]. *Sciences de la Société, 50/51*, 279–295.

Joule, R. V., Bernard, F., & Halimi-Falkowicz, S. (2008). Promoting ecocitizenship: In favor of binding communication. *International Scientific Journal for Alternative Energy and Ecology, 6*, 214–218.

Joule, R. V., Girandola, F., & Bernard, F. (2007). How can people be induced to willingly change their behavior? The path from persuasive communication to binding communication. *Social & Personality Psychology Compass, 1*, 493–505.

Joule, R. V., Py, J., & Bernard, F. (2004). Qui dit quoi, à qui, en lui faisant faire quoi ? Vers une communication engageante [Who say what to whom and by making him do what? Toward a binding communication]. In M. Bromberg & A. Trognon (Eds.), *Psychologie sociale et communication* (pp. 205–218). Paris, France: Dunod.

Karniol, R., & Ross, M. (1996). The motivational impact of temporal focus: Thinking about the future and the past. *Annual Review of Psychology, 47*, 593–620.

Kortenkamp, K. V., & Moore, C. F. (2006). Time, uncertainty, and individual differences in decisions to cooperate in resource dilemmas. *Personality and Social Psychology Bulletin, 32*, 603–615.

Lewin, K. (1942). Time perspective and morale. In G. Watson (Ed.), *Civilian morale* (pp. 48–70). Boston, MA: Houghton Mifflin.

Lindsay, J. J., & Strathman, A. (1997). Predictors of recycling behavior: An application of a modified health belief model. *Journal of Applied Social Psychology, 27*, 1799–1823.

Moser, G. (2009). *Psychologie environnementale: les relations homme-environnement* [Environmental psychology: Man-environment relations]. Bruxelles, Belgium: De Boeck.

Ostrom, E. (1990). *Governing the commons: The evolution of institutions for collective action*. Cambridge, MA: Cambridge University Press.

Perloff, R. M. (2003). *The dynamics of persuasion*. Mahwah, NJ: Erlbaum.

Platt, J. (1973). Social traps. *American Psychologist, 28*, 641–651.

Stern, P. C. (2005). Understanding individuals' environmentally significant behavior. *Environmental Law Reporter, 35*, 10785–10790.

Strathman, A., Gleicher, F., Boninger, D. S., & Scott Edwards, C. (1994). The consideration of future consequences: Weighing immediate and distant outcomes of behavior. *Journal of Personality and Social Psychology, 66*, 742–752.

Strathman, A., & Joireman, J. (2005). A brief history of time. In A. Strathman & J. Joireman (Eds.), *Understanding behavior in the context of time* (pp. 3–7). Mahwah, NJ: Erlbaum.

Trommsdorff, G. (1983). Future orientation and socialization. *International Journal of Psychology, 18*, 381–406.

Tversky, A., & Kahneman, D. (1974). Judgement under uncertainity: Heurstics and biases. *Science, 185*, 1124–1130.

Zimbardo, P. G., & Boyd, J. N. (1999). Putting time in perspective: A valid, reliable individual-differences metric. *Journal of Personality and Social Psychology, 77*, 1271–1288.

The Role of Individual Differences in Promoting Pro-Environmental Behavior

Dóra Medvés and Judit Kovács

Institute of Psychology, University of Debrecen, Hungary

Abstract

This pilot study explores the effects of using different aspects of environmental protection and different levels of perceived criticality on *judgements* about the efficacy of social marketing. Specifically, four alternatives of a message related to promoting the reduction of packaging in domestic waste are introduced to a group of Hungarian school-leavers. On the basis of the findings related to the willingness to behave pro-environmentally, we predict that individual differences play an important role in evaluating pro-environmental messages. Consistent with our expectations, significant results were revealed: Pro-self people indicated messages, emphasizing anthropocentric aspects of environmental protection and high levels of perceived criticality more effective than pro-socials, who did not find less effective messages with totally different aspects. Finally, connectedness to nature also had effects on decisions.

Key words: connectedness to nature, environmental attitudes, perceived criticality, social marketing, social value orientation

Introduction

One of the most important social issues today is to solve the problem of global environmental change. During the last decades, environmental problems were highlighted not only by natural scientists but also by social and behavioral scientists. Research has to explore how humans can be encouraged to behave in a cooperative way with thousands of unknown others in a vast social dilemma situation (e.g., Van Vugt, 2001). As behavior in all kinds of social dilemmas can also depend on individual differences (Weber, Kopelman, & Messick,

2004), it is worth considering whether some specific individual features also influence the evaluation of messages encouraging pro-environmental behavior.

Thus, the aim of our study was to investigate whether various informational messages are evaluated differently by people with different kinds of individual characteristics. Findings from the research can contribute to the development of more efficient campaigns promoting pro-environmental behavior, by encouraging people to behave this way.

Characteristics of Social Marketing

The definition of social marketing refers to "the design, implementation, and control of programs calculated to influence the acceptability of social ideas and involving considerations of product planning, pricing, communication, distribution, and marketing research" (Kotler & Zaltman, 1971, p. 5). Social marketing provides an appropriate framework for the promotion of environmental protection and many other socially relevant issues.

Advertising for pro-environmental behavior can be achieved in several ways thanks to the development of social marketing. Steg and Vlek (2009) present an integrative model for promoting this behavior. They state that intervention strategies aimed at altering a certain kind of behavior can be invented after identifying the factors determining that kind of behavior. These intervention strategies can be categorized in several ways. One of them differentiates between structural and informational strategies (Messick & Brewer, 1983). The former tries to alter circumstances of the choices, while the latter can result in changes in perceptions, motivations, knowledge, and norms; furthermore it is more effective in the case of convenient and inexpensive behaviors, when the external pressure on behaving pro-environmentally is weak.

In our research – where the specific focus is on promoting the reduction of packaging in domestic waste – the informational strategy is an adequate framework, because the required behavior can be achieved at a relatively low cost, if the actors are well informed.

Influences of Social Value Orientation and Perceived Criticality on Pro-Environmental Behavior

The problem of waste separation and disposal, like many other environmental problems, can be defined as an immense social dilemma. In order to encourage individuals' pro-environmental behavior, factors influencing decisions in social dilemma situations have to be considered. From the multitude of influencing factors (Ostrom, 2000) we have chosen differences in social value orientation and perceived criticality of contributions.

The construct of social value orientation was primarily developed to evaluate behavior in the prisoners' dilemma and it was only later extended to social dilemmas (Eek & Gärling, 2008). Social dilemmas can be characterized by the existence of conflict between collective and the individual interests and empirical evidence has established that people differ in their preferences about the outcomes for themselves and an unknown other, in these situations (e.g., Messick & McClintock, 1968). At least three categories can be distinguished by these preferences: Cooperator (maximizing the overall outcome), individualist (maximizing the personal outcome), or competitor (maximizing the difference between the other person's and the personal outcome). Further distinctions can be made (and will be used in the present study as well): *Pro-socials* (cooperators) perceive situations in terms of *collective welfare*, while *pro-selfs* (individualists and competitors) emphasize their *own interests*. The role of social value orientation in social dilemmas has been established in many studies. For instance, pro-socials tend to behave in a more cooperative way than pro-selfs and they also pay more attention than pro-selfs to the long-term collective interest (e.g., Kramer, McClintock, & Messick, 1986). Both of these characteristics are important for pro-environmental behavior.

Perceived criticality is also one of the factors affecting participants' decisions in social dilemma situations. By definition, perceived criticality refers to "one's perception about one's criticality in the provision of public good" (Chen, Au, & Komorita, 1996, p. 39). The notion was conceived by Kerr (1989), who relied on Bandura's concept of self-efficacy (1977). According to experimental evidence (Chen et al., 1996; De Cremer & van Dijk, 2002), the higher the perceived criticality, the more people contribute in a social dilemma situation.

In the field of marketing research, Perceived Consumer Effectiveness is a similar construct. It is used to measure individuals' beliefs whether their actions make a difference in solving a problem. The results of Laskova's (2007) research showed that, for the case of green behavior, perceived consumer effectiveness has predictive impact.

Environmental Attitudes

Attitudes, together with subjective norms and perceived behavioral control, are significant determinants of behavior (mediated by behavioral intention according to the Theory of Planned Behavior (Ajzen, 1991), which is a widely-used model for the attitude-behavior relationship. The Theory of Planned Behavior is often used to predict the probability of pro-environmental behavior. For instance, Kaiser (2006), who included a moral-related component in his examination, found that conservation behavior determined by intention can be best predicted by people's attitude under the influence of moral norms. Of course, the connection can also be formulated in another way: Involvement in pro-environmental behavior, even if it was first evoked by external incentives, can develop commitment to environmentalism (Pallak, Cook, & Sullivan, 1980).

To analyze the role of attitude in pro-environmental behavior, the notion of environmental attitude was developed. Environmental attitudes can be defined as "the collection of beliefs, affects, and behavioral intentions a person holds regarding environmentally related activities or issues" (Schultz, Shriver, Tabanico, & Kazian, 2004, p. 31). Research on environmental attitudes is widespread. For example, Milfont and Duckitt (2004, 2006) examined the structure of environmental attitudes and created a questionnaire to measure

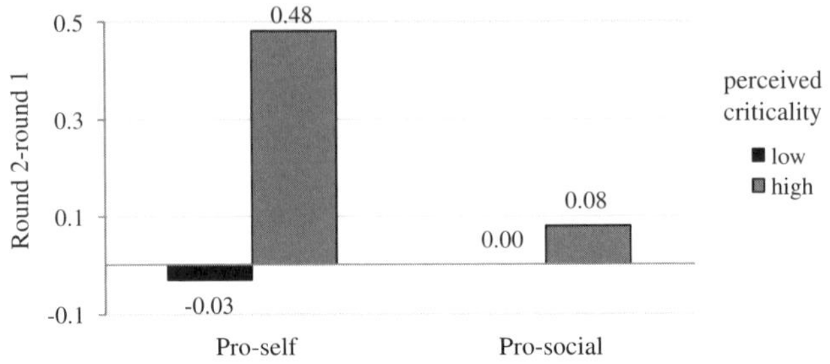

Figure 1. The interactive effect of social value orientation and perceived criticality on cooperation. (Reprinted from J. Kovács & D. Medvés, 2009, © 2009 Alkalmazott Pszichológia. Reproduced with permission.)

it. Both the extended (*Environmental Attitude Inventory*) and the short forms of their questionnaire are valid and reliable measures of environmental attitudes (Milfont & Duckitt, 2010).

A Preliminary Study Exploring the Role of Social Value Orientation, Environmental Attitudes, and Perceived Criticality in Pro-Environmental Behavior

Because pro-environmental behavior is usually conceptualized and examined as a social dilemma situation (e.g., Van Vugt, 2001), the authors (Kovács & Medvés, 2009) conducted a study to examine the effect of social value orientation and perceived criticality on behavior in a step-level social dilemma situation. A between-subject design was applied to evaluate the connection between social value orientation and perceived criticality. The results (Figure 1) demonstrated that people with pro-self social value orientation increased their cooperation in the second round, when they learned that the common good had been provided barely any above the necessary level of functioning by the group in the first round, while pro-socials did not adjust their behavior to the feedback ($F = 2.797$; $df = 1$; $p < .1$).

Results derived from the short form of the *Environmental Attitude Inventory* showed that pro-socials were less likely to support anthropocentric aspects of protection of the environment than pro-selfs ($Z = -1.365$; $p < .1$).

Connectedness to Nature and Pro-Environmental Behavior

According to Schultz (2002, p. 67), connectedness to nature demonstrates "the extent to which an individual includes nature within his/her cognitive representation of self."

Previous studies established that connectedness to nature is related to numerous environmentally relevant concepts. For instance, Schultz (2001) found that the differences in the extent of the interconnectedness of the self and nature are attached to different types of environmental concerns. Additionally, connectedness to nature is not only correlated with biospheric environmental concerns but also with self-reported pro-environmental behavior, a result that was also confirmed by Davis, Green, and Reed (2009).

We can therefore state that connectedness to nature plays an important role in human-environment issues.

The Study

Results of our previous research provided evidence that social value orientation and perceived criticality of contributions are essential factors in predicting pro-environmental behavior. Thus, the aim of our current study was to explore whether these constructs (expanded by the factor of connectedness to nature) also play significant roles in making judgements about the effectiveness of pro-environmental messages related to packaging in domestic waste, by testing a target group of Hungarian school-leavers.

Participants

Ninety-five Hungarian school-leavers (47 male and 48 female) from the Tóth Árpád Secondary Grammar School, Debrecen, took part in the experiment. Participants were 17.87 years old on average (SD = 0.58). The questionnaire was filled out during school time in the classroom; participants were not compensated for participation, but received credit for their contribution. The process lasted approximately 30 minutes.

Materials

Measurement of Social Value Orientation

To distinguish between people with various kinds of social value orientation, participants filled in the triple-dominance measurement of social values, which is a nine-item decomposed game questionnaire (from Van Lange, Otten, De Bruin, & Joireman, 1997). During the procedure of allocation tasks, respondents decide between three alternatives. They decide whether they want to maximize the allocated points in total (cooperative choice), maximize the number of their own points (individualistic choice), or maximize the difference in points between the other person and themselves (competitive choice). If a participant chooses a certain alternative consistently at least six times, s/he can be classified into the respective social value orientation category.

Environmental Attitude Inventory

To examine participants' environmental attitudes, we adapted two scales of the *Environmental Attitude Inventory* (developed by Milfont & Duckitt, 2006). The scale of *Conservation Motivated by Anthropocentric Concern* measures the extent to which people support anthropocentric concerns for human welfare in connection with environmental protection. On the contrary, the scale of *Ecocentric Concern* estimates the extent of people's nostalgic concerns about environmental damage. The reason for applying these scales of the *Environmental Attitude Inventory* is that, in our preliminary study, people with different types of social value orientation supported different aspects of environmental protection. In both cases, the ten-item version of the scales was used.

Inclusion of Nature in Self Scale

To measure the perceived relationship between self and nature, our questionnaire contained the *Inclusion of Nature in Self* scale adopted by Schultz (2001). In this item, participants are asked to indicate the extent of their interconnectedness with nature using circles labeled "nature" and "self" that overlapped to various degrees.

Developing Pro-Environmental Messages

Pro-environmental messages were created that promoted the reduction of packaging in domestic waste. In a short statement, we attracted participants' attention to the massive extent of domestic waste that comes from packaging (see Appendix A). To develop these messages, we acquired accurate information about the proportion of different types of wastes from the Hungarian Waste Management Plan related to the specific region the participants live in.

Towards the end of the statement two dichotomous variables appeared: The level of perceived criticality and the aspect of environmental attitudes. In the case of low perceived criticality, a significant result of the behavior can be achieved by the collaboration of 80% of the inhabitants of the region, while in the case of high perceived criticality, this proportion is only 20%. In the case of ecocentrically focused communication, the natural environment could benefit from the reduction of domestic waste, whilst in the case of anthropocentrically focused communication, humans could benefit (see Appendix A).

Thus, four different alternatives of the message were created by the two dichotomous variables emphasizing the following aspects of reducing packaging in domestic waste:

- Anthropolow: Focus on anthropocentric concerns and low perceived criticality;
- Anthropohigh: Focus on anthropocentric concerns and high perceived criticality;
- Ecolow: Focus on ecocentric concerns and low perceived criticality;
- Ecohigh: Focus on ecocentric concerns and high perceived criticality.

The messages were evaluated by two different methods. First, participants made six paired comparisons on a seven-point scale, where 2 messages out of the 4 were always compared by their effectiveness. Then, all types of messages were presented simultaneously and participants ranked them by their effectiveness.

Results

Anthropocentric and Ecocentric Aspects of Environmental Attitudes and Messages

Results of the reliability analysis of anthropocentric and ecocentric scales of the *Environmental Attitude Inventory* suggest that both scales are reliable ($\alpha = 0.62$ and $\alpha = 0.87$, respectively). The reliability of the scales did not increase by deleting items; thus, we used the original 10-item versions.

Before conducting further analyses of the messages, it is worth considering whether the score on specific attitude scales and the evaluation of messages with specific "eco" or "anthropo" focus are correlated. Significant correlations would serve as a positive manipulation check. To test these correlations, evaluations from the three different comparisons for a single message were totalled (higher scores represent greater effectiveness) and correlated with the Ecocentric Attitude Scale and the Anthropocentric Attitude Scale. There was a significant correlation of the Ecocentric Attitude Scale, both with the evaluation of Ecohigh ($r = .24$; $p < .05$) and Ecolow ($r = .38$; $p < .05$) messages. In the case of the Anthropocentric Attitude Scale, a significant correlation was found only with the evaluation of Anthropohigh messages ($r = .31$; $p < .05$).

From the results of the correlation analysis, we can conclude that the Anthropocentric and the Ecocentric Scale of *Environmental Attitude Inventory* correspond with anthropocenric and ecocentric aspects of messages promoting packaging waste reduction.

Consistency of the Evaluation of Messages

Whereas effectiveness of messages was evaluated in two different ways (paired comparison and ranking), consistency of evaluations was tested. While the independent variables in the messages were structured in an easily perceivable way, the method of ranking evoked manifest aspects and conscious elaboration in the evaluation process. The paired comparisons (less conscious measurement) can verify the results of the ranking task.

To test the consistency of the messages' evaluation, we compared the rankings from the two methods. Total congruence was considered as consistent response, whereas the least difference was considered as a non-consistent response. Based on the results of independent sample *t*-tests, we can state that consistent participants' attitudes did not differ from non-consistent ones', either in anthropocentric or in ecocentric aspects. However, there was significant difference between their connectedness to nature ($t(72) = 2.39$; $p < .05$). Participants less connected to nature had more consistent responses.

Effects of Social Value Orientation and Connectedness to Nature

Based on the results from the preliminary study, we expected pro-self people to be more sensitive to messages with high perceived criticality and Anthropo focus. Regarding connectedness to nature, from the work of Davis et al. (2009), we expected people with high connectedness to nature to appreciate ecocentric aspects more.

To test our expectations, the impact of social value orientation and connectedness to nature was examined with univariate analysis on the evaluation of the messages from the ranking task. The interactive effect of the two independent variables (social value orientation and connectedness to nature) was not significant, but each of them had significant effects.

Social value orientation exerted its effect in the case of Anthropohigh messages ($F(1) = 4.39$; $p < .05$). As Figure 2 demonstrates, the message emphasizing anthropocentric aspects of environmental attitudes and high level of perceived criticality is considered more effective by pro-self than by pro-social people.

In the case of the Ecolow message, social value orientation also proved to be a significant determinant of evaluation ($F(1) = 5.47$; $p < .05$). It is noteworthy that this does not mean that pro-social people find the Ecolow message especially attractive, but they do not really distinguish between Anthropohigh and Ecolow messages (see Figures 2 and 3).

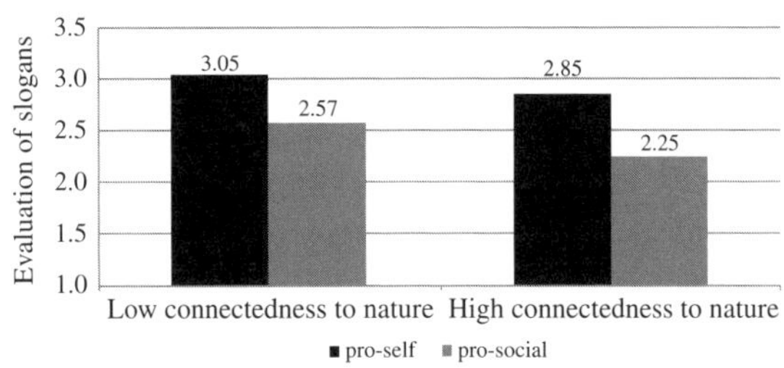

Figure 2. Evaluation of Anthropohigh message.

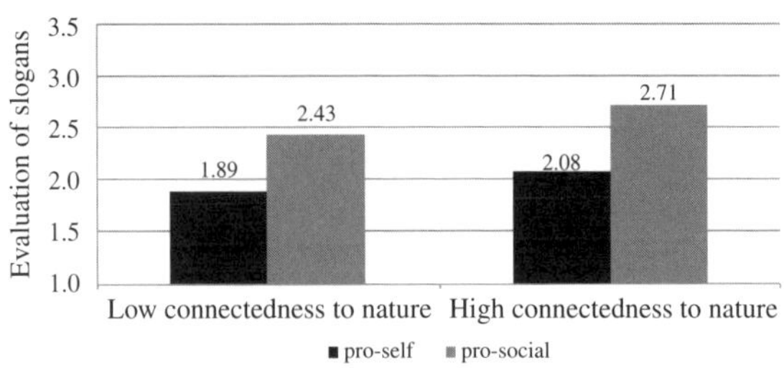

Figure 3. Evaluation of Ecolow message.

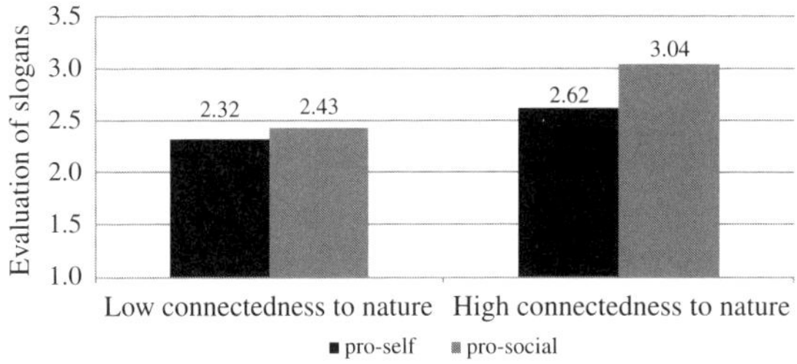

Figure 4. Evaluation of Ecohigh message.

To analyze the effect of connectedness to nature, the continuous variable was transformed to a categorical one (the scale was divided at score 4: 1–4 values were categorized as low, and 5–7 values as high connectedness to nature). According to the results of the univariate analysis, people with high connectedness to nature judged the Ecohigh message as being more effective ($F(1) = 3.34$; $p < .1$) than did people with low connectedness to nature (see Figure 4).

General Discussion

The aim of our study was to evaluate the role of individual differences (social value orientation and connectedness to nature) and characteristics of the means of social marketing (aspect of environmental protection and perceived criticality) in judgements about the effectiveness of pro-environmental messages. In order to test our hypotheses, we conducted questionnaire research to measure individual deviations in the perception of the messages' effectiveness and the interaction of the personal and situational variables.

Our results confirm that different types of social value orientation play significant roles in evaluating the persuasiveness of messages encouraging pro-environmental behavior. Significant dissimilarities between pro-self and pro-social people were demonstrated in their reactions to manipulated variables in the messages. While messages emphasizing anthropocentric aspects of environmental protection and high perceived criticality of pro-environmental actions were evaluated as being more effective by pro-selfs, pro-socials did not judge messages with ecocentric and low perceived criticality as being less effective than Anthropohigh ones. These results correspond with previous findings that pro-self people find pro-environmental behavior acceptable if they can benefit from it (Van Vugt, Meertens, & Van Lange, 1995), or when the criticality of their behavior is emphasized (Kovács & Medvés, 2009). In contrast, pro-socials are likely to prefer pro-environmental behavior, even if it requires more effort (Van Vugt et al., 1995), and behave cooperatively when their actions do not have considerable criticality (Kovács & Medvés, 2009). We conclude that using Anthropohigh messages in promoting pro-environmental attitudes is to be highly recommended: This makes the activity more attractive in the eyes of pro-self people and at the same time it does not encourage pro-socials to refrain from the activity.

In the present study, an additional variable was examined. Because previous findings have established the importance of connectedness to nature in pro-environmental behavior (e.g., Schultz, 2001; Davis et al., 2009), our aim was to study the effect of this concept on decisions about pro-environmental messages. It influenced participants' judgements: People more connected to nature evaluated messages in a less consistent way, and they found messages emphasizing ecocentric aspects and a high level of perceived criticality more effective than less connected people. Both results require some explanation, because they are not trivial. At this point – since we lack empirical evidence – we can rely only on speculative arguments. Perhaps more connected people are so heavily committed to environmentally related topics that they cannot distinguish between the messages, as in the case of paired comparisons. However, when they are forced to distinguish in the ranking task, they emphasize the most important aspect for themselves, namely the ecocentric one, and they enrich the ecocentric focus with the motivating message of high criticality. Making connected people (who most probably are more active in pro-environmental campaigns) aware of the superior effectiveness of Anthropohigh messages is worth considering, in order to make these campaigns more efficient.

The results from such a small study sample should be treated with caution: To generalize the results, more evidence would be needed from other samples (age, school, nation) and with other activities. Despite these limitations, our results can contribute to make social marketing more effective by developing tailored messages emphasizing the importance of pro-environmental behavior.

To summarize the main findings of our research, we can say that pro-environmental messages emphasizing anthropocentric concerns and high criticality of contributions are evaluated as being especially effective by pro-self people, and pro-socials do not regard them as being demotivating. Consequently, using Anthropohigh messages in promoting pro-environmental attitudes is highly recommended. Moreover, highly connected people consider messages with an ecocentric focus and high criticality of contributions most effective. As pro-environmental activists are more probably highly connected people, making them aware of the attractiveness of Anthropohigh messages is recommended.

Acknowledgments

The work is supported by the TÁMOP-4.2.2/B-10/1-2010-0024 project. The project is co-financed by the European Union and the European Social Fund.

References

Ajzen, I. (1991). The theory of planned behavior. *Organizational Behavior and Human Decision Processes, 50*, 179–211.

Bandura, A. (1977). Self-efficacy: Toward a unifying theory of behavioral change. *Psychological Review, 84*, 191–215.

Chen, X-P., Au, W. T., & Komorita, S. S. (1996). Sequential choice in a step-level public goods dilemma: The effects of criticality and uncertainty. *Organizational Behavior and Human Decision Processes, 65*, 37–47.

Davis, J. L., Green, J. D., & Reed, A. (2009). Interdependence with the environment: Commitment, interconnectedness, and environmental behavior. *Journal of Environmental Psychology, 29*, 173–180.

De Cremer, D., & Van Dijk, E. (2002). Perceived criticality and contributions in public good dilemmas: A matter of feeling responsible to all? *Group Process & Intergroup Relations, 5*, 319–332.

Eek, D., & Gärling, T. (2008). A new look at the theory of social value orientations: Prosocials neither maximize joint outcomes nor minimize outcome differences but prefer equal outcomes. In A. Biel, D. Eek, T. Gärling, & M. Gustafsson (Eds.), *New issues and paradigms in research on social dilemmas* (pp. 10–26). New York, NY: Springer.

Kaiser, F. G. (2006). A moral extension of the theory of planned behavior: Norms and anticipated feelings of regret in conservationism. *Personality and Individual Differences, 41*, 71–81.

Kerr, N. L. (1989). Illusions of efficacy: The effects of group size on perceived efficacy in social dilemmas. *Journal of Experimental Social Psychology, 25*, 287–313.

Kotler, P., & Zaltman, G (1971). Social marketing: An approach to planned social change. *Journal of Marketing, 35*, 3–12.

Kovács, J., & Medvés, D. (2009). Közös ügyek?! Van-e jelentősége? [Common affairs?! Are they critical?]. *Alkalmazott Pszichológia, 3*, 51–74.

Kramer, R. M., McClintock, C. G., & Messick, D. M. (1986). Social values and cooperative response to a simulated resource conservation crisis. *Journal of Personality, 54*, 576–592.

Laskova, A. (2007). Perceived consumer effectiveness and environmental concerns. *Proceedings of the 13th Asia Pacific Management Conference* (pp. 206–209). Melbourne, Australia.

Messick, D. M., & Brewer, M. B. (1983). Solving social dilemmas: A review. In L. Wheeler & O. Shaver (Eds.), *Review of personality and social psychology* (Vol. 4, pp. 11–44). Beverly Hills, CA: Sage.

Messick, D. M., & McClintock, C. G. (1968). Motivational bases of choice in experimental games. *Journal of Experimental Social Psychology, 4*, 1–25.

Milfont, T. L., & Duckitt, J. (2004). The structure of environmental attitudes: A first- and second-order confirmatory factor analysis. *Journal of Environmental Psychology, 24*, 289–303.

Milfont, T. L., & Duckitt, J. (2006). Preservation and utilization: Understanding the structure of environmental attitudes. *Medio Ambiente y Comportamiento Humano, 7*, 29–50.

Milfont, T. L., & Duckitt, J. (2010). The environmental attitudes inventory: A valid and reliable measure to assess the structure of environmental attitudes. *Journal of Environmental Psychology, 30*, 80–94.

Ostrom, E. (2000). Collective action and the evolution of social norms. *Journal of Economic Perspectives, 14*, 137–158.

Pallak, M. S., Cook, D. A., & Sullivan, J. J. (1980). Commitment and energy conservation. In L. Bickman (Ed.), *Applied social psychology annual* (pp. 235–254). Beverly Hills, CA: Sage.

Schultz, P. W. (2001). Assessing the structure of environmental concern: Concern for the self, other people, and the biosphere. *Journal of Environmental Psychology, 21*, 327–339.

Schultz, P. W. (2002). Inclusion with nature: The psychology of human-nature relations. In P. Schmuck & W. P. Schultz (Eds.), *Psychology of sustainable development* (pp. 61–79). Dordrecht, Netherlands: Kluwer Academic.

Schultz, P. W., Shriver, C., Tabanico, J. J., & Khazian, A. M. (2004). Implicit connections with nature. *Journal of Environmental Psychology, 24*, 31–42.

Steg, L., & Vlek, C. (2009). Encouraging pro-environmental behavior: An integrative review and research agenda. *Journal of Environmental Psychology, 29*, 309–317.

Van Lange, P. A. M., Otten, W., De Bruin, E. N. M., & Joireman, J. A. (1997). Development of prosocial, individualistic, and competitive orientations: Theory and preliminary evidence. *Journal of Personality and Social Psychology, 73*, 733–746.

Van Vugt, M. (2001). Community identification moderating the impact of financial incentives in a natural social dilemma. *Personality and Social Psychology Bulletin, 27*, 1440–1449.

Van Vugt, M., Meertens, R. M., & Van Lange, P. A. M. (1995). Car versus public transportation? The role of social value orientations in a real-life social dilemma. *Journal of Applied Social Psychology, 25*, 258–278.

Weber, J. M., Kopelman, S., & Messick, D. M. (2004). A conceptual review of decision making in social dilemmas: Applying a logic of appropriateness. *Personality and Social Psychology Review, 8*, 281–307.

Appendix A – The Message With Two Independent Variables

The amount of packaging in domestic waste is 55 kg/person/year in Hungary. For the 1.5 million inhabitants of our region, this number is 85,749 tons/year. In order to achieve significant results, *1.2 million (low perceived criticality) / 300 thousand (high perceived criticality)* people have to choose products with environmentally friendly packaging. By reducing packaging in domestic waste, enlarging landfills in the region becomes unnecessary, thus a considerable amount of *money can be saved (anthropocentric focus) / land can be preserved for nature (ecocentric focus).*

Mental Representations of Nature

The Importance of Well-Being

Kathrin Röderer and Renate Cervinka

Institute of Environmental Health, Center for Public Health, Medical University of Vienna, Austria

Abstract

This qualitative study was aimed at exploring mental representations of nature. The qualitative content analysis was based on 133 responses to three open-ended questions in a paper and pencil questionnaire. Aspects of nature emerged as dominant mental representations of nature. They ranged from wild nature to cultivated landscapes. Well-being was the second most prominent category of mental representations of nature. Mental recreation turned out to be the main purpose for spending time in nature. Altogether, positively connotated associations clearly prevailed. These three distinct findings led to recommendations for the design of urban green space. Urban green has to provide opportunities to positively experience nature within walking distance. Nature experiences should be possible in cultivated areas and in "wild spots" in different stages of proliferation, tailored to peoples' individual needs. These natural areas will foster personal well-being as well as pro-conservation behaviour for a sustainable future.

Key words: definition of nature, qualitative content analysis, urban design, urban green space, well-being

Introduction

In times of global change, the urbanisation and density of human habitats increase continuously, while natural free spaces decline. In urban areas, we face a loss of open land and unspoilt natural areas where people can experience nature and derive benefits from spending time in natural environments. Empirical evidence highlights the positive effects of experiencing nature on health, well-being, and restoration (e.g., Abraham, Sommerhalder, Bollinger-Salzmann, & Abel, 2007; Groenewegen, Van den Berg, De Vries, & Verheij, 2006; Hartig, Evans, Jamner, Davis, & Gärling, 2003; Hartig, Kaiser, & Bowler, 2001;

Health Council of the Netherlands, 2004). To support personal well-being as well as pro-sustainable behaviour, green spaces for restoration must be provided in densely populated areas. These open and green spaces should be offered close to peoples' homes or be accessible by sustainable mobility.

How should these green places for restoration in urban areas be designed to optimally foster health and well-being? How can these areas promote peoples' quality of life in an optimal fashion? To adequately design near-natural spaces according to peoples' needs, it is essential to know what mental conceptions of nature people have. This paper aims at exploring mental representations of nature and thereby detecting key factors for the design of urban green spaces. Thus, quality of life, health, and well-being will be fostered.

Definitions and Mental Representations of Nature

How can the term "nature" be defined? Nature as the natural environment is the sum of all organic and inorganic elements. Its scale ranges from the micro-level (single leaves, a blossom, a grain of sand) to the macro-level (woodlands, oceans) and finally to the all-encompassing cosmos (Gifford, 2007). Wohlwill (1983) notes that the natural environment is "the vast domain of organic and inorganic matter that is not a product of human activity or intervention" (p. 7). But the term "nature" refers not only to natural environments in the sense of Hellpach (1939). Nature as a socio-cultural construct is more than that; it also relates to experienced, cultivated, and shaped nature, as well as to images of nature (Flade, 2010). Nature is a symbol for (1) life itself, vitality, growth, and change (the cycles of living and dying), for (2) continuity or enduringness (stability, timelessness, universality), for (3) an elemental force (innocent purity, cleansing power), and for (4) mystery or spirituality (Knopf, 1987).

Krömker (2004) asked students which word best represented the term "nature." The most common responses were terms such as forests, trees, plants, animals, water, and air. They were categorised as "natural elements" and represented the physical natural environment. Responses like "life" or "livelihood" were also elicited often, as were terms from the category "tranquility and recreation." Other categories were "aesthetics" and "entireness." Generally, nature had a positive connotation. People did not think primarily of risks or threats. The perception of nature is not only a psychological, but also a socio-cultural phenomenon (Knopf, 1987). Images of nature are very important in conservation psychology. The starting point of conservational efforts is not nature as the natural physical environment, but rather images of nature.

Nature is often seen in contrast to culture. Clayton and Opotow (2003), for instance, define nature as an environment in which human influence is minimal or not obvious. In this sense, nature is everything not made by humans, whereas culture refers to the man-made environment. But this contrastive definition of nature and culture seems to be very artificial and not helpful for the discussion (see e.g., Kaplan & Kaplan, 1989). In real life, nature and culture are closely interwoven. In Central Europe, "wilderness" areas are rare.

Nature is no longer wild and unaffected but a man-made environment. Kaplan and Kaplan (1989, p. 2) state that "nature includes parks and open spaces, meadows and abandoned fields, street trees and backyard gardens." Another classification, proposed by the Health Council of the Netherlands (2004), named six types of nature: Urban nature, agricultural nature, production forest, traditional rural nature, natural forests, and wild nature.

The theoretical considerations related to the definition of nature given above are drawn predominantly from studies that follow a quantitative approach. But what do people really associate with nature, what do they think and feel about it? What are their motivations for spending time in nature or for interacting with it? The aim of this study was to clarify these questions by exploring peoples' mental representations of nature.

Method

Participants and Setting

A sample of 133 participants (52.6% female) was surveyed with a pencil and paper questionnaire. The age of the participants ranged from 16 to 74, with a mean age of 34.9 years and a median of 28 years. In terms of education, 6.8% of the participants had completed elementary school, 13.5% junior high school, 45.9% senior high school, 20.3% had graduated from university or college, and 12.8% had other levels of education. All participants were fluent in German and lived in the metropolitan area of Vienna (2.3 million inhabitants). Thanks to its wide range of cultural attractions, excellent infrastructure, a clean environment and a high percentage of open and green space (45.6% within city limits), Vienna has recently been recognised as the best city worldwide with respect to quality of living (Mercer, 2010).

Procedure

Participants responded to three open questions. The questions were generated on the basis of a literature review and prior quantitative and qualitative research. They covered cognitive, motivational, and affective aspects of "nature" as an overall assessment of nature, analogous to the three components of attitudes (Aronson, Wilson, & Akert, 2008). The three questions were:

1. What comes to your mind when you think of nature?
2. For which purpose do you spend time in nature in your leisure time?
3. Which emotions do you experience while spending time in nature?

Analysis

Responses were analysed using qualitative content analysis (Berg, 2003; Mayring, 2007). The data set was screened separately by five investigators (three psychologists and two

Table 1. Examples of how actual responses became a category

Actual responses	Topic	Category
"beautiful landscape with meadows, forests and mountains" "forest, grassland, fields, wide landscape" "this stereotype of a flowering meadow with nearby forest on a beautiful sunny day"	*Landscape*	Aspects of nature (everything that has to do with nature itself and its appearance)
"animals, plants, trees" "mighty beech trees, birds, deer" "flora and fauna"	*Creatures*	
"nature = untouched area" "sparse intrusion of man" "no concrete, few people"	*Unspoilt nature*	

senior psychology students), in order to (1) identify the relevant topics. Based on these topics, (2) categories were formed, and (3) cross-checked with the aid of the remaining topics. (4) Classifications of answers were discussed with the aim of reaching a consensus between the raters and (5) necessary revisions were made until all responses could be classified into topics and categories. After achieving sufficient consistency, (6) the entire data set was reanalysed and responses were assigned to the respective categories.

Responses incorporating multiple topics across categories were counted in the relevant categories. Multiple references to the same topic within one response were classified only once per topic.

Table 1 shows how responses became topics and then categories. Responses that were repeated very often as single words (e.g., "green") generated their own topic.

Most answers were given as single words; very few respondents gave longer statements or complete sentences. The following results are the condensed findings of the analysis. In order to underscore the validity of the results and to illustrate the findings, quotations from the original responses are added in quotation marks.

Results

Table 2 presents the results for the first question: "What comes to your mind when you think of nature?" For clarity, infrequent responses were subsumed under the category Miscellaneous.

Table 2 shows the four categories that were sufficient to summarise peoples' thoughts about nature. Aspects of nature, including parts of landscapes, as well as plants and

Table 2. Categories, topics, number of responses and examples that emerged in the analysis of Question 1

Category	Topic	Number of responses	Examples
Aspects of nature	*Landscape*	52	"mountains," "forests," "meadows"
	Creatures	52	"animals and trees," "plants," "the little owl in front of my window," "vegetables"
	Green	25	"green"
	Unspoilt nature	21	"untouched," "virgin forest," "no concrete, few people"
	Living space	19	"a habitat for animals and humans," "climate," "we could not live without nature"
	Sensations	18	"birds twittering," "the smell of mushrooms"
	Water	14	"water," "rivers and lakes," "swimming lakes"
	Force of nature	12	"force of nature, incredible power, natural phenomena," "weather, storm"
	Sun	11	"sun"
	Landscaped by humans	10	"parks and garden," "farming," "the bike lane to work"
Well-being	*Restoration and health*	25	"recreation and relaxation," "serenity," "to be comfortable," "healthy"
	Fresh air	23	"pure, fresh air"
	Tranquility	16	"tranquility"
	Positive emotions	11	"happiness," "high spirits"
Activity	*Activity/sports*	21	"to take a walk," "sportive activities," "hiking, swimming," "exercise in nature"
Leisure time	*Leisure time*	14	"leisure time," "holidays," "being away from everyday life"
Miscellaneous	*Freedom*	9	"freedom"
	Sustainability	6	"sustainability"
	Environmental destruction	5	"environmental destruction"
	Spirituality	5	"spirituality"

animals, were mentioned most often. Restoration and health followed. Most responses appeared in the category Aspects of nature. Actual natural features or the appearance of nature were subsumed in this category. *Landscape* can be seen as a geographical term that consisted of landscape-related features. *Creatures* contained all living things, such as

Table 3. Categories, topics, number of responses and examples that emerged in the analysis of Question 2

Category	Topic	Number of responses	Examples
Aspects of nature	*Restoration/ relaxation*	72	"relaxation, recreation," "to rest"
	Getting new strength and energy	20	"getting new energy," "to recharge my batteries," "it gives me energy"
	Well-being	13	"to clear the mind," "to increase my well-being," "because it feels good"
	Positive emotions	9	"to rejoice in nature," "positive mood," "having fun"
	Tranquility	9	"tranquility"
	Being away/ compensation/ balance	9	"distance to the city," "compensation for work"
	Enjoying nature	7	"to enjoy the surrounding nature," "it is beautiful," "to witness the changing seasons"
	Liberation	3	"to experience liberation"
Physical recreation	*Activity*	78	"going for a walk," "sports," "swimming, cycling"
	Adventure	12	"positive diversion," "to observe and marvel," "to have new experiences"
	Fresh air	10	"getting fresh air," or "to enjoy the fresh air"
Other purposes	*Contemplation*	8	"to think about life and knowledge," "to contemplate"
	Social context	8	"to meet friends for a picnic," or "to play with the kids"
	Spirituality	5	"peace of mind," or "to escape into the 'real' world"
	Utilitarian behaviour	5	"my parents live there," "if my route leads me through it"
	Sensations	3	"to listen to birds"

Table 4. Categories, topics, number of responses and examples that emerged in the analysis of Question 3

Category	Topic	Number of responses	Examples
POSITIVE EMOTIONS	*Recreation/ restoration*	60	"recreation," "relaxation"
	Tranquility	34	"to calm down," "it has calming effects on me," "peace of mind"
	Joy	29	"gladness"
	Freedom	28	"being free" or "liberation"
	Happiness	26	"happiness"
	Satisfaction	20	"satisfaction"
	Well-being/health	13	"feeling comfortable," "the feeling of doing something good for my health"
	Unspecified positive emotions	12	"I enjoy it," "esteem for nature"
	Getting new strength and energy	12	"getting new strength and energy"
	Balance/ compensation	10	"equilibrium," "being centred"
	Fascination/ curiosity	8	"excitement," "amazement"
	Serenity	6	"serenity"
	Security	3	"security"
	Amorousness	3	"romance," "love"
	Pureness	3	"pureness"
	Peace	2	"peace"
	Optimism	2	"optimism"
	Thankfulness	2	"thankfulness"
	Connectedness to nature	2	"connectedness to nature"
	Inspiration	1	"inspiration"
NEGATIVE EMOTIONS	*Anxiety*	6	"concern," "astonishment and anxiety, if one realises what nature can create, what can happen"
	Sadness	3	"sadness"
	Loneliness	3	"loneliness"
	Boredom	2	"boredom"
	Melancholy	1	"melancholy"
NEUTRAL/ UNDEFINED	*Activation*	7	"positive strain (hiking)," "athleticism," "exercise"
	Spirituality	5	"yearning," "self-discovery," "gratitude"
	Childhood memories	3	"childhood memories"

plants and animals. *Green* consisted of all responses mentioning the colour "green." *Unspoilt nature* included aspects of wilderness not influenced by humans. *Living space* encompassed responses related to the surrounding living space, life itself and to sustenance provided by the environment. *Sensations* consisted of actual natural sensations, such as smelling or hearing, and *Water* consisted of water-related features of the natural environment. *Force of nature* contained responses referring to the power of nature, or to the weather as a manifestation of this power. *Landscaped by humans* included environments designed or shaped by humans.

The second most important category was WELL-BEING. This category subsumed topics related to well-being. *Restoration and health* included restoration- and health-related topics; in *Fresh air*, the character of freshness was highlighted. *Positive emotions* contained all the positive emotions that were mentioned.

Responses related to physical and sporting activities or leisure-time related aspects were infrequent.

Table 3 presents the categories that emerged from the question: "For which purpose do you spend time in nature in your leisure time?" Three categories were derived, one related to physical recreation, one related to mental recreation and one consisting of all other purposes.

Most responses could be subsumed under the motivational category MENTAL RECREATION, although activity – as a form of physical recreation – was mentioned most often (78 times). It was clear from the number of statements that recreation was the most important reason for spending time in nature. MENTAL RECREATION subsumed issues related to restoration, energy, well-being, physical and mental health, positive feelings and emotions while spending time in nature, tranquility, balance, as well as positive experiences had in nature. PHYSICAL RECREATION consisted of the topics connected with physical recreation, such as physical activities, adventure, and experiences in nature. Other responses mentioned contemplation, social interaction, spirituality, utilitarian behaviour, and natural sensations.

Table 4 illustrates the responses to the question: "Which emotions do you experience while spending time in nature?" The responses to this question could be classified into three categories. Most of the topics in this category were frequently named directly. Therefore, further examples are only given for topics that contain more information than the wording itself.

Positive emotions in connection with nature were reported most frequently, including recreation, tranquility, joy, freedom, and happiness. However, negative emotions were also reported in connection with exposure to nature, for example, anxiety or sadness, as well as emotional states of unclear valence such as activation, spirituality, or childhood memories.

Discussion and Conclusions

This qualitative study was aimed at exploring peoples' mental representations of nature. The results suggest that redefining theoretical concepts of nature is necessary. Additionally, recommendations for the design of urban green spaces can be derived from our analysis.

Most responses involving thoughts about nature could be subsumed under the category ASPECTS OF NATURE (mentioned in total 234 times). Also, well-being and recreation played a major role in respondents' mental representation of nature (mentioned a total of 75 times). The categories ACTIVITY and LEISURE TIME can also be associated with well-being. When asked for their reasons for spending time in nature, most respondents mentioned physical and mental recreation. This is in line with observations made by Knopf (1987), who pointed out that peoples' perceptions of natural environments depend on former experiences with these environments. It is not the natural physical elements, but the mental representations of nature and the "set of symbols people hang on them" (Knopf, 1987, p. 786). Knopf does not take restorative experiences into account; however, based on our findings, they turned out to be very important. People who have had restorative experiences in nature will think of restoration-related natural features in the future. But the images of nature related to well-being in our study do not only represent former experiences in nature; they also simultaneously represent ongoing restoration-related needs and expectations.

Obviously, many people experience nature as a restorative environment, as the 142 responses in the category MENTAL RECREATION of Question 2 show. Therefore, people have positive associations with and emotions towards nature, as can be deduced from the 276 responses to the third question in the category POSITIVE EMOTIONS. Kals, Schumacher, and Montada (1999) highlight the importance of past and present experiences in nature as powerful factors for the formation of an emotional affinity towards nature. This positive emotional affinity, in turn, is a very important predictor of pro-conservation behaviour and therefore essential for establishing a sustainable lifestyle. Therefore, the relevance of positive experiences in nature cannot be overestimated. To sum up, there is a relationship between physical, spatial, and experiential aspects of nature that are related to psychological restoration, good health and well-being, and also to environmental protection.

Positive experiences in nature are revealed in the mental representations of nature and constitute an important facet in the perception of nature. Despite extensive media coverage, neither environmental destruction nor environmental risks were mentioned frequently in the responses. Positive associations with nature clearly prevailed. In line with Krömker (2004), we found "tranquility and recreation" played a major role in peoples' spontaneous associations with nature in highly industrialised regions. Krömker ascertained that the perception of nature is culturally dependent. The categories in her study were based on a macro-level analysis across cultures. We, on the other hand, identified categories on a micro-level, as we surveyed mental representations of Austrian participants in detail. We do not dispute the cultural dependency of perceptions of nature but, at the same time, we postulate an additional interrelationship with the development and state of industrialisation in a society.

The importance of well-being found in our study is relevant to the definition of nature in several disciplines. Nature can be experienced in different ways (Flade, 2010), for instance, in connection with restoration as revealed in the present study. According to our findings, this connection with restoration is so prominent that it made its way into mental representations of nature. Nature elicits thoughts and feelings of well-being. Well-being in relation to the environment in general was already mentioned by Stokols and Altman (1987) in their definition of environmental psychology as "the study of human behaviour and well-being in relation to the socio-physical environment" (p. 1). There is broad empirical evidence for the positive connection between nature and well-being, but in previous scientific studies on nature, well-being was not especially highlighted. Because of its importance, we urge including considerations related to well-being in psychological approaches to the design and evaluation of open and green spaces in urban areas.

Moreover, our findings reveal that participants perceive nature comprehensively, related both to natural as well as to man-made features. Participants' mental representations include a broad scope of associations ranging from natural to cultivated landscapes. This finding empirically supports Flade's (2010) recent proposition that nature is more than the mere physical environment. These findings are in line with the transformations landscapes have undergone in Central Europe in past centuries. Very few wild and unaffected natural environments can be found in highly industrialised areas. Most of the Austrian landscape is culturally shaped, especially in the metropolitan area of Vienna. The distinction between nature and culture is not only artificial but also not helpful, as Kaplan and Kaplan (1989) point out. Moreover, the theoretical definitions of nature as wilderness and unaffected environment do not adequately display peoples' actual perceptions of nature. Our findings update this definition. We propose considering well-being in psychological definitions of nature: Nature comprises all animate and inanimate objects. It encompasses both unaffected as well as cultivated manifestations. Further, nature has a material and a symbolic character. Contents related to well-being are closely associated with nature; they are represented in subjective conceptions of nature.

The topics and categories that emerged in our study lead to clear recommendations for the design of urban green space. As Hunziker (1995) states, "an ideal landscape is [...] not only an optimum combination of aesthetically relevant (visual) landscape elements, it also has to give an optimum satisfaction of other interests" (p. 409). These interests lie, according to this study, in restoration and health and in mental as well as physical well-being (e.g., sports/activity). Therefore, an ideal urban green space should provide users with the opportunity to have positive experiences and to relax and recuperate, but also to be physically active. At the same time, urban green spaces should be composed of natural elements that incorporate a certain savagery or "wildness," to provide access to impressions of wild nature close to peoples' homes.

The qualitative methodological approach proved to be well-suited to the investigation of mental representations of nature. The open questions we used in our study can easily be included into questionnaires devoted to diverse topics. One limitation of our approach was that, using written responses, it was not possible to clarify the meaning of responses by further inquiry. Other limitations refer to the study sample. The participants' ages were not normally distributed; the majority of participants were younger than 30 years. Their educational level was higher than average. However, the sample in this study consisted mainly of subjects with university degrees and thus potential future decision makers. This may also be relevant for future design decisions.

To conclude, sustainable use of local urban green space is fostered by tailor-made design solutions based on users' needs and expectations. The results from our qualitative analysis highlight the need for urban green spaces that are associated with wild nature on one hand and well-being on the other hand. Therefore, we recommend introducing areas supporting these needs within walking distance of peoples' homes whenever possible. This could be arranged within parks or realised on waste land. An urban brownfield, for instance, could deliberately be protected for wildness and be linked with other brownfields in different stages of proliferation. The need for wilderness may be met in this way. On the other hand, areas should be designed that support needs for restoration. In such recreational areas, strict regulations for protection from various urban stressors such as noise, crowding, or vandalism, should be implemented. With the presence of such spaces, sustainable use of urban open and green spaces could be supported and unsustainable leisure time mobility could be counteracted.

Acknowledgments

This chapter has benefited from the comments of Sandrine Depeau, Giuseppe Carrus, and Terry Hartig. Its preparation was supported by the students of the Faculty of Psychology at the University of Vienna. The authors are grateful for methodological support from Stephanie Hiesberger and Michael Kundi.

References

Abraham, A., Sommerhalder, K., Bollinger-Salzmann, H., & Abel, T. (2007). *Landschaft und Gesundheit. Das Potential einer Verbindung zweier Konzepte* [Landscape and health. The potential of linking two concepts]. Bern, Switzerland: Universität Bern.

Aronson, E., Wilson, T. D., & Akert, R. M. (2008). *Social psychology* (6th ed.). Upper Saddle River, NJ: Prentice Hall.

Berg, B. L. (2003). *Qualitative research methods for the social sciences* (5th ed.). Boston, MA: Allyn and Bacon.

Clayton, S. & Opotow, S. (Eds.). (2003). *Identity and the natural environment. The psychological significance of nature*. Cambridge, MA: The MIT Press.

Flade, A. (2010). *Natur psychologisch betrachtet* [Nature from a psychological perspective]. Bern, Switzerland: Huber.

Gifford, R. (2007). *Environmental psychology: Principles and practice* (4th ed.). Colville, WA: Optimal Books.

Groenewegen, P. P., Van den Berg, A. E., De Vries, S., & Verheij, R. A. (2006). Vitamin G: Effects of green space on health, well-being, and social safety. *BMC Public Health, 6*, 149–158.

Hartig, T., Evans, G. W., Jamner, L. D., Davis, D. S., & Gärling, T. (2003). Tracking restoration in natural and urban field settings. *Journal of Environmental Psychology, 23*, 109–123.

Hartig, T., Kaiser, F. G., & Bowler, P. A. (2001). Psychological restoration in nature as a positive motivation for ecological behavior. *Environment & Behavior, 33*, 590–607.

Health Council of the Netherlands. (2004). *Nature and health: The influence of nature on social, psychological and physical well-being*. The Hague, Netherlands: Health Council of the Netherlands.

Hellpach, W. (1939). *Geopsyche* (5th ed.). Leipzig, Germany: Engelmann.

Hunziker, M. (1995). The spontaneous reafforestation in abandoned agricultural lands: Perception and aesthetic assessment by locals and tourists. *Landscape and Urban Planning, 31*, 399–410.

Kals, E., Schumacher, D., & Montada, L. (1999). Emotional affinity toward nature as a motivational basis to protect nature. *Environment & Behavior, 31*, 178–202.

Kaplan, R., & Kaplan, S. (1989). *The experience of nature. A psychological perspective*. Cambridge, MA: Cambridge University Press.

Knopf, R. C. (1987). Human behavior, cognition, and affect in the natural environment. In D. Stokols & I. Altman (Eds.), *Handbook of environmental psychology* (Vol. 1). New York, NY: Wiley.

Krömker, D. (2004). *Naturbilder, Klimaschutz und Kultur* [Images of nature, climate protection, and culture]. Weinheim, Germany: Beltz.

Mayring, Ph. (2007). *Qualitative Inhaltsanalyse. Grundlagen und Techniken* (9th ed.) [Qualitative content analysis. Basics and techniques]. Weinheim, Germany: Beltz UTB.

Mercer. (2010). Quality of living survey. Retrieved from http://www.mercer.com/press-releases/quality-of-living-report-2010

Stokols, D., & Altman, I. (1987). Introduction. In D. Stokols & I. Altman (Eds.), *Handbook of environmental psychology*. New York, NY: Wiley.

Wohlwill, J. F. (1983). The concept of nature. A psychologist's view. In I. Altman & J. F. Wohlwill (Eds.), *Human behavior and environment. Behavior and the natural environment* (Vol. 6). New York, NY: Plenum.

Biodiversity, Perceived Restorativeness, and Benefits of Nature

A Study on the Psychological Processes and Outcomes of On-Site Experiences in Urban and Peri-Urban Green Areas in Italy

Massimiliano Scopelliti,[1,3] Giuseppe Carrus,[2,3] Francesca Cini,[2,3] Stefano Mastandrea,[2,3] Francesco Ferrini,[4] Raffaele Lafortezza,[6] Mariagrazia Agrimi,[5] Fabio Salbitano,[4] Giovanni Sanesi,[6] and Paolo Semenzato[7]

[1]Libera Università Maria Ss. Assunta (LUMSA), Italy
[2]Department of Cultural and Educational Studies, University of Roma Tre, Italy
[3]CIRPA – Center for Inter-University Research on Environmental Psychology, Italy
[4]University of Florence, Italy
[5]Tuscia University, Italy
[6]University of Bari, Italy
[7]University of Padua, Italy

Abstract

Research on restorative environments has frequently showed a more positive impact of natural vs. built environments on human well being. However, the restorative potential of different typologies of nature has been less investigated. A still open issue is the role of biodiversity in the restoration process. This study compared five typologies of urban green spaces possessing increasing levels of biodiversity: An urban plaza with green elements, an urban park, a pinewood, a botanical garden, and a peri-urban natural area. One hundred and twenty-five residents of Padua, Italy, filled in a questionnaire that measured length and frequency of visits to the green spaces, activities performed there, perceived restorativeness, and affective qualities of the place, self-reported benefits of the visit. Results showed a positive relation between biodiversity in the settings, perceived restorative properties, and self-reported benefits. Perceived restorativeness and affective qualities also mediate the relation between exposure to nature and self-reported benefits. Theoretical and practical implications are discussed.

Key words: affective qualities, biodiversity, place-experience, psychological benefits, restorative environments

Introduction

Restorative environments are settings capable of promoting, and not merely allowing, the recovery of psychological well-being, defined through stress reduction, increase in positive emotions, and renewal of cognitive resources (Hartig, 2004). The literature on restorative environments has assumed an evolutionary perspective as a theoretical framework (the *Biophilia* Hypothesis, Wilson, 1984), postulating that human beings, who have evolved in natural environments, have developed an innate tendency to positively respond to nature as a consequence of an adaptation process. This positive response also includes psychological restoration, conceived by Ulrich (1983) and Kaplan and Kaplan (1989) in terms of stress reduction (*Stress Reduction Theory* – SRT) and recovery of directed attention (*Attention Restoration Theory* – ART), respectively. Several studies have provided empirical evidence supporting both SRT (e.g., Hartig, Mang, & Evans, 1991; Ulrich et al., 1991) and ART (e.g., Berto, 2005; Staats, Kievet, & Hartig, 2003). Kaplan (1995) has also suggested an integrative framework in this respect, explaining how stress and directed attention may be related. Recent studies have provided empirical support for this relationship. Laumann, Gärling, and Stormark (2003) showed that mentally fatigued people perform better in a cognitively demanding task after viewing a natural vs. urban video. In a field study, Hartig, Evans, Jamner, Davis, and Gärling (2003) found better physiological, emotional and cognitive outcomes for people taking a walk in a natural area than in an urban environment. The study suggests that physiological and attention restoration can complement each other, even if it is difficult to establish which process mediates the other.

Empirical research on restorative environments has consistently shown the benefits of human-nature transactions in several domains. The role of wilderness in promoting psychological well-being has been explored for some time (e.g., Kaplan & Talbot, 1983): People in wild environments can distance themselves from everyday life, reflect on their goals from a different perspective, experience relaxation and peace. With reference to everyday environments, Ulrich (1984) showed that a view onto a natural vs. built scene through a window in a hospital promotes a better and faster recovery in surgery patients. Similar findings were found also in residential settings, university dormitories (Tennessen & Cimprich, 1995) and workplaces (Heerwagen & Orians, 1986). In urban settings, where distressing places and situations are widespread, the beneficial effects of green areas have also been identified: Nature emerged as a resource for the promotion of physical activity and psychological well-being (Ulrich & Addoms, 1981), and positive social interaction (e.g., Sullivan, Kuo, & DePooter, 2004).

Measuring the Restorative Potential of Environments

ART identifies four environmental properties promoting restoration: *Being away* implies a change of scenery and/or experience compared to daily routines; *fascination* refers to the capability of aesthetically pleasant environments to catch a person's attention without mental effort; *extent* refers to the properties of connectedness and scope in environments, because all elements are coherently related to one another, and promise to engage the mind beyond what is being immediately perceived; *compatibility* has to do with the level of perceived congruence between the characteristics of the environment and people's needs and inclinations. To measure these properties, Hartig, Korpela, Evans, and Gärling (1997) developed the Perceived Restorativeness Scale (PRS), which consists of 26 items measuring the four restorative properties proposed by ART. Through the use of this and other similar tools, and concurrent measures of physiological and cognitive restoration, the empirical literature has consistently revealed more positive responses to natural vs. built settings (Berto, 2005; Hartig et al., 2003; Laumann et al., 2003; Purcell, Peron, & Berto, 2001; Staats et al., 2003; van den Berg, Koole, & van der Wulp, 2003). More recently, Scopelliti & Giuliani (2004) emphasised that many studies on restorative environments conceived people-environment transactions leading to restoration in terms of visual perception of the environment – often analysed through a photo or video presentation – whereas the role of overall experience in the environment (e.g., the multi-sensorial character of perception itself, individual feelings, activities performed, etc.) was less considered. Scott and Canter (1997) have shown empirically that people conceptualise the environment differently, depending on whether the focus is on the visual perception or on the experience in the place. It is questionable whether mere visual perception can adequately represent the complexity of the process leading to restoration in a real environment. In this respect, Staats and Hartig (2004) analysed the role of social interaction in restorative experience, and found that – when perceived safety is controlled for – the presence of other people may diminish the benefits of natural environments. Scopelliti and Giuliani (2005) considered both the activity performed and social interaction as potential moderators of positive outcomes for the elderly. They found an effect of social interaction on the restorative potential of natural environments, which was perceived to be higher when people were alone. The authors suggest that social interaction in natural environments may represent a source of distraction from the relationship with the environment, which is restorative in itself. Thus, it is clear that the role of variables moderating restorative nature experience needs further investigation.

Open Issues in Research on Restorative Environments

In the study of restorative environments, nature has often been considered as an undifferentiated typology and contrasted to built environments. Little attention has been paid to the analysis of the restorative potential of *different* kinds of natural environments. Fuller, Irvine, Devine-Wright, Warren, and Gaston (2007) found that biodiversity can increase the psychological benefits associated with experiencing green areas. This preliminary result is still compatible with an evolutionary perspective, because biodiversity plays a fundamental role in life support and for an ecosystem's continuity (e.g., Wilson, 1999). It is then possible to hypothesise that human beings developed positive responses to this specific feature of natural settings in the course of evolution. Therefore, it seems to be interesting to better understand if – and how – there can be a relationship between biodiversity and psychological restoration.

A further issue is to understand how various experiences in restorative environments, in terms of length and frequency of exposure to nature, and activities performed there, can affect restoration.

Because the literature on restorative environments has mainly considered restoration as an outcome (or effect) promoted by people-environment transactions, it is important to gain more insight into the role of psychological processes, such as perceptions of and affect toward the environment, which might, in turn, lead to psychological benefits.

A better understanding of the environmental properties and psychological mechanisms promoting restoration can also support more sustainable and health-oriented planning and management of everyday environments.

The Study

This study aimed at analysing the relationships between ecological-structural and psychological variables, namely biodiversity and restoration, in urban and peri-urban green areas, and at examining the role of some experience variables (length and frequency of visits, activities performed) in promoting restoration. A further aim was to better understand what psychological mechanisms – if any – intervene in the process linking exposure to nature to restoration. The study was conducted in the city and the peri-urban area of Padua, Italy. Five settings with a different level of biodiversity were identified as a context for the study. A photo of each of these five settings is shown in Figure 1.

Aims and Hypotheses

Three main aims and related hypotheses were identified for this study.

The first aim was to analyse the relationships between the level of biodiversity and the restorative potential of environments. It was hypothesised that:

H1: The level of biodiversity positively affects the perception of restorative properties and perceived benefits.

The second aim was to analyse the effect of experience variables, referring to both activities performed and exposure to the environment (length and frequency of visits), on the restorative potential of environments. On the basis of studies on social interaction in restorative environments, it was hypothesised that:

H2: Environmentally-oriented vs. socially-oriented activities in restorative environments increase the perception of restorative properties and perceived benefits.

In accordance with the literature on wilderness, which stresses the evolution through time of people-environment transactions and benefits, it was hypothesised that:

H3: The length and the frequency of visits in restorative environments positively affect the level of perceived benefits.

The third aim was to understand the role of psychological variables (perceived restorativeness and affective quality of the environment) in the restoration process. The mediation of these variables in the relationship between exposure to nature (length and frequency of visits) and perceived well-being was tested (Baron & Kenny, 1986; Evans & Lepore, 1997).

Methods

Selection of Environments

Five environmental typologies in the urban and peri-urban area of Padua, with different levels of biodiversity, were identified according to forestry science parameters. The environments were: The mixed built-natural square *Prato della Valle*, the urban park *Parco Iris*, the pinewood, the botanical garden, and the peri-urban green area *Colli Euganei*. The level of biodiversity ranged from a minimum at *Prato della Valle* up to a maximum at *Colli Euganei*.

Participants, Procedure, and Measures

An opportunistic sample of 124 respondents (62 females; mean age 37.2 y; $SD = 15.1$) was contacted at the sites. Participants were almost equally distributed across the five settings ($n = 23, 26, 26, 24, 25$ for square, urban park, pinewood, botanical garden and peri-urban area, respectively). Subjects were invited to take part in the study and asked to fill in a paper-and-pencil questionnaire, referring to their experience in the environment. The questionnaire was arranged as follows:

Section 1: open-ended, multiple-choice, and Likert-type scale questions on exposure to the environment (length and frequency of visits, crowding), main activity performed

Figure 1. Pictures representing the five sites: (a) urban square; (b) urban park; (c) pinewood; (d) botanical garden; (e) peri-urban green area. (Photographs: the authors)

(socialising, walking, contemplation, or physical activity), and socio-demographic data. These measures were adapted from a previous study (Lafortezza, Carrus, Sanesi, & Davies, 2009).

Section 2: Italian version of the PRS (Pasini, Berto, Scopelliti, & Carrus, 2009), measuring the restorative properties of the environments on a 5-step Likert scale (0–4). Examples of items are: "*Spending time here gives me a break from my day-to-day routine*" (being-away); "*It is chaotic here*" (extent); "*This place is fascinating*" (fascination); "*I can do things I like here*" (compatibility).

Section 3: a measure of the affective quality of environments, based on the Russell and Pratt (1980) model, in which the bipolar adjectives unpleasant-pleasant, distressing-relaxing, sleepy-arousing, gloomy-exciting, were used (higher scores correspond to the presence of positive qualities). These measures were taken from the Italian version of the affective qualities scale (Perugini, Bonnes, Aiello, & Ercolani, 2002).

Section 4: a measure of psychological and physical benefits experienced in the environment on a 5-step Likert scale (0–4). The items used were "*How much do you feel psychologically better than usual?*" and "*How much do you feel physically better than usual?*" The items were adapted from Lafortezza et al. (2009).

Results

Effects of Biodiversity Levels on the Restorative Potential of the Settings

The level of biodiversity, ranging across the five environmental typologies considered, was used as an independent factor and each of the four restorative components were considered separately as dependent variables. The potential effect of the perceived level of crowding in the settings was controlled for as covariate, and for this reason we chose an ANCOVA model to test the hypotheses; a significant effect of the covariate was detected ($F_{(4,\ 119)} = 26.54$, $p < .001$). Coherently with hypothesis 1, results showed a significant effect of the level of biodiversity on each component: *Being away* ($F_{(4,\ 118)} = 12.25$, $p < .001$, $\eta^2 = .29$), *extent* ($F_{(4,\ 118)} = 12.97$, $p < .001$, $\eta^2 = .31$), *fascination* ($F_{(4,\ 118)} = 22.04$, $p < .001$, $\eta^2 = .43$), *compatibility* ($F_{(4,\ 118)} = 24.80$, $p < .001$, $\eta^2 = .46$) as well as on an average aggregate *restorativeness* score ($F_{(4,\ 118)} = 25.55$, $p < .001$, $\eta^2 = .46$). A significant effect of the level of biodiversity was also detected for psychological ($F_{(4,\ 118)} = 10.36$, $p < .001$, $\eta^2 = .26$) and physical ($F_{(4,\ 118)} = 9.51$, $p < .001$, $\eta^2 = .24$) benefits reported by the respondents. Figure 2 shows a general trend in which the mean scores of the dependent variables increase as settings shift from a lower level to a higher level of biodiversity. In the histograms, significant differences between the settings are indicated by colours for each of the dependent variables, at the level of $p < .05$. In the case of *being away*, the mixed colour for *Parco Iris* indicates a non-significant difference with both the urban square *Prato della Valle*, on the one hand, and the pinewood and the botanical garden, on the other hand.

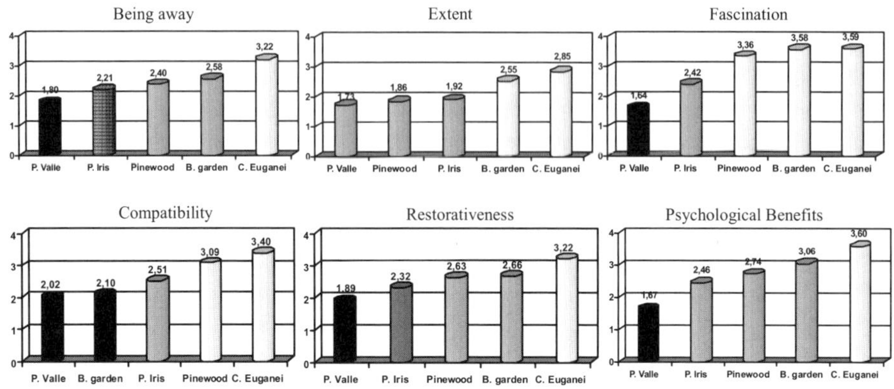

Figure 2. Mean scores of restorative properties and psychological benefits across the five environments.

Effects of Experience Variables on the Restorative Potential of the Environment: Activity, Length, and Frequency of Visits

To assess the effects of the various activities carried out during the experience upon the perceived restorative potential and self reported benefits, we distinguished between four groups of respondents, according to the main activity they reported while in the setting. These activities were socialising with other people ($N = 16$), walking ($N = 43$), contemplating the place ($N = 39$), and physical exercise ($N = 26$), and they can be considered as indicators of increasing levels of personal involvement with the setting.

Coherently with hypothesis 2, results showed a significant effect of the main activity on three of the restorative components: *Being away* ($F_{(3,\ 120)} = 7.45, p < .001, \eta^2 = .16$), *fascination* ($F_{(3,\ 120)} = 9.33$, $p < .001$, $\eta^2 = .19$) and *compatibility* ($F_{(3,\ 120)} = 6.42$, $p < .001$, $\eta^2 = .14$). No significant effect of the main activity emerged on *extent* ($F_{(3,\ 120)} = 3.83$, *ns*). A significant effect of the main activity also emerged on the aggregate score of *restorativeness* ($F_{(3,\ 120)} = 10.43, p < .001, \eta^2 = .21$) and on the psychological ($F_{(3,\ 120)} = 7.59$, $p < .001$, $\eta^2 = .16$) and physical ($F_{(3,\ 120)} = 18.17$, $p < .001$, $\eta^2 = .31$) benefits reported by the respondents.

Figure 3 shows a general trend in which the mean scores of the dependent variables increase when respondents shifted from an activity that implied a lower level of involvement with the setting (such as socialising) towards activities implying a higher level of involvement (such as walking, contemplating the place, and performing physical exercise). Again, the colours in the histograms indicate significant differences between the activities for each of the dependent variables, at the level of $p < .05$.

A hierarchical multiple regression analysis was also performed to assess the role of length and frequency of visits to the environment upon the self-reported benefits. In a first block, experience variables related to exposure to the environment (length and frequency

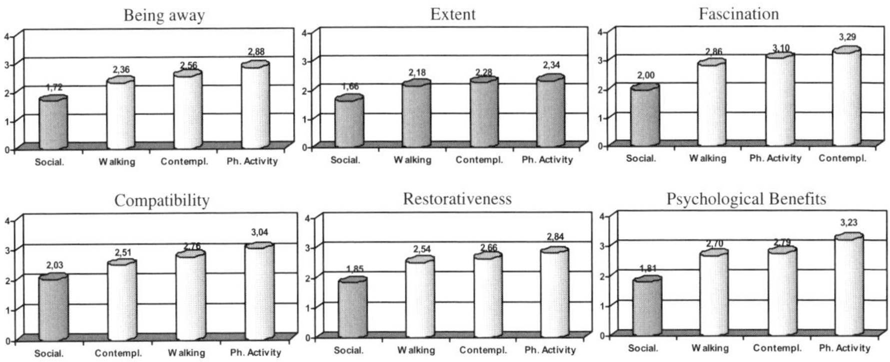

Figure 3. Mean scores of restorative properties and psychological benefits for different activities in the environment.

of visits to the setting) were inserted as predictors. In a second block, the psychological variables referring to the perceived properties of the environment were added to the model (perceived restorativeness and affective qualities of the place), using an aggregate score of self-reported psychological and physical benefits as the criterion. Prior to the final test, bivariate correlations among predictors were calculated, to check for possible multicollinearity. This eventuality was ruled out, since the maximum correlation value was $r = .53$ (between the affective qualities *pleasant* and *relaxing*).

In line with hypothesis 3, as shown in Table 1, the predictors at step 1 accounted for a significant proportion of variance in the criterion ($R^2 = .113$, $F_{(2,\ 121)} = 7.68$, $p < .01$). Both length and frequency of visits emerged as significant positive predictors of self-reported benefits: ($\beta = .228$ and .241, respectively; $p < .01$). The addition of predictors at step 2 significantly increased the proportion of explained variance ($R^2 = .513$, $F_{(5,\ 118)} = 24.85$, $p < .001$). After controlling for the variables in block 2, length and frequency of visits were no longer significant, while perceived *restorativeness* ($\beta = .410$, $p < .001$) and the affective qualities *arousing* ($\beta = .215$, $p < .01$) and *relaxing* ($\beta = .158$, $p < .05$) positively predicted self-reported benefits.

These results suggest a mediating role of the perceived characteristics of the environment, namely perceived restorativeness and affective qualities, upon the relationship between exposure to the environment, expressed in terms of length and frequency of visits, and the positive outcomes reported by the participants. The results of mediation analyses are reported in the next paragraph.

Mediation Effects of Perceived Environmental Characteristics on the Relationship Between Exposure to Nature and Self-Reported Benefits

To test the mediation model, multiple regression and Sobel test were performed. Figure 4 shows how the relationship between length of the visit in the natural setting and

Table 1. Hierarchical multiple regression analysis: predictors of perceived benefits

		β	t	p
Model 1: Exposure to the environment variables				
Length of visit		.228	2.666	.009
Frequency of visits		.241	2.816	.006
Model 2: Psychological variables				
Length of visit		.105	1.570	.119
Frequency of visits		.122	1.853	.066
Restorativeness		.420	4.891	.000
Affective quality: Arousing		.215	2.788	.006
Affective quality: Relaxing		.158	1.981	.049
Model fit	R	R^2	F	p
Model 1	.336	.113	7.68	.001
Model 2	.716	.513	24.85	.000

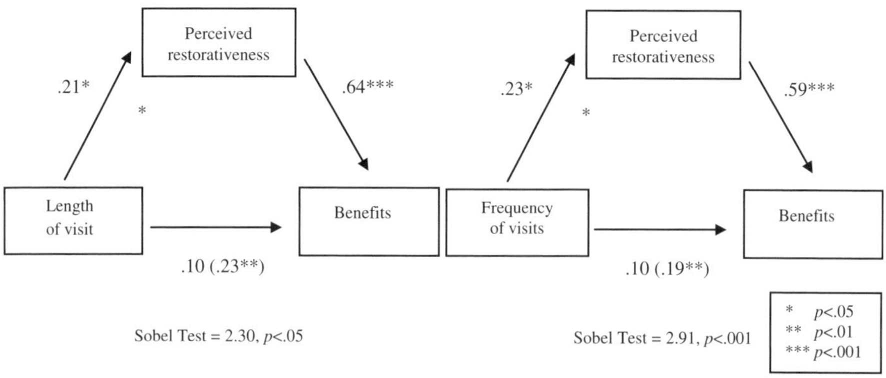

Figure 4. Mediation analysis: the role of perceived restorativeness.

self-reported benefits was fully mediated by perceived restorativeness (Sobel test: 2.30, $p < .05$): The relationship ($\beta = .23, p < .01$) between length of visit and self-reported benefits was no longer significant when controlling for perceived restorativeness. This, in turn, positively predicted self-reported benefits ($\beta = .64, p < .001$). Similarly, the relationship between frequency of visits to the natural setting and self-reported benefits was fully mediated by perceived restorativeness (Sobel test: 2.91, $p < .001$), with a significant predictor-criterion relationship ($\beta = .19$, $p < .01$) dropping down to non-significance after

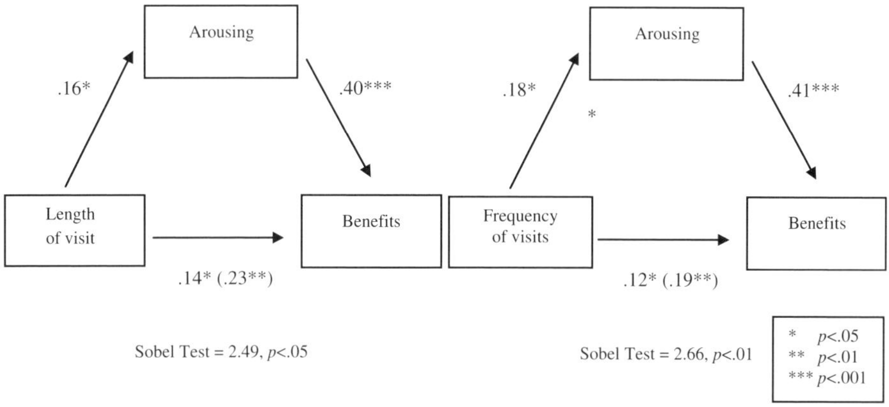

Figure 5. Mediation analysis: the role of the affective quality arousing.

controlling for the effect of the mediator, which, in turn, significantly predicted the criterion ($\beta = .59$, $p < .001$). In other words, spending more time in contact with nature leads individuals to better acknowledge its restorative properties, and this, in turn, leads them to perceive more beneficial outcomes.

A further set of regressions was performed to test for the mediating role of affective qualities of the place on the relationship between exposure to nature and self-reported benefits. Figure 5 shows how the relationship between length of the visit in the natural setting and self-reported benefits is partially mediated by the *arousing* affective quality of the environment (Sobel test: 2.49, $p < .05$): The significant relationship ($\beta = .23$, $p < .01$) between length of visit and self-reported benefits decreased to $\beta = .14$, $p < .05$ after controlling for the *arousing* quality, and this, in turn, positively predicted self-reported benefits ($\beta = .40$, $p < .001$). Likewise, the relationship between frequency of visits to the natural setting and self-reported benefits is partially mediated by the *arousing* affective quality of the environment (Sobel test: 2.66, $p < .01$), with a substantial decrease in the predictor-criterion association after controlling for the effect of the mediator (from $\beta = .19$, $p < .01$ to $\beta = .12$, $p < .05$), which in turn significantly predicts the criterion ($\beta = .41$, $p < .001$). Again, these results suggest that spending more time in contact with nature causes individuals to perceive it as more beneficial, and this can partly be attributed to an increased perception of the presence of arousing elements in the environment.

The same test, performed for the *relaxing* affective quality, did not show a mediation pattern, because the predictor-mediator link was not significant for either the length or the frequency of visits. Therefore, only a direct effect of the *relaxing* quality of the environment on experienced benefits – as it emerged from the multiple regression analysis – was found.

Discussion

This study addressed some undeveloped issues in the literature on restorative environments. Previous research often considered the broad natural vs. built environments distinction, demonstrating the higher restorativeness of the former over the latter (Berto, 2005; Hartig et al., 1991, 2003; Laumann et al., 2003; Tennessen & Cimprich, 1995; Ulrich et al., 1991). In the present study, we analysed the role of the level of biodiversity, which differentially characterises various typologies of urban nature, and the moderating effect of place-experience variables (activities performed and exposure to the environment) in the perception of the restorative properties of the environment and associated benefits.

Our comparisons showed an increase in the restorative properties and benefits experienced by respondents as a function of the level of biodiversity: From setting with the lowest biodiversity (*Prato della Valle*) to those with the highest (*Colli Euganei*), with the middle-level settings (*Parco Iris*, the pinewood, the botanical garden) in between. This general trend is in line with our hypothesis and previous findings (Fuller et al., 2007), suggesting a relationship between ecological-structural properties of environments and psychological processes.

At a general level, the amount of contact with nature was found to affect its positive outcomes: Respondents who spent a longer time and repeatedly visited natural settings reported more benefits. These results indicate that an evolution of the experience over time occurs, yielding deeper restoration and thus confirming previous findings (Kaplan & Talbot, 1983).

In line with previous research showing more restorative outcomes for people in solitude (Scopelliti & Giuliani, 2005; Staats & Hartig, 2004), we also found a moderating effect of the activities performed in the environment on the perception of the restorative properties and experienced benefits: People involved in environmentally-oriented activities showed a more positive experience than people involved in socialising activities.

In terms of the psychological processes promoting restoration, our analyses showed that the relationship between exposure to the natural environment, expressed in terms of length and frequency of visits, and experienced benefits is mediated by the perception of the restorative properties and the arousing affective quality of the environment. In other words, the more, longer and deeper the transactions with the environment are, the more people are able to perceive its positive qualities, and these perceptions, in turn, promote positive outcomes. A different pattern emerged for the affective quality *relaxing*, which is not linked to the amount of exposure and might thus represent an immediately perceivable feature of restorative settings.

Conclusions

The study has demonstrated the role of biodiversity in promoting benefits for people spending time in urban natural environments. This suggests that the richness of biodiversity in these environments should be preserved, not only for the sake of nature itself, but also for the positive outcomes for residents. This finding can thus encourage decision makers to develop more sustainable policies in urban settings, and it also addresses the paradox that characterises current urban development. On the one hand, as van den Berg et al. (2007) suggest, the compact city provides ecological advantages in terms of carbon emissions, and therefore urban sprawl might be seen as a negative phenomenon that needs to be buffered. At the same time, the global phenomenon of urbanisation and increasing building density in cities often reduces urban residential quality and negatively affects people's well-being. Therefore, urban planning strategies that promote the presence of larger natural spaces rich in biodiversity could contribute to improving residents' health status.

Our findings also support the hypothesis that various experiences in urban natural environments may promote benefits to different extents. These results might be helpful for a healthier management of urban and peri-urban natural settings: The arrangement and layout should provide a positive involvement through differentiation of areas, accessibility and variety of activities. This would invite people to spend more time in the environment and would offer them opportunities for a more intensive interaction with nature.

Our study also has some shortcomings: The reliability of the results from this relatively small sample should be confirmed in future research with larger samples, and also in different geographical areas and with different populations. For example, epidemiological studies have suggested the plausibility of a link between green space availability and well-being at a population level (e.g., Mitchell & Popham, 2008). Future studies should address this question more directly, for instance, through longitudinal field studies. These might make it possible to test the impact of changes in exposure to biodiversity-rich green spaces, for example, as a consequence of residential mobility or urban development plans.

Despite these shortcomings, the present findings provide the basis for new theoretical insights and also offer practical suggestions for more sustainable management and use of urban natural environments.

References

Baron, R. M., & Kenny, D. A. (1986). The moderator-mediator variable distinction in social psychological research: Conceptual, strategic, and statistical considerations. *Journal of Personality and Social Psychology, 51*, 1173–1182.

Berto, R. (2005). Exposure to restorative environments helps restore attentional capacity. *Journal of Environmental Psychology, 25*, 249–259.

Evans, G. W., & Lepore, S. J. (1997). Moderating and mediating process in environment-behavior research. In G. T. Moore & R. W. Marans (Eds.), *Advances in Environment, Behavior and Design* (Vol. 4, pp. 255–285). New York, NY: Plenum Press.

Fuller, R. A., Irvine, K. N., Devine-Wright, P., Warren, P. H., & Gaston, K. J. (2007). Psychological benefits of greenspaces increase with biodiversity. *Biology Letters, 3*, 390–394.

Hartig, T. (2004). Restorative environments. In C. Spielberger (Ed.), *Encyclopaedia of Applied Psychology* (Vol. 3, pp. 273–278). San Diego, CA: Elsevier/Academic Press.

Hartig, T., Evans, G. W., Jamner, L. D., Davis, D. S., & Gärling, T. (2003). Tracking restoration in natural and urban field settings. *Journal of Environmental Psychology, 23*, 109–123.

Hartig, T., Korpela, K., Evans, G. W., & Gärling, T. (1997). A measure of restorative quality in environments. *Scandinavian Housing and Planning Research, 14*, 175–194.

Hartig, T., Mang, M., & Evans, G. W. (1991). Restorative effects of natural environment experiences. *Environment and Behavior, 23*, 3–26.

Heerwagen, J. H., & Orians, G. H. (1986). Adaptations to windowlessness: A study of the use of visual decor in windowed and windowless offices. *Environment and Behavior, 18*, 623–639.

Kaplan, S. (1995). The restorative benefit of nature: Toward an integrative framework. *Journal of Environmental Psychology, 15*, 169–182.

Kaplan, R., & Kaplan, S. (1989). *The experience of nature: A psychological perspective*. New York, NY: Cambridge University Press.

Kaplan, S., & Talbot, J. F. (1983). Psychological benefits of a wilderness experience. In I. Altman & J. F. Wohlwill (Eds.), *Behaviour and the natural environment* (pp. 163–203). New York, NY: Plenum Press.

Lafortezza, R., Carrus, G., Sanesi, G., & Davies, C. (2009). Benefits and well-being perceived by people visiting green spaces in periods of heat stress. *Urban Forestry & Urban Greening, 8*, 97–108.

Laumann, K., Gärling, T., & Stormark, K. M. (2003). Selective attention and heart rate responses to natural and urban environments. *Journal of Environmental Psychology, 23*, 125–134.

Mitchell, R., & Popham, F. (2008). Effect of exposure to natural environment on health inequalities: An observational population study. *The Lancet, 372*, 1655–1660.

Pasini, M., Berto, R., Scopelliti, M., & Carrus, G. (2009). Measuring the restorative value of the environment: Contribution to the validation of the Italian version of the Perceived Restorativeness Scale. *Bollettino di Psicologia Applicata, 257*, 3–11.

Perugini, M., Bonnes, M., Aiello, A., & Ercolani, A. P. (2002). Il modello circonflesso delle qualità affettive dei luoghi. Sviluppo di uno strumento valutativo italiano [The affective quality of places' circumplex model: Development of an Italian valutative tool]. *Testing Psicometria Metodologia, 9*, 131–152.

Purcell, T., Peron, E., & Berto, R. (2001). Why do preferences differ between scene types? *Environment and Behavior, 33*, 93–106.

Russell, J. A., & Pratt, G. (1980). A description of the affective quality attributed to environments. *Journal of Personality and Social Psychology, 38*, 311–322.

Scopelliti, M., & Giuliani, M. V. (2004). Choosing restorative environments across the lifespan: A matter of place experience. *Journal of Environmental Psychology, 24*, 423–437.

Scopelliti, M., & Giuliani, M. V. (2005). Restorative environments in later life. An approach to well-being from the perspective of environmental psychology. *Journal of Housing for the Elderly, 19*, 203–226.

Scott, M. J., & Canter, D. V. (1997). Picture or place? A multiple sorting study of landscape. *Journal of Environmental Psychology, 17*, 263–281.

Staats, H., & Hartig, T. (2004). Alone or with a friend: A social context for psychological restoration and environmental preferences. *Journal of Environmental Psychology, 24*, 199–211.

Staats, H., Kievet, A., & Hartig, T. (2003). Where to recover from attentional fatigue: An expectancy-value analysis of environmental preference. *Journal of Environmental Psychology, 23*, 147–157.

Sullivan, W. C., Kuo, F. E., & DePooter, S. F. (2004). The fruit of urban nature. Vital neighborhood spaces. *Environment and Behavior, 36*, 678–700.

Tennessen, C., & Cimprich, G. (1995). Views to nature: Effects on attention. *Journal of Environmental Psychology, 15*, 77–85.

Ulrich, R. S. (1983). Aesthetic and affective response to natural environment. In I. Altman & J. F. Wohlwill (Eds.), *Behaviour and the natural environment* (pp. 85–125). New York, NY: Plenum Press.

Ulrich, R. S. (1984). View through a window may influence recovery from surgery. *Science, 224*, 420–421.

Ulrich, R. S., & Addoms, D. L. (1981). Psychological and recreational benefits of a residential park. *Journal of Leisure Research, 13*, 43–65.

Ulrich, R. S., Simons, R. F., Losito, B. D., Fiorito, E., Miles, M. A., & Zelson, M. (1991). Stress recovery during exposure to natural and urban environments. *Journal of Environmental Psychology, 11*, 201–230.

van den Berg, A., Hartig, T., & Staats, H. (2007). Preference for nature in urbanized societies: Stress, restoration, and the pursuit of sustainability. *Journal of Social Issues, 63*, 79–96.

van den Berg, A. E., Koole, S. L., & van der Wulp, N. Y. (2003). Environmental preference and restoration: (How) are they related? *Journal of Environmental Psychology, 23*, 135–146.

Wilson, E. O. (1984). *Biophilia*. Cambridge, MA: Harvard University Press.

Wilson, E. O. (1999). *The diversity of life*. New York, NY: W. W. Norton.

Place Making and Urban Design

The Meaning of Place

Spontaneous Memorials in a Changing Urban Landscape

Mats Lieberg

Department of Landscape Architecture, Swedish University of Agricultural Sciences, Alnarp, Sweden

Abstract

Spontaneous memorials trigger questions about the meaning of place and the relationship between the public and private domain. Based on two case studies and a workshop within an interdisciplinary research project in Sweden, this chapter discusses the use of spontaneous memorials in the urban environment and how they function in an everyday context. The issue is not simply what phenomenon memorials *are*, but what they *do* with us. On-site observations and interviews with bereaved individuals and various professionals such as landscape architects and designers are used to examine how urban memorials acquire life and meaning, and sometimes also conflicts, through the actions of various groups. The results indicate that spontaneous memorials support multiple functions and agendas and sometimes also challenge social, cultural, and ideological notions of what a proper place of death should comprise. Most often they have a memorial purpose by being "monuments" in the public space, but sometimes they also open the way for future cultural and political actions, through events held on-site. These research findings will help us understand diverse conceptions of memorial places, enabling us to visualise and realise the various ideas of meaning and memorialisation in the contemporary urban landscape.

Key words: meaning of place, public space, spontaneous memorials, sustainable planning, urban landscape

Spontaneous Memorials

Spontaneous memorials are places in the urban landscape where people place fresh flowers, lighted candles, and photos, for example, at the site of a fatal motor accident, a murder or a disaster. These places have become more common, especially in Northern Europe. They are recognised as being important private expressions in a secularised world. They represent a shift from an institutional to an individual notion of death – the individualisation of death (Petersson, 2010).

This chapter focuses on the use of spontaneous memorials in the urban environment and how they function in an everyday context. The research questions try to answer what kind of phenomenon memorials *are*, and what they *do* with us. These symbolically charged sites, whether spontaneously created or deliberately planned, often aim at speaking to our feelings. They form part of the lives of individuals and families and of the collective consciousness. They are hedged off by rituals and symbolic patterns of behaviour – something that can occasionally lead to conflicts and tensions between various groups and interests in society. Memorial sites and objects not only act as messengers and narrators of history, but also engender life, meaning, and activities in their vicinity.

Interviews and conversations with bereaved individuals and with various professionals such as landscape architects and designers are used in this contribution to examine how urban memorials acquire life and meaning through the actions of various groups. This requires analysis of the intentions of initiators and designers, and also of the everyday use of the sites. Particular attention is devoted here to spontaneously created memorials, a type of everyday, unofficial monument to commemorate people who died in an accident or a disaster. At such sites, private grief and loss at the death of a relative or close friend is expressed in public, with the help of letters, objects, items of clothing, and other material objects that create a particular closeness between the observer and the memorial. An additional question is whether these places can also help to develop and enrich the contemporary urban environment.

Early and Recent Studies

Spontaneous memorials are reasonably well documented in academic disciplines such as ethnology, sociology, geography, and religious studies. In these studies, existential, psychological, and phenomenological perspectives are associated with imparting a sense of meaning to the place of a death (Henzel, 1989; Nieminen Kristofersson, 2006; Petersson, 2009a). The impact of political, religious, cultural, and social structures in the production of memorial sites has also been addressed (Doss, 2002; Hartig & Dunn, 1998; Kong, 1990; Nieminen Kristofersson, 2006; Petersson, 2004, 2009b). Other studies have focused on the dissatisfaction of the bereaved with the burial customs or formalities and regulations of conventional cemeteries (Clark & Franzmann, 2006; Hartig & Dunn, 1998).

The increase in spontaneous memorials has sometimes been interpreted as a reaction to death as an invisible phenomenon in modern Western society (Haney, Leimer, & Lowery,

1997), or to a decline in public belief in institutions such as the church or religious organisations (Clark & Cheshire, 2003/2004; Clark & Franzmann, 2006). The mixing of various cultures and traditions, due to immigration and new settlements, and modes of drawing on cultures and religions inherent to a certain geographical location, has been the subject of several studies (Barrera, 1997; Clark & Cheshire, 2003/2004; Everett, 2002; Petersson, 2009a). Studies of roadside memorials in the US (Barrera, 1997), the UK (Monger, 1997), Ireland (MacConville & McQuillan, 2009) and Europe in general (Köstlin, 1999) report early examples of spontaneous memorials for victims of motor vehicle accidents and traditional practices similar to present-day memorials.

According to many researchers (for instance Monger, 1997; Walter, 1996), it is difficult to know how old or recent the phenomenon of creating spontaneous memorials is. Although it seems to have developed since the mid-1980s, occasional earlier cases and similar regional customs can be found (Petersson, 2010). The phenomenon of spontaneous memorials in Sweden has been little studied to date (exceptions: Ehn, 2007; Nieminen Kristofersson, 2006; Petersson, 2009a). However, British (for instance Monger, 1997), Northern American (Everett, 2002; among many others) and Australian studies on roadside memorials (see, e.g., Clark & Franzmann, 2006) have been influential in the research field of spontaneous memorials.

The proliferation of spontaneous memorials that we see today can be linked to the theoretical conceptualisation of place (Canter, 1997; Gustafson, 2001; Relph, 1976; Tuan, 1977) and to the expanding focus in post-modern society on individuality and privatisation (Clark & Franzmann, 2006). This is also recognised in other practices connected with death, such as burial ceremonies (Gustafsson, 2003), newspaper death announcements (Dahlgren, 2000), and the design of memorial stones (Gustafsson, 2003). Of special interest in the Swedish context is the deregulation, in the twentieth century, of some of the spatial institutions of the nineteenth century, for instance the cemetery.[1]

Space of Appearance

Earlier research indicates that some degree of turbulence often arises around memorials and monuments (Ehn, 2007; Hartig & Dunn, 1998; Nieminen Kristofersson, 2006; Petersson, 2009a). They act as a type of moral signpost intended to interpret the past, but also pointing forward to how the world should be. In her book, *The Human Condition*, the philosopher Hanna Arendt (1958) pointed out the multiple meanings and significance of memorials in terms of invoking the future. They offer a shared place and shared matter that people can use and relate to in different ways – whether this be recognition or opposition. According to Arendt (*ibid.*) they create an opening, a "space of appearance," where something interesting and, in its most extreme interpretation, political, becomes visible and

[1] In 1990, Sweden adopted a new burial law giving the owner of a grave site greater freedom to decide about the design of the memorial stone and the site (Begravningslag, 1990: 1144).

takes shape. At the actual site of an event, it becomes possible for individuals to test their relationship to history, but also to consider their relationship to themselves and to the other people they meet at the site, and the social collective to which they belong. In this way, a memorial becomes a site that emphasises the meaning of place, as it can be mapped around and between the three poles of self, others, and environment (Gustafson, 2001). Thus memorials, together with rituals and ceremonies, deal with much more than simple remembrance – they also form the basis for progressive action.

Methodology

The empirical results were developed during 2007 and 2009 through case studies within an interdisciplinary research project at the Department of Landscape Architecture, Swedish University of Agricultural Sciences.[2] A combination of quantitative and qualitative methods were used for data collection and data analysis (Bryman, 2008). The study included on-site observations and photo documentation of ten memorial places, and forty-four semi-structured in-depth interviews with twenty-five memorial users, six relatives, two taxi drivers and two bus drivers, two professional artists, three landscape architects, and four representatives of the municipal authority and the Swedish Roads Administration. The interviews revolved around opinions on, and experiences of, memorial sites and ritual places, including conflicts and contradictory attitudes and actions. The main focus of the on-site observations was to learn more about the behaviour of visitors to the memorial, as well as about the interaction between the site and the surrounding area. In addition, observations were made on the interaction between actors in the design process and the contribution of visitors to this process. Archive documents and material from the media and the internet were also used, along with a certain amount of public data and statistics. In a workshop situation, where the knowledge from the interviews and observations was confronted with design expertise, the research results were transformed into tangible design proposals on how spontaneous memorials could be developed in contemporary urban landscapes.

A Spontaneous Memorial and a Public Monument

The first case study focused on a public monument unveiled on 29 October 2009, in remembrance of the sixty-three young people who had died ten years earlier in a fire at a youth club

2 The research project "Designing Places for Memory and Meaning in the Contemporary Urban Landscape" received three years of funding (2007–2009) from the Swedish research council Formas, followed by a one-year debriefing period during 2010 (Lieberg, 2006). The leader of the project is urban sociologist Professor Mats Lieberg; other participants are sociologist Ann-Britt Sörensen, landscape architect Professor Carola Wingren, artist and designer Anna Petersson and landscape architect Sabina Jallow. The project also has a reference group consisting of theologians, artists, architects, landscape architects, sociologists, and ethnologists.

Figure 1. The spontaneous memorial (left) and the public monument (right), in remembrance of 63 young people killed in a fire at a youth club in Gothenburg in 1999. (Photographs: the author)

in Gothenburg, Sweden. The positioning of this monument has its origins in the spontaneous memorial that immediately appeared in the car park of the premises at Backaplan, where the catastrophic fire occurred. The authorities in Gothenburg were very accommodating and understanding at the time and, after a few months, they granted permission for a temporary memorial to be set up in the car park. This memorial took the form of a plain blue bench where relatives could place flowers, candles, photos, and other mementos to honour those who had died (Figure 1, left). A decision was also made, in consultation with the relatives, to rebuild the burned-out youth club and allow it to become part of the memorial site.

Memorial objects that had been left at the site were collected and transferred to the Gothenburg City Museum. Flowers and plant material were composted and the compost was later used to plant two memorial trees at the site beside the public monument. From all the candles left at the site, the nuns at Vadstena Convent created sixty-three new candles that now form part of a display at the restored memorial hall, together with objects found at the site, for example, burned shoes, clothing and a mobile phone that had belonged to one of the victims. According to both Gustafson (2001) and Gibson (2004), material things such as candles or photographs of the deceased have an existential dimension, in that they not only mediate between self, others, and the environment, but also act as materialised compensators for our experience of the relationships between these states.

A year after the fire, a joint relatives' association, BrandOffrens Anhöriga (BOA), was created for all those bereaved by the fire. During the first year, the work of this association was characterised by strong tensions between different religious groups. BOA has since been active both in decisions concerning the rebuilding of the burnt-out youth club into a memorial hall and in choices relating to the public monument and the memorial site situated in front of the memorial hall. In the ten-year period that elapsed from the fire to the final unveiling of the permanent public monument, a design process evolved that included Gothenburg City Council and the bereaved (Figure 2). The final decision on choosing the artist Claes Hake's design was taken by Gothenburg City Council, with the consent of BOA.

The black monument, a polished dark kyanite stone slab bearing the names and ages of all the young victims engraved in gold, the red paving, the memorial trees and honeysuckle, and the flat natural stone that reflects the light of the sky (Figure 1, right) are all important memorial objects (Wingren, 2009) which, together with the objects in the rebuilt memorial hall, create the space of appearance referred to by Arendt (1958). They all bear witness to the fact that the disaster affected not only the nearest and dearest, but also all of Gothenburg. All the visitors to the site are reminded of their own relationship to similar events, whether now or in the future. While the engraved names of the victims bear witness to their ethnic diversity, the inclusion of their ages underlines the universal loss caused by their death. The sixty-three names on the stone and the sixty-three candles inside the memorial hall together represent the individual grief and loss shared by the relatives, but they also link the memorial site indoors (the place of death) with the new outdoor memorial. By providing a memory of the past, they also open up a way to evoke the future.

For the relatives, the unveiling of the new monument and memorial hall was a great relief after ten years of hope and expectation. As one of our interviewees put it, "we can now go on and think about the future." The fact that it took almost ten years for the permanent public monument to be completed meant that there was time for discussions and reflection on the appearance of the final design and also on the way in which this tragic event could be used in education and prevention. Today, young people and school classes from all over Sweden, and even from other countries, come on study visits to the site, to learn how rapidly a catastrophic fire can develop. According to the representatives of the relatives' association, the monument and memorial hall provide an important complement to the private graves located in various graveyards, through being a site of collective grieving where the families and those left behind come together to remember their own loss in the disaster, but also that of their friends and neighbours. People visit the actual graves more as private individuals, or as a family, to remember their nearest and dearest. Thus the private and the shared appear to be two complementary processes, each with its own rituals and ceremonies.

The monument at Backaplan in Gothenburg and the rituals and ceremonies enacted there are thus a reflection of something much more complex than a memorial. They create the foundation for creative and progressive actions that help the relatives go on with their lives, while also comprising an important component of accident prevention work with children and young people.

The Role of the City Council

The City Council of Gothenburg was tremendously aware of its design actions from the beginning of the process. By supporting the relatives and friends of the victims, and by being part of the mourning process, the municipal authorities were able to observe how the situation developed, not from above or outside, but by letting public servants and professionals step right into the process. Through this approach, the City Council provided room for the spontaneous memorial that arose, while at the same time ensuring that these activities did not disrupt or disturb other citizens.

Through its initiative with the blue bench during the early stages, the City Council made it possible for the bereaved to deliberate and communicate a variety of social meanings and competing historical narratives. In this way, the City Council managed and protected the design actions of private individuals. This was possible due to an organised emergency plan, with a structure for providing pastoral care for citizens, but also for the civil servants who formed part of the emergency team. Figure 2 shows how civil servants from the city and those grieving the deaths were able to interact at the specific place.

Examples of design actions by the City Council included cooperating with the media, providing adequate information about the situation, putting the emergency team together and formulating their instructions, keeping the site tidy and illuminating it, directing the placement of the flowers, renovating the hall, taking care of candle ends, flowers, and cuddly toys, organising the site for the permanent monument, choosing the artist, and organising the ceremony for the ten-year memorial day. Examples of design actions by citizens included placing flowers, forming the relatives' association, locating victims, relatives and friends, interacting with the City Council, to mention but a few. Thus together, everyone

Figure 2. Diagram showing how civil servants from the city and mourning citizens interacted at the specific site of the Gothenburg disaster. (© 2010 Sabina Jallow. Reproduced with permission.)

involved, and perhaps also everyone deciding to stay out of the process, performed actions and designed a memorial place.

The planning of the public monument reaffirmed a sense of community and made it possible for social tensions to be expressed. The facts that the monument was not designed too soon after the incident and that the bereaved participated in the decision-making process are probably the main reasons for the warm welcome that the final monument received.

By providing information and care for those involved, the City Council was able to avoid mass psychoses, counterattacks, and frustration and instead, guide the trauma in a more positive direction, possibly increasing acceptance between the various ethnicities, professions, and ages. Thus, the disaster resulted in a community integration project. The controlling authority played a bridging role as a generator and catalyst.

The memorial site, including memorial objects and rituals in the memorial hall and at the final monument, can be interpreted as a "passage landscape" (Wingren, 2009), a place that simultaneously contains several states of velocities, directions, and presence. The memorial place and the actions helped all the actors, including the bereaved, to switch from one state of mind to another on their way back to normal life. In this respect, the spontaneous memorial, as well as the final monument, can be regarded as tools used by planners, designers, and others involved to bridge the gap between the space of death and the space of life.

Roadside Memorials

The second case study within our project focused on roadside memorials, which are individualised memorials at the site of road accidents where the bereaved design and organise their own rituals and activities (Petersson, 2010). The study comprised semi-structured, in-depth interviews with family members of the victims, professional drivers and employees at the Swedish Union of Transportation.[3]

One important theme expressed by the family members interviewed concerned their visits to the fatal accident site. One informant said that he places flowers at the accident site in order to reflect upon, understand, and finally integrate the death of his son into his life. He described his regular visits to the site as something that has developed into "a little ritual" (Petersson, 2010).

Relatives bereaved by traffic accidents also spoke about the practice of bringing home items such as fallen leaves found at the accident site, or splinters of glass from the crashed car, to keep as material reminders with a special connection to the deceased. Another related practice mentioned by these informants was the placing of material objects from the deceased's personal or social life, such as pillows or sports equipment, as well as letters

3 The results from this case study were recently published in a doctoral thesis by Anna Petersson (2010) at the Department of Architecture and the Built Environment, Lund Institute of Technology, Sweden.

or gifts addressed to the victims, at the accident site. For the bereaved, these practices were closely associated with a need and a desire to continue to care for the deceased person.

Interviews with representatives of the Swedish Road Administration confirmed that there are a number of conflicts and tensions about the roadside memorials. These informants expressed great concern over the recent increase in roadside memorials on Swedish roads. When the authorities decided to remove these memorials, because they regarded them as formally illegal constructions located on state-owned or private property, the bereaved were offended and accused the authorities of intruding on their private grieving process. The conflict resulted in a campaign organised by the National Motorists' Association to introduce standard red crosses to replace the spontaneous roadside memorials along some stretches of road. However, this too attracted the displeasure of many of the bereaved, who regarded the actual site of the traffic accident as an important and meaningful memorial to which they wished to return in order to remember their loved one.

To further investigate the meaning of physical places and material objects in the types of encounters described above, Petersson (2010) used the concept of place as a "field of care," introduced by the cultural geographer Tuan (1977). According to Tuan, fields of care may emerge in places with which one has repeated contact and where one creates interpersonal networks, such as the home, a public building, a supermarket, a street corner, a marketplace, a neighbourhood, a park, a town, and so on. Repetition and habit, as well as personal acquaintance with the physical and material environment, are essential for a place to evoke affection in its users, which is also a sign that a place is a field of care (Tuan, 1977). Tuan's reflections were used by Petersson (2010) to shed light on the fact that, for some of the persons bereaved by traffic accidents, the presence of the deceased could be felt at the site of the accident rather than at their grave. Several of the bereaved also spoke of other memorial places which they visited to remember the dead, such as a special glade for picking mushrooms, a favourite restaurant or summer cottage, places which are very much in line with fields of care in Tuan's sense, namely, environments and rituals with deep meaning for the bereaved and the deceased, but perhaps insignificant to others.

Memorials at Sites of an Accident or Catastrophe

In an additional investigation, linked to the previous case study, we explored the large spontaneous memorial that developed in a car park in front of a supermarket in the city of Landskrona during 2009. On-site observations and individual interviews with visitors at the memorial place were carried out, in order to obtain information about the incident and the public reactions. The study is about an elderly woman who was attacked by a young man from an immigrant background. The man was enraged because the woman had told him not to park his car in front of the entrance of the supermarket and he pushed the woman so hard that she fell backwards and broke her neck. On the day after the incident, a spontaneous memorial with flowers and lighted candles was created by local citizens in front of the supermarket. The death of the woman released pent-up anger among the citizens of Landskrona that had been building up during years of increased violence and segregation in the city. On a small stone left at the site someone wrote: "Why this violence?" A second memorial place was later created by the citizens at a public square in front of the town hall. At the time of the incident, the situation in Landskrona was already a tense, due to conflicts between native Swedes and immigrants. The incident and the memorial site became a symbolic place, a space of appearance, where the initial fatal event developed into a public political manifestation against the current situation in Landskrona. The Landskrona case is an example of public space being turned into a memorial through actions and rituals held on-site.

Workshop on Urban Memorial Places

In order to analyse and reflect upon the amassed research material, the research group invited a group of professionals to a one-day workshop, as part of the research project. The aim of the workshop was to find out if the research results could be transformed into tangible design proposals on how spontaneous memorials might be developed and integrated into the contemporary urban landscape. The participants comprised twelve professional artists, landscape architects, cemetery architects, and a psychologist of religion, who were invited to work on the experimental task of creating common urban memorial places without connections to a specific event or person (Figure 3).

The research material gathered in this workshop consisted of images and sound recordings, as well as notes, drawings, and three-dimensional sketch models concerned with how designers and planners can gain knowledge from spontaneous memorials. This includes, for instance, respecting the need of the bereaved to visit the site of the fatality, to spontaneously leave things there, and to continually re-visit the site.

Figure 3. Workshop situation and sketch model of a public memorial park with places for both remembering and forgetting, created by the artist Aniela Wojdyno and the landscape architect Anders Jönsson. (Photographs: the author)

A conclusion that can be drawn from the design suggestions presented in the workshop is that it is difficult to consciously create a memorial without having a specific event, person, or place as the focal point. However, by creating optimal conditions, planners and designers can offer people the opportunity and the tools to adopt a place for creating their own remembrance and meaning in their everyday life. If the site is good enough, it can even function as a meaningful memorial for a number of people simultaneously. A sign that a place has become this meaningful is when people make up their own name for it. This touches in general upon the phenomenological (Relph, 1976) and psychological (Canter, 1997) conceptualisation of place and, specifically, upon Tuan's (1977) concept of place as "a field of care" – a place to which one returns for mutual contact and for solitude and contemplation.

Conclusions

Spontaneous memorials support multiple functions and agendas, that is, quiet reflection as well as active re-creation. Most often they have a memorial purpose by being "monuments" in the public space. Sometimes they initiate conflicts and turn everyday life upside down by pointing to its temporary nature and fragility, but they can also open the way for future cultural and political actions, through events held on-site. Then public space becomes a memorial in itself.

When authorities have to deal with spontaneous memorials, it is essential for them to reflect upon the many ways people have of expressing grief. It is particularly important to recognise the need of some bereaved people, for instance, those who have lost a child, to design a memorial site according to their own wishes. Spontaneous memorials generate reflections on the sacred and profane and trigger questions about the meaning of place and the relationship between the public and private domain. They can be characterised as a kind of individualised occupation of public space – probably the reason why some people feel offended by, while others are attracted to them.

The meaning of place is often achieved with the help of narrative objects, for instance engraved monuments, sculptured stones, and specific plants or materials that have the capacity to evoke associations to a specific place, person, state, or event. In the same way, fresh flowers and lighted candles function as narrative objects, by creating a connection to death and grief, hence informing passers-by that this is a place where a person's life was lost.

The design and location of spontaneous memorials can sometimes challenge social, cultural, and ideological notions of what a proper place of death should look like. Memorial objects and "passage landscapes" are tools that can be used by designers and planners to bridge the gap between the space of life and the space of death. This knowledge can also be applied to the planning and design of the urban environment in general, by paying attention to the existence and importance of spontaneous and informal activities in the urban environment. Hereby, research on spontaneous memorials can contribute to more general empirical and theoretical discussions regarding the role and meaning of place in the contemporary urban landscape.

References

Arendt, H. (1958). *The human condition*. Chicago, IL: The University of Chicago Press.

Barrera, A. (1997). Mexican-American roadside crosses in Starr County. In J. S. Graham (Ed.), *Hecho en Tejas: Texas-Mexican folk arts and crafts* (pp. 278–292). Denton, TX: University of North Texas Press.

Begravningslag (Burial law). (1990: 1144). Retrieved from http://www.notisum.se/RNP/sls/lag/19901144.htm

Bryman, A. (2008). *Social research methods*. New York, NY: Oxford University Press.

Canter, D. (1997). The facets of place. In G. T. Moore & R. W. Marans (Eds.), *Advances in environment, behaviour and design* (Vol. 4, pp. 109–147). New York, NY: Plenum.

Clark, J., & Cheshire, A. (2003/2004). RIP by the roadside: A comparative study of roadside memorials in New South Wales, Australia, and Texas, United States. *Omega – Journal of Death and Dying, 48*, 229–248.

Clark, J., & Franzmann, M. (2006). Authority from grief, presence and place in the making of roadside memorials. *Death Studies, 30*, 579–599.

Dahlgren, C. (2000). *När döden skiljer oss åt ... Anonymitet och individualisering i dödsannonser: 1945–1999* [When death separates us ... Anonymity and individualization in obituary notices: 1945–1999]. Stockholm, Sweden: Databokförlaget AB.

Doss, E. (2002). Death, art and memory in the public sphere: The visual and material culture of grief in contemporary America. *Mortality, 7*, 63–82.

Ehn, B. (2007). Hos mig kommer du alltid finnas kvar. Monumentaliseringens uttrycksformer [You will always be with me. Expressions of remembrance]. In J. Frykman & B. Ehn (Eds.), *Minnesmärken. Att tolka det förflutna och besvärja framtiden* (pp. 341–381). Stockholm, Sweden: Carlsons.

Everett, H. (2002). *Roadside crosses in contemporary memorial culture*. Denton, TX: University of North Texas Press.

Gibson, M. (2004). Melancholy objects. *Mortality, 9*, 285–299.

Gustafson, P. (2001). Meanings of place: Everyday experience and theoretical conceptualizations. *Journal of Environmental Psychology, 21*, 5–16.

Gustafsson, G. (2003). *När det sociala kapitalet växlas in: om begravningar och deltagandet i begravningar* [Redeeming social capital: On funerals and participation in funerals]. Lund Studies in Sociology of Religion, 4. Lund: Centrum för teologi och religionsvetenskap, Lunds Universitet.

Haney, C. A., Leimer, C., & Lowery, J. (1997). Spontaneous memorialisation: Violent death and emerging mourning ritual. *Omega – Journal of Death and Dying, 35*, 159–171.

Hartig, K. V., & Dunn, K. M. (1998). Roadside memorials: Interpreting new deathscapes in Newcastle, New South Wales. *Australian Geographical Studies, 36*, 5–20.

Henzel, C. (1989). Crosses in the roadside landscape of north eastern Mexico. *Journal of Cultural Geography, 11*, 93–106.

Kong, L. (1990). Geography and religion; trends and prospects. *Progress in Human Geography, 14*, 355–371.

Köstlin, K. (1999). Roadside memorials. The cross in European culture. In A. Gustavsson & M. S. Montez (Eds.), *Folk religion – Continuity and change.* (pp. 277–285) Lisbon: Instituto de Sociologia e Etnologia das Religioes.

Lieberg, M. (2006). *Designing places for memory and meaning in the contemporary urban landscape* (Unpublished reports). Alnarp, Sweden: Department of Landscape Architecture, Swedish University of Agricultural Sciences.

MacConville, U., & McQuillan, R. (2009). Remembering the dead: Roadside memorials in Ireland. In A. Kasher (Ed.), *Dying, assisted death and mourning* (pp. 135–155). Amsterdam and New York: Rodopi.
Monger, G. (1997). Modern wayside shrines. *Folklore, 108*, 113–114.
Nieminen Kristofersson, T. (2006). Från olycksplats till minnesplats [From accident site to memorial]. In T. Lundén (Ed.), *Katastrof! Olyckans geografi och antropologi* (pp. 269–280). Stockholm, Sweden: Svenska sällskapet för antropologi och geografi.
Petersson, A. (2004). *The presence of the absent. Memorials and places of ritual* (Doctoral dissertation). Lund Institute of Technology, Sweden.
Petersson, A. (2009a). Swedish offerkast and recent roadside memorials. *Folklore, 120*, 75–91.
Petersson, A. (2009b). The altar of the dead: A temporal space for memory and meaning in the contemporary urban landscape. In S. Bergmann, P. Scott, M. Jansdotter, & H. Bedford (Eds.), *Nature, space and the sacred: Interdisciplinary perspectives* (pp. 131–44). Surrey: Ashgate.
Petersson, A. (2010). *Representing the absent. A study of memoralisation and its material culture in spontaneous and official memorial places in contemporary Sweden* (Doctoral dissertation). Lund Institute of Technology, Sweden.
Relph, E. (1976). *Place and placelessness*. London, UK: Pion.
Tuan, Y.-F. (1977). *Space and place: The perspective of experience*. Minneapolis, MN: University of Minnesota Press.
Walter, T. (1996). Funeral flowers: A response to Drury. *Folklore, 107*, 106–107.
Wingren, C. (2009). *En landskapsarkitekts konstnärliga praktik – kunskapsutveckling via en självbiografisk studie* [The artistic practise of a landscape architect – Knowledge development through a self biographical study] (Doctoral dissertation). Swedish University of Agricultural Sciences, Alnarp, Sweden.

Effective Common Place Design Implied by Behavioural Evidence

Fresh Perspectives on Facility Management Research Promoting Environmentally Friendly University Campuses

Fahed A. Khasawneh,[1] Akikazu Kato,[1] and Gen Taniguchi[2]

[1]Graduate School of Engineering, Mie University, Tsu City, Japan
[2]Graduate School of Engineering, Nagoya University, Nagoya, Japan

Abstract

Quality and economy are the goals of higher education institutions; applying facility management guarantees achieving such goals. Initiatives to create efficient buildings on campuses are a valuable contribution to efforts aiming to protect the global environment. Based on qualitative methods, this study focuses on analysing actual patterns of use within common place of dining facilities. Data from five dining facilities at three university campuses have been studied. The results showed that the table layout within a dining facility is one of the primary factors that influenced its success as a common place; also, anchored tables were favoured by more diners. We conclude that the dining facilities should incorporate more seating alternatives that correspond to the groups of users served. Additionally, the Campus facility managers should consider the behaviours of frequent users of the facility and design spaces that meet their expectations. Creating common place in dining facilities to promote social integration leads to sustainability.

Key words: common place, facility management, place-making, sustainability, university campus

Introduction

The goal of facility management on a university campus is to increase effectiveness. Adopting sustainability and understanding the needs of the campus community might be the smartest way to achieve it. Creating better campus environments that run parallel to academic excellence enables universities to be more successful in recruiting and retaining distinguished students and faculty. Based on the concept of place-making in campus planning and design, introduced by Dober (2003), understanding the physical and social components of campus common place is important to create liveable and memorable places.

A closer look at the users in a campus dining hall reveals that eating is a by-product of other activities and interactions. Dober (1996) elaborates on this concept, explaining that the changes in the type and appearance of food services are some of the most significant changes taking place in campus centres these days. Many researchers have pointed out that the design of a dining facility is of the utmost importance. Kimes and Robson (2004) suggest that, generally, users prefer anchored tables in dining facilities rather than unanchored ones. Anchored tables are those that are found near a permanent architectural feature (such as a window, wall, partition, or column) that provides more privacy and helps to establish and maintain personal territory, allowing for both more comfort and more opportunities for interaction and extended stay times. In addition, Klassen, Trybus, and Kumar (2005) explain that planning food services on campus is important, because it affects the quality of life on campus.

The APPA Thought Leaders' Series, edited by Lunday (2006), identified resource scarcity and affordability, sustainability, energy, and environmental resource management as major issues affecting the facilities management profession within the context of the university campus. The International Sustainable Campus Network, ISCN (2009) urges universities to adopt sustainability principles in a comprehensive manner. Three sustainable principles are promoted in the ISCN sustainable campus charter: Firstly, to demonstrate respect for nature, secondly, to ensure long-term sustainable campus development and, thirdly, to align the organization's core mission with sustainable development. The campus planning and the design of individual buildings along with their management needs to reflect the adoption of sustainability, furthermore, the university curricula and the on-campus community should promote the application of sustainability principles.

Research Objectives and Methodology

The main objective of this study is two-fold. First, it focuses on systematically studying the place-making qualities, on both the physical and social level, that encourage campus users to use the campus common places, especially those within the dining facilities, in a manner that promotes informal interaction. The second goal is to gain better understanding of the actual behaviours and patterns of use of dining facilities from the viewpoint of "common place," aiming, in the long run, to create sustainable environments that make use of available space effectively and efficiently.

The methodology of the research was more qualitative in nature, making use of the techniques of visual documentation, observation, behavioural mapping, and video recording. Studying the physical features of the dining facility common place required the use of casual observation surveys. In these surveys, the physical features of each relevant common place, in addition to the actual patterns of use, were explored and documented by photography. These data were later combined, as recommended by Sommer and Sommer (2002), with the observer's first-visit personal feelings and impressions. The study of actual behaviours in a dining facility common place made use of observation sessions conducted by video recording. These recordings were used to plot place-centred maps (see Sommer & Sommer, 2002). These maps allowed us to show where users sat most often and the patterns of seat selection. Collier and Collier (1986) favoured using video recording in observation sessions that involved crowded settings, as is the case in dining facilities.

Case Studies

Based on preliminary observations, five dining facilities were chosen for the study. The video recording sessions were conducted for two hours during lunch times. The case studies included, first, the Toyohashi University of Technology (TUT) restaurant (two surveys performed on 26 and 27 April 2007). Second, at Nagoya University Higashiyama Campus, the study investigated two dining facilities: The North-Coop restaurant (two surveys performed on 10 and 11 May 2007) and the Forest restaurant (two surveys performed on 17 and 18 May 2007). Finally, two other facilities were studied at Mie University: Cafeteria 1 (one survey performed on 25 October 2007) and Cafeteria 2 (one survey performed on 26 October 2007). More information about the selected case studies is presented in the table below (Table 1).

Table 1. Basic information from the selected dining facilities case studies

		Seat no.		Number of users (entrance)			Number of users (exit)			Rate of seat occupancy			
	Area (m^2)	Total	Observed	Peak time	Max (in)	Sum (in)	Peak time	Max (out)	Sum (out)	Peak time	Average (%)	Max (%)	Min (%)
TUT	600.16	380	276	12:40–12:50	197	588	12:50–13:00	139	653	12:50	22.7	54.0	1.1
North-coop	536.14	390	160	11:50–12:00	151	1095	12:10–12:20	172	1162	12:15	47.0	76.3	20.0
Forest	416.42	220	166	12:00–12:10	119	979	12:20–12:30	116	918	12:20	51.7	83.1	10.8
Cafeteria 1	536.14	358	223	12:00–12:10	227	770	12:40–12:50	150	666	12:25	27.2	52.9	7.6
Cafeteria 2	416.42	324	249	12:00–12:10	123	920	12:40–12:50	171	999	12:25	34.4	61.0	10.4
Average	501.06	334	215	–	163	870	–	150	880	–	36.6	65.5	10.0

The table shows that the average rate of seat occupancy was highest (51.7%) in Forest restaurant, compared to other case studies, which indicates the success of its layout in allowing more users to dine, even though it focuses less on providing shared tables. Also, this facility operated at more than four times its total capacity within the two hours of the survey.

The Forest restaurant, which is the newest dining facility at Nagoya University, was seen to be unique. More attention has been paid to the design elements, including the variety of table layouts, colours and décor elements, and lighting scheme. In addition, the data gathered there was more complete; therefore, the results from Forest restaurant will be used to demonstrate the following general findings.

General Patterns of Use

The video recordings were analysed by plotting the users of the dining halls every five minutes, to create behavioural maps; the aim was to understand the general patterns of use.

An analysis of the resulting plotted data from the five dining facilities case studies clearly demonstrated that the use of the dining halls could be understood by dividing the patterns of use into three phases related to time and the congestion seen in the halls. These phases were the pre-peak, peak, and post-peak periods. Each phase exhibited several common aspects of use, including the preference for certain seat locations within the dining hall, the table choice mode, the number of users in general and the number of users sharing tables, and differences in procedures to achieve privacy and preserve personal space.

In the pre-peak phase, use of the dining hall was generally stress-free: There were more empty tables than the number of users entering the facility, the choice of table was based on personal preferences, and it was seen that, in the early stage, users preferred to sit in the more accessible parts of the dining hall, closer to the cashiers. Priority was given to sitting at anchored tables. As time passed, more users tended to sit in the other parts and at shared tables. As for the table choice mode, users in this phase tried to place themselves at unoccupied tables, to avoid sharing tables with others, most often choosing to sit at the corner seats of tables. Small groups preferred single tables and single users preferred to use bar-like seating.

In the peak period, the number of users in the dining hall increased, the users spread to almost all tables located in the hall, and table preference and table choice mode followed a seek-and-hunt strategy (looking for empty seats and then moving to occupy them). Sharing was observed, even at small tables. Proximity to others was tolerated, especially at large shared tables and at bar-like seating. It is important to mention that, even in the most crowded situations, the shared tables were only fully occupied in relatively few cases. It was noticed that many groups respected the borders of small tables used to form the shared tables as being territorial markers that guaranteed the integrity of their personal zones.

During the post-peak phase, there were fewer users than during the peak phase, but generally more than during the pre-peak phase. More users sat at anchored tables, and users were still spread over a large proportion of the dining hall tables. As for table choice mode, tables were selected according to personal preferences and perceived value of tables, with users favouring anchored tables near windows. Groups of users first selected unoccupied tables, occupying either corner seats of the table. Users avoided sharing tables with others when possible.

Forest Restaurant Case Study

The Forest restaurant survey performed on 18 May 2007 was chosen for further analysis because it was conducted in a more rigorous manner, and because the frequent users behaved more naturally, since they were accustomed to the presence of cameras and observers on the second day of the survey.

Behavioural Maps

The behavioural maps that were created clarified the patterns of use described earlier within the context of this particular case; for instance, it was observed that, during the early pre-peak phase, users preferred to sit near the southeast wall of the dining hall, due to ease of access. Also, during the peak phase, sharing was seen, even at small tables, but especially at six-seat tables, which were frequently shared by two groups simultaneously (Figure 1). Sharing tables with others was avoided during the post-peak phase, yet when sharing occurred, groups placed themselves on either corner seats of the table.

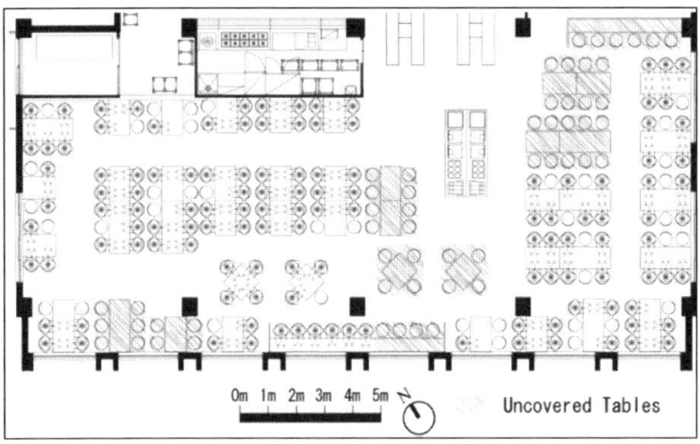

Figure 1. The peak phase as seen in the dining hall at 12:35 p.m.

Structured Analysis

A table-by-table analysis was conducted, focusing on:

i. *The group formation of users:* The numbers of group members, gender, and the total number of users.
ii. *Table profile:* Anchored or free tables; each table was given a code, to make analysis easier (Figure 2).
iii. *Activities:* A distinction was made between the meal duration, which we defined as the time during which food was eaten, and interaction duration, which we defined as the sustained interaction (more than two minutes) between members of a group after finishing the meal.
iv. *Patterns of users:* In terms of belonging to "eat and go" or "eat and stay" patterns. The users who stayed after they had finished eating were considered to belong to the "eat and stay" pattern.

Single users and more groups ranging in size from two-user to five-user groups preferred to sit at anchored tables. This observation was most frequent for single users and two-user groups; the single users mostly sat at bar-like seating, ate quickly, and left, creating space for subsequent single users to select this type of seating. Many two-user groups sat at 4-seat and six-seat anchored tables, denying other larger groups the opportunity to make use of such tables. It can be said that as the group size decreased the tendency of such small groups to select anchored tables increased. In fact, more six-user groups selected free tables, although many six-seat anchored tables were made especially to accommodate such groups. Likewise, for large groups, more users chose free tables, and this behaviour corresponded to the size of tables.

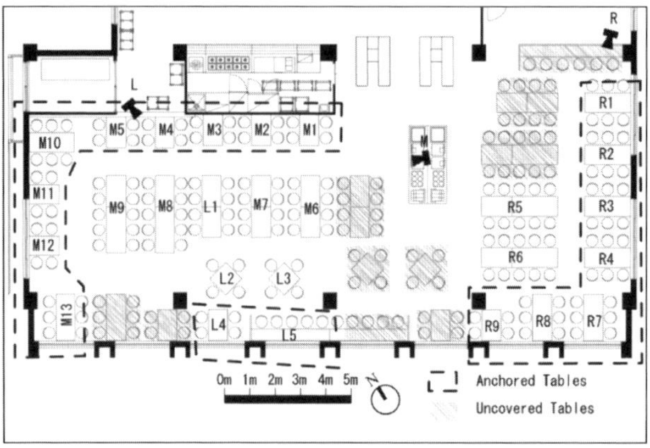

Figure 2. Tables codes used during structured analysis.

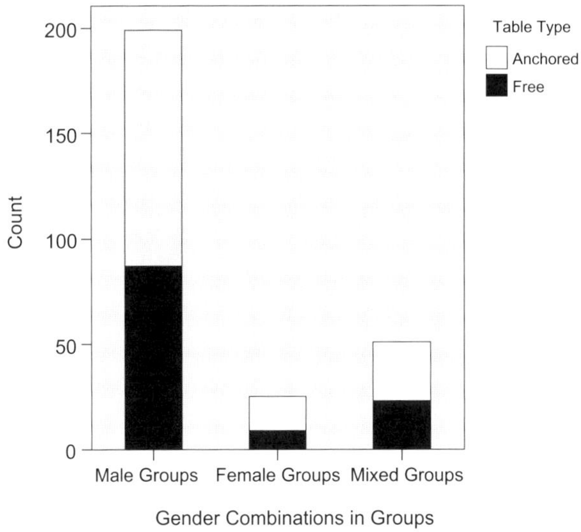

Figure 3. Relationship of groups' gender to table type selection.

Generally, all groups, regardless of gender combinations, preferred to sit at anchored tables. This may be related to the high value of such tables and the comfort they provide to users. However, the gender effect was clear, especially for female groups, who chose anchored tables twice as often as free tables. It is important that the ratio of female users to male users was 1:4. Also, the difference of table type selection was less apparent for mixed groups (Figure 3).

Small Group Ecology

This analysis uses knowledge based on small group ecology, according to Sommer (1969), to analyse the behaviours of groups of users. The aim is to understand the mechanisms and dynamics of users' behaviours while selecting seats and while interacting with the physical and social features and with other users. Each group of users was tracked and studied in relation to the available table layout. Two examples are discussed in the following paragraph.

Single users avoided sitting directly across from other single users, although close proximity while sitting side-by-side was accepted in crowded conditions. In one case, a single user had no choice but to sit across from other users as the other side was fully occupied; he chose a seat in front of a two-user group and avoided sitting across from other single users.

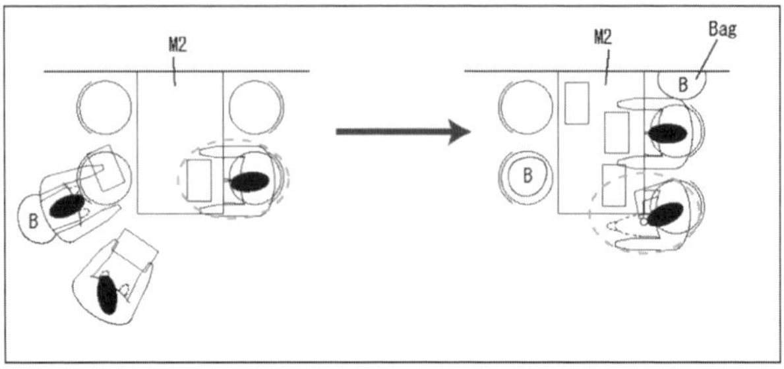

Figure 4. A single user's territoriality invasion and reaction.

Users tried to keep their personal zone intact and reacted to any invasion. A single user was sitting at a 4-seat table, and his position on an aisle seat conveyed his desire to control the table and keep intruders away. Due to crowded conditions, a two user group decided to share table with him. A female user sat on the seat beside the single user and moved her seat towards him. The single user's posture and gestures, in addition to the action of moving his seat to the opposite corner of table, showed his annoyance and discomfort; he finished eating quickly and left (Figure 4).

Discussion

In reference to dining facility common place, Whyte (2001) explains that food is one of the major attractions in public places. Milshtein (1999) stresses the importance of providing more seating alternatives and food serving methods to create successful places.

Swanquist (1999) argues that campus dining facilities should be flexible and multi-functional and act as hybrid food service locations that are comfortable and lively places to hang out with friends. According to Deasy and Lasswell (1990), the users of dining facilities fall into two categories. First, there are those who like to eat alone and quickly; second, there are those who want to eat in groups and consider dining to be a social event. A successful dining facility should provide for both alternatives. A common place, according to Komatsu and Kato (1994), includes all non-specialized space that is connected to and adjacent to specialized spaces that are designed to fulfil the formal purpose of any facility. Strange and Banning (2001) stress that educational environments need to possess certain elements to succeed: A feeling of security and belonging, mechanisms to induce students to engage in campus life, and a sense of community. Scott-Webber (2004) elaborates that certain behaviours of potential campus users is induced and affected by the university's physical features and institutional nature. The ISCN (2009) second principle in the sustainable campus charter calls for campus-wide master planning and target setting that includes environmental and social goals to ensure long-term sustainable campus development, where social integration is of utmost importance to ensure user diversity and to create indoor and outdoor space for social exchange and shared learning.

Features of Prominent Place-Making Elements

The findings of the research corroborate many of the previously mentioned research literature; the analysis of the case studies showed that prominent place-making elements may include:

i. *Clear and strong definition of the dining process*: Providing dining facilities that follow a clear service process that is reflected by internal planning and design with wide movement and service aisles.
ii. *Multiplicity of seat and table configurations and layouts:* Table layouts should try to cater to the demands of various potential users by providing places for those who intend to eat and go and separate places for those who intend to eat and then stay for interactions. The dining facilities should try to provide the highest number of anchored tables, which, as we have seen throughout this research findings, most users preferred.
iii. *Providing proper lighting and colour schemes and other complimentary elements that provide the users with a feeling of comfort and relaxation.*

Frequent Activities Pertaining to the Common Place

The activities pertaining to the common place were more prominent during the relaxed and stress-free time periods, including:

i. *Multiplicity of activities:* The user considers the dining facility to be a comfortable place to practise a wide range of activities, besides eating, in a planned manner.
ii. *Meeting others:* The dining facility is used as a meeting place; one form of such meetings are lunch meetings. In other cases, visitors make use of the available tables and seats to hold a discussion, as the main activity; here food and beverages are not consumed.
iii. *Spontaneous meetings:* The users arrive separately with the intention of enjoying a meal, yet due to the nature of the dining facility as a gathering place, they end up meeting friends.

iv. *Following daily dining habits and rituals:* Users on a campus develop affiliations with certain groups and even develop attachment to certain seats or tables. This is reflected in many rituals and habits that make the dining event more relaxing and enjoyable.
v. *Hanging out:* The dining place is seen more like a lounge or a place to hang out and pass time. Users keep the table for a long time and their activities include a wider range of purposeless activities.

Conclusion

The creation of effective dining facilities as common place should be the primary goal in campus planning, design, and management. It will help, in the long run, to create an environmentally friendly campus that applies and promotes sustainability principles and, in particular, social integration, by creating common place for social exchange and shared learning. The concept of place-making is essential for creating liveable common areas, the focus being on both physical and social components. One of the most important place-making elements within this context is providing more anchored tables as a base for table layouts in the dining facility. Institutional dining facilities are especially crowded during lunch hours; this limits comfortable seating choices and reduces users' control over personal territory. Users will seek anchored tables that provide more control and make it easier to defend territory and personal space, both of which help to provide a comfortable environment amid crowding. Having more control in an environment provides more freedom and comfort which induces more interaction. Providing anchored tables is an essential response to the users' needs for social interaction and secure territory. In addition, multiplicity of choice should be incorporated into the planning and design of dining facilities. Table configurations should take into account the profile of the groups served, the actual patterns of use, and the needs of its frequent users. Shared tables proved to be unsuccessful and unpopular while, on the other hand, bar-like seating was popular, especially among single users. Two-seat tables were also seen to be necessary in the dining hall, as two-user groups constituted a sizable proportion of the groups. The facility manager of a campus should consider the actual uses and behaviours of frequent users when designing effective dining facilities. The priority must be to provide more control and/or interventions in the environment – this kind of consideration is what leads to a successful common place.

References

Collier, J., & Collier, M. (1986). *Visual anthropology: Photography as a research method.* Albuquerque, New Mexico: University of New Mexico Press.

Deasy, C. M., & Lasswell, T. E. (1990). *Designing places for people: A handbook on human behavior for architects, designers and facility managers*. New York, NY: Whitney Library of Design.

Dober, R. P. (1996). *Campus architecture: Building in the groves of academe*. New York, NY: McGraw-Hill.

Dober, R. P. (2003). *Campus design*. Ann Arbor, MI: Society for College and University Planning (SCUP).

International Sustainable Campus Network, ISCN. (2009). *What is the ISCN? Introduction and Analysis.* Retrieved from http://www.international-sustainable-campus-network.org/index.php?id=74

Kimes, S., & Robson, S. (2004). The impact of restaurant table characteristics on meal duration and spending. *Cornell Hotel and Restaurant Administration Quarterly, 45*, 333–346. Retrieved from http://cqx.sagepub.com/

Klassen, K. J., Trybus, E., & Kumar, A. (2005). Planning food services for a campus setting. *Hospitality Management, 24*, 579–609. Retrieved from http://www.sciencedirect.com/

Komatsu, H., & Kato, A. (1994). Study on effectiveness of layout of seats with POE and the concept "affordance". In A. Seidel (Ed.), *EDRA 25. Banking on Design?* (pp. 124–130). San Antonio, TX: Environmental Design Research Association (EDRA).

Lunday, E. (2006). *University facilities respond to the changing landscape of higher education: Including the top ten critical facilities issues* [APPA Thought Leaders Series]. Alexandria, VA: APPA and CFaR. Retrieved from http://www.appa.org/files/pdfs/appa40a_thoughtleaders_3.pdf

Milshtein, A. (1999). Bye bye cafeteria … hello restaurant-style dining. *College Planning and Management, 2*(11), 46–49. Retrieved from http://www.peterli.com/cpm/

Scott-Webber, L. (2004). *In Sync: Environmental behavior research and the design of learning spaces*. Ann Arbor, MI: Society for College and University Planning.

Sommer, R. (1969). *Personal space: The behavioral basis of design*. Englewood Cliffs, NJ: Prentice-Hall.

Sommer, R., & Sommer, B. (2002). *A practical guide to behavioral research: Tools and techniques*. New York, NY: Oxford University Press.

Strange, C. C., & Banning, J. H. (2001). *Educating by design: Creating campus learning environments that work*. San Francisco, CA: Jossey-Bass.

Swanquist, B. (1999). A new trend in dining design. *College Planning and Management, 2*(4), 34–36. Retrieved from www.peterli.com/cpm/

Whyte, W. H. (2001). *The social life of small urban spaces*. New York, NY: Project for Public Places Inc (PPS).

Responding to Complex Urban Challenges With Integrated Research in Education

Case Study of a Master Programme in Urban Design

Ombretta Romice,[1] Michael Mehaffy,[1] Sergio Porta,[1] and Kevin Thwaites[2]

[1]Urban Design Studies Unit, Department of Architecture, University of Strathclyde, Glasgow, United Kingdom
[2]Department of Landscape, University of Sheffield, United Kingdom

Abstract

It is widely recognized that the complexity of urban challenges today demands new strategies and skills for urban designers. In many cases, these skills do not yet exist, and may even be specific to new kinds of projects. Thus, an important new skill must be the ability of practitioners to perform rigorous, effective project-based research – to "learn to learn" in a complex, challenging environment. In response to this need, we present and analyse a methodology for teaching urban design that gives students project-based research skills, while at the same time contributing to an ongoing professional research agenda. This integrated research methodology is currently in development at the University of Strathclyde Urban Design Studies Unit (Glasgow, UK).

Key words: adaptability, complexity, teaching, urban design

Introduction

Urban design professionals are now facing an unprecedented convergence of challenges related to human habitats, including new economic pressures, social conflicts from immigration, and growing environmental damage. These challenges are complicated by an often accelerating scale and pace of urban development, leaving the professions poorly prepared for the results of this rapid urbanisation: Unbalanced growth, chaotic patterns of use, and growing fragmentation of the essential connective structure of the city. It is widely recognised that these challenges must be addressed with a new generation of more precise tools.

Scale, pace, and development economics no longer allow costly social and environmental remediation. Professionals need to *understand in a new way* the problems and requirements of development, and identify effective, locally-calibrated solutions for them – solutions that may rely more upon incremental processes and local capacity-building, and less upon large-scale remediation.

We believe this means, among other things, that urban design needs an effective new form of real-time, project-based research. Such a project-based research methodology must gather evidence and understand how to act strategically on this evidence, often leveraging modest resources. It needs to move away from the currently conventional, standardised tools in urban planning and develop new, more contextual ones. New does not necessarily mean a complete break from the past; rather, it means *more appropriate* for unique or unprecedented problems – that is, adaptable and manageable under the pressure of scale, and less reliant on mainstream planning frameworks.

This comes at a time when a number of authors have expressed concerns about the theoretical foundations of urban design. For example, Cuthbert (2007, p. 177) reviewed discourse and advancements in the field over the last 50 years and denounced a lack of "any concerted attempt within the discipline to link the material creation or 'designing' of urban space and form to fundamental societal processes." He criticized urban design as self-referential, lacking theoretical frameworks appropriate for today's development scenarios, and failing to link to related disciplines of sociology, geography, politics, economics, psychology, amongst others. Cuthbert warned that urban design lacks a clear object around which to formulate needed theories. At the same time, he argued for the legitimacy of urban design as an independent discipline, no longer in need of finding a compromise between architecture and planning.

We report here how we have sought to respond to these challenges by embedding research activities into mainstream teaching, involving postgraduate students in the further development and application of new urban design methodologies and strategies (including project-specific ones), in a Project-Based Learning (PBL) format. In addition to defining learning and teaching objectives, we make explicit, within the studio setting, the research context, the research problem, and the theoretical tools that students will assist to refine and apply in response to a real-world project scenario. As we will discuss below, we conclude that this approach provides clear benefits, both for the quality of student learning, and for the research and development of effective new urban design tools and methodologies. Whilst the UK research and teaching system is the specific context for this research, we believe the principles and outcomes are likely to be applicable in many other contexts.

The Research Hypothesis: A Framework

Our integrated research curriculum was, in turn, dependent on a research hypothesis that we believed was likely to respond to the larger challenges of urban design as outlined above. The genesis of this hypothesis was a book edited by some of the authors, called *Urban Sustainability Through Environmental Design* (Thwaites, Porta, Romice, & Greaves, 2007), in which we reviewed current best practice, and pointed out remaining shortcomings in producing places that were sustainable from a social, economic, and environmental point of view, and that could perform adequately under the dynamic requirements of urban development today. Whilst we acknowledged enormous advances in practice in recent years, we concluded that the discipline still lacks a robust theoretical framework sufficient to allow credible analytical research.

A central theme of this book was the urgent need to re-introduce a more incremental, time-conscious perspective into contemporary urban planning and design. Parallel research in urban morphology demonstrated that urban environments whose form had evolved through time under the structuring influence of social processes, rather than being delivered in pre-packaged form by professional planning and design processes, tended to be socially more diverse, experientially fulfilling, and sustainable (e.g., Alexander, 1967; Frank & Stevens, 2007; Habraken, 1998).

In response, we developed a framework design hypothesis to be evaluated in our research. This hypothesis had two related components, which we termed "Plot-Based Urbanism" (PBU) and "Urban Seeding," (US), which we describe in more detail below (Porta & Romice, 2010). They both embody ideas that are well-established in the literature, but which we structure into a new kind of context-specific design methodology. It was this methodology that we wished to evaluate in the research with our students. In this sense, research was occurring at three distinct levels: One, the evaluation of the design methodologies themselves; two, the evaluation of the teaching methodologies, in the context of PBL; and three, the design research that the students were performing for the specific project.

Briefly, PUB recognises again the fundamental role of the plot in generating the complex spatial structure of ordinary urban fabrics. The methodology adopts strategies to influence how the plot is shaped, its size and geometry, its relationship with the street and the street hierarchy, how it forms up street fronts and, eventually, urban blocks. Moreover, it engages with the ways that all this informs human activities and urban functions, and the ways that property, usage, and control are maintained.

The second, related methodology. "Urban Seeding" seeks to support the processes by which incremental changes adapt to social needs through strategic and catalytic interventions. If PBU provides the *essential structures*, US supports their completion and modification in time. It provides the capacity-building tools with which these modifications will be made, in response to dynamic social needs and conditions.

We chose to evaluate this framework hypothesis by assigning students to apply the methodology to actual projects, using a "research and development" format. In this way, we could test the efficacy of the elements of the methodology on an actual project. In particular, we were assessing and refining the framework methodology; the students were conducting their own project-based research; and lastly, we were assessing and refining the research-based curriculum model itself.

We felt that as long as these distinct activities and objectives were kept clear, this multi-stranded, integrated approach was, in fact, a benefit: The conclusions from one form of research would also help to inform the other. At the same time, the students would benefit from the rich environment of PBL in two complementary forms: They would perform hands-on research work on an actual urban project, and they would assist in the larger research project to develop the framework and its tools.

Research Into Practice: Towards a Critical Synthesis of Theories and Tools in Urban Design

Our first step to make the proposed design studio operational as an integrated education and research experience was to conduct (and to develop, as a curriculum for students) a critical assessment of mainstream theories of urban design, to define the theoretical foundations we would use, the questions we would ask, and the tools we would adopt. There is, of course, a well-established body of influential and widely implemented theories on what "good" urban form should be (e.g., Lynch, 1984), and a well-established set of best practices to organise the complexity of places and cities, to achieve such normative goals (e.g., Bentley, Alcock, Martin, McGlynn, & Smith, 1985; Llewelyn-Davies, 2000).

But alongside this body is an equally well-established critical discourse, informed by a rich history of empirical observation and planning theory, arguing that the heavily planned is not *per se* better than the non (or the *less*) planned, and indeed is often distinctly inferior (e.g., Alexander, 1965; Jacobs, 1961). This history of awareness was informed by anthropological observation of the quality of life of people and groups living in the new highly planned and innovative environments of the 20th century (e.g., Coleman, 1985; Towers, 2000).

Such criticisms are today widely acknowledged, to the point that UN Habitat has recently launched a wide reflection on the need for and the character of a new urban planning discipline altogether, to face the challenges of the 21st century (Farmer et al., 2006). Indeed, these criticisms have in large part fuelled an era of "post-modern" design that has sought to recover the complexity that had apparently been lost. Yet, no credible authors of whom we are aware have been able to show that the complexity that, say, Jane Jacobs argued for so persuasively, has been achieved today, even within the most context-sensitive products of the planning culture. Clearly, more reform of practice is needed – and more effective methodologies that learn from these lessons.

The complexity intended was by no means a spontaneous chaos, as opposed to order; rather, it was (in Jacobs' own term) *organized complexity*, which has a *structural quality*, one which affords the exercise of both planning by experts and forms of participation by stakeholders, acting as catalysts for local management and adaptation of the urban environment, within the parameters set through planning and formal participation. There are echoes of trends in other fields, including (as Jacobs herself noted) biology and other "sciences of complexity." Within anthropology and sociology, the most relevant concept may be "capacity-building," or its variant "asset-based community development" (see e.g., Kretzmann & McKnight, 1993), where the goal is not to impose a top-down solution, but to find leverage points within the existing set of conditions that will catalyze a desired transformation.

These principles, which we clarified and adopted within our own theoretical review, laid the foundation for the further development and clarification of the proposed design methodology we called "Urban Seeding" (i.e., capacity-building through strategic and catalytic design intervention), which operates to transform the spatial structures of PLU. We took this design hypothesis to the inaugural studio, and with the students we set out to clarify in our own minds *what kind* of planning was best suited to set the *right spatial structure* for future change and adaptation.

Researching the Design Methodology of PBU in Higher Education: The Strathclyde Prototype

We developed the studio in our Masters in Urban Design programme as an inter-disciplinary course, offered to students from disciplines such as planning, engineering, sociology, and geography. We structured the course according to the PBU framework, using the ideas and structures discovered and illustrated so far. Our dual aim, as mentioned, was to research these theoretical principles through application of the educational practices they imply, while simultaneously providing skills in project research to students, within a specific context of PBL. Again, the students learned to research their own project, and they simultaneously assisted us with the research of the curriculum as a whole, and the specific tools being developed and refined.

An essential goal of the course has been to equip students with effective tools of three types: Tools for the structural analysis of the urban fabric, its spaces and measures; tools for the structural analysis of the city in terms of connections and centralities; and tools for the design of the essential underlying ordinary structure of places, such that this structure is capable of changing in time, of maintaining its ordinary character, and of allowing extraordinary and novel features to be embedded where necessary. Since students are not experts in the subject field, they provide an effective test of the practicality and accessibility of the tools for a variety of users. The feedback they generate as users contributes to the refinement and further development of the theoretical models being applied, and generates material for measurable research outputs at international levels.

Again, it is critical that this process occurs in the context of a real project within an immersive, PBL environment. The context of the Masters project is commissioned by the City Council of Glasgow, in partnership with local stakeholders. This allows moving the studio permanently out of the university and into the study area, with a number of benefits in terms of motivation and cohesiveness of the students' understanding of the place and involvement of the community (Figure 1). Typically, the study area is a portion of the city needing a structured analysis, long-term development frameworks, and Detailed Masterplans and design solutions. This provides a unique learning and teaching opportunity, grounded in a real-world scenario, and the opportunity for students to perform useful empirical research within the developing framework of new theoretical perspectives and their application.

Students conduct on-site analyses of various kinds and derive both "objective" (observed) and "subjective" (perceived) qualities of space, to understand the area, and the links between social, economic, and civic characteristics and the formal fabric of

Figure 1. Public consultation took place over nine months and involved stakeholders and local groups in each phase of the work. (Source: courtesy of the University of Strathclyde, Glasgow, UK)

the place. Then they develop long-term plans for the improvement of the area, formulating economic, social, and physical scenarios. These transformations are recorded in Foundation Masterplans, documents which specify new street networks, nodes, and areas of influence at local, neighbourhood, and metropolitan levels, and changes of density to support facilities redistribution (Figure 2).

In parallel, students work out a morphometric analysis of urban blocks of the area and of areas with similar characteristics around the city, representing well-established urban types compatible to those in the area under investigation, to understand the structurally essential properties which make up such urban forms and their proposed transformations. In doing so, the students collect fundamental formal parameters that they can then use to generate urban types that are well-adapted to the indigenous ones and their context, whilst at the same time allowing the addition of a degree of change and innovation. This morphometric analysis focuses on the detailed study of the formal structural properties of plots in relation to street-fronts and street types (Figure 3).

All parameters derived from the morphometric analysis are then collected and structured within the Local Urban Code (LUC) (Figure 4), a guide to accompany the development of the Detailed Masterplans that will follow. The spatial structural principles of PBU then guide the refinement of the long-term strategic frameworks within Foundation Masterplans, which specify the main planning decisions through two main factors: 1) changes in the type and structure of street networks and 2) the distribution of densities to accommodate changes in use and in population distribution. This step is very important, because it specifies the structural characteristics for the Detailed Masterplans. It indicates, on a specific street, the degree of publicness of a street front, and with it the number of plots,

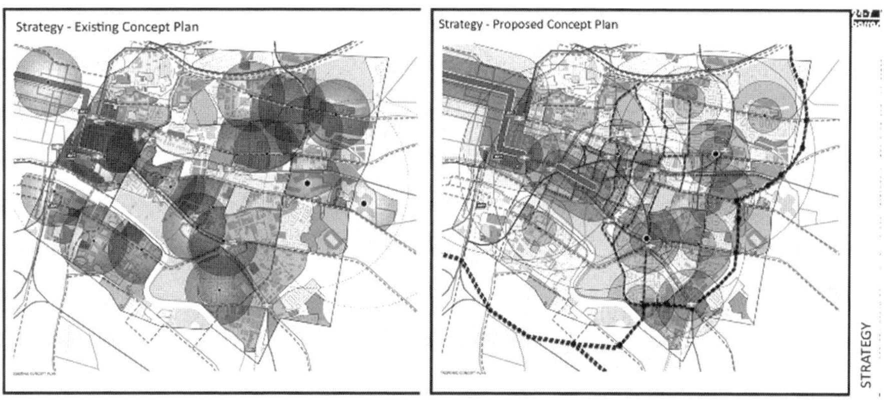

Figure 2. Foundation Masterplans stem from the combination of the study of current conditions (represented as an overlay of densities, infrastructure, services with their area of influence, barriers) and the strategy to overcome all issues identified, also represented using the same information. (Source: courtesy of the University of Strathclyde, Glasgow, UK)

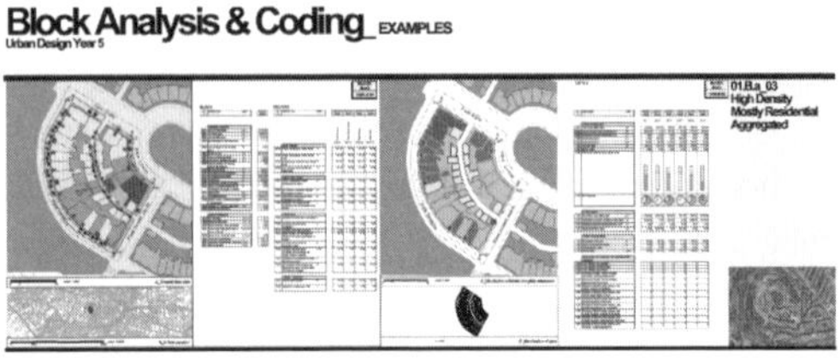

Figure 3. The morphometric analysis of urban form. (Source: courtesy of the University of Strathclyde, Glasgow, UK)

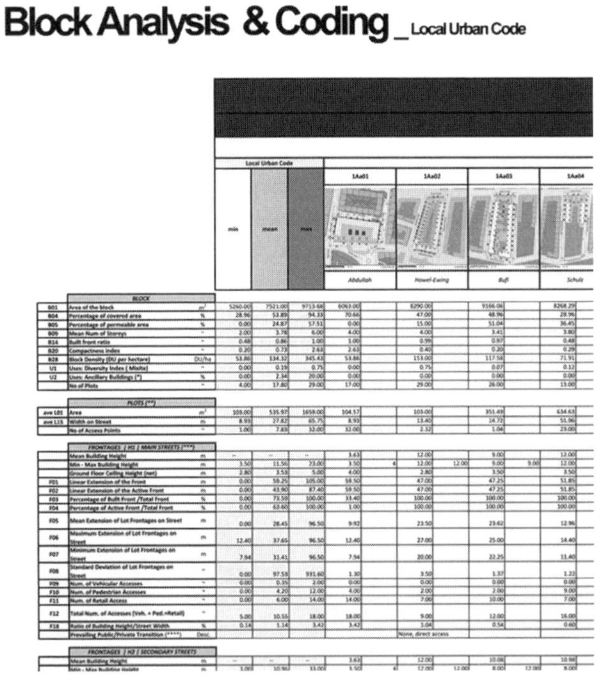

Figure 4. The Local Urban Code (LUC) is a combination of all information gathered from the morphometric analysis and used as a design code for detailed masterplanning. (Source: courtesy of the University of Strathclyde, Glasgow, UK)

heights, setbacks, proportion of public versus private activity, number of entrances, etc., characteristic of a *typical street of that kind.*

It is important to note here that these indications are typical, and structurally essential, but that they by no means impose any specific architectural style on the project. This should be left to the work of inhabitants, the architects who will develop the Masterplan, and other stakeholders. This approach develops a fundamental principle of PBU: A focus on the *structure* of spaces, which should leave the largest possible room for personal articulations, arrangements, and adaptations in time.

By combining Foundation Masterplans and LUC, students produce the Final Masterplans, which are comprehensive design documents at scale 1:1,000, tackling infrastructural, service, and detailed design decisions. They include the built form, differentiating between ordinary and extraordinary fabric; the public realm, and special events, that is, specialist areas, public landmarks, etc. Our LUC puts great emphasis on the relation between street types (and fronts) and plots. Within the blocks which derive, plots are to be seen as the units of design; they will vary in their degree of autonomy within a frontage, in accordance to the publicness of such frontage.

The Final Masterplans contain *specifications on ordinary fabric* such as localised densities, building types with their relative access, scenes (i.e., fronts behaving exceptionally to mark a special character, even within the ordinary fabric), location of services and facilities in mixed types; on *special events,* such as their location, footprint, and main/secondary access; and on the *public realm*, such as the detailed design of roads per type, including signage and circulation patterns (main, secondary, tertiary, local, lane, etc.), and landscaping (Figure 5). More subjective characteristics, such as visual links and community significance, are taken into account, and tested with local groups.

Finally, these Final Masterplans are developed by students acting as architects who design individual elements in accordance to the specification contained in the LUC. Thus, students design individual buildings within individual plots that together form up street fronts and blocks that are diverse but at the same time share fundamental characteristics derived from the LUC. The thesis pursued here is that when the structure of a place is

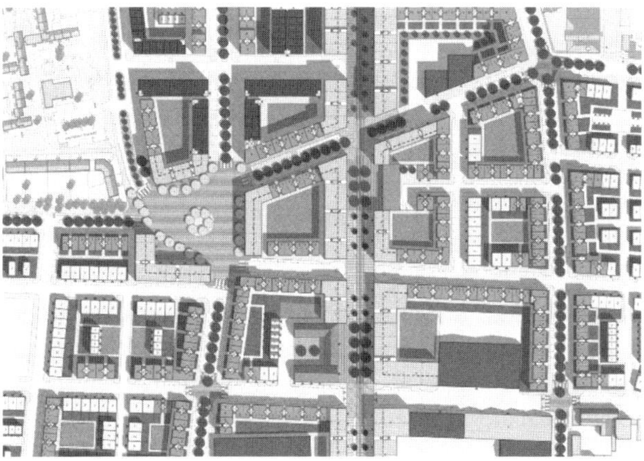

Figure 5. The detailed final Masterplan is a flexible document, style neutral but structurally very precise. (Source: courtesy of the University of Strathclyde, Glasgow, UK)

appropriate to its location and function, its character can be designed in several ways, but its essence will work.

Conclusions

Armstrong (1999) identified three research potentials of design studio teaching: 1) education research to pioneer and pilot new approaches to teaching, 2) consideration of the design studio and its outputs as theoretical creative work, 3) the studio as part of a larger research programme. The experience described in this chapter fits all three points. The third point is of particular interest to us, since the development of new design methods is an ongoing endeavour of the Urban Design Studies Unit. We are now expanding this research-integrated curriculum further, since it has so far produced a number of significant benefits:

Students have:

- *Gained control of their learning process*: The explicit aim of the course, combined with the structured approach through each step of the design work, allowed them to consistently assess their performance;
- *Engaged creatively*: Whilst they mainly work around given structural elements, they also add to the list through their morphological analyses and help develop a process to integrate such elements into a replicable and useable design approach;
- *Been empowered*: With a clear role, they have tested research hypotheses and influenced their refining, moving away from a traditional teacher-learner relationship to a more engaged and responsible one, where they are not just recipients of pre-existing knowledge but active collaborators in the development of new knowledge.

There are, of course, dangers in such a combined research and education model, and in particular, there is a real danger of confusion. The variety of subjects of research – the evolving tools on the one hand, and the projects themselves on the other – must be kept distinct. The fast-track research and development of the tools themselves can also confuse students, who may be more comfortable learning to use well-established, easily identified tools.

And yet, as we noted at the outset, urban environments and their challenges are increasingly dynamic, and we are more than ever convinced that students are best-served by learning a dynamic, evidence-based, *research*-based approach in response. Most particularly, we believe students are well-served to understand that their job as designers is not to impose rigid design templates, but to facilitate the kinds of collaborative, structural processes that will achieve healthier urban growth over time. In this process, research is an indispensable dynamic methodology that must be aimed at the adaptive urban subject, and simultaneously at the adaptive processes of design itself.

Details of this work and specific research outputs are available on our UDSU website, at http://www.udsu-strath.com. We will continue to update this site with additional results.

References

Alexander, C. (1965). A city is not a tree. *Architectural Forum*. Republished on rudi.net. Retrieved from http://www.rudi.net/pages/8755

Alexander, C. (1967). The city as a mechanism for sustaining human contact. In W. R. Ewald (Ed.), *Environment for man* (pp. 60–102). Bloomington, IN: Indiana University Press.

Armstrong, H. (1999). Design studios as research: An emerging paradigm for landscape architecture. *Landscape Review, 5*, 5–25.

Bentley, I., Alcock, A., Martin, P., McGlynn, S., & Smith, G. (1985). *Responsive environments*. London, UK: The Architectural Press.

Coleman, A. (1985). *Utopia on trial*. London, UK: Hilary Shipman.

Cuthbert, A. (2007). Urban design: Requiem for an era – review and critique of the last 50 years. *URBAN DESIGN International, 12*, 177–223.

Farmer, P., Froimovic, M., Hague, C., Harridge, C., Narang, S., & Shishido, R., . . . Vogell, J. (2006). *Reinventing planning: A new governance paradigm for managing human settlements*. Retrieved from http://www.globalplannersnetwork.org/pdf/reinventingplanningenglish.pdf

Frank, K., & Stevens, Q. (2007). *Loose space: Possibility and diversity in urban life*. London, UK: Routledge.

Habraken, N. J. (1998). *The structure of the ordinary: Form and control in the built environment*. Cambridge, MA: MIT Press.

Jacobs, J. (1961). *The death and life of great American cities*. New York, NY: Random House.

Kretzmann, J., & McKnight, J. (1993). *Building communities from the inside out: A path toward finding and mobilizing a community's assets*. Evanston, IL: Center for Urban Affairs and Policy Research, Northwestern University.

Llewelyn-Davies. (2000). *Urban design compendium*. London, UK: English Partnerships and The Housing Corporation.

Lynch, K. (1984). *Good city form*. Cambridge, MA and London, UK: MIT Press.

Porta, S., & Romice, O. (2010). *Plot-based urbanism*. Paper commissioned by Architecture and Design Scotland. Retrieved from http://www.udsu-strath.com/wp-content/uploads/2011/11/Plot-Based-Urbanism_10_OR-SP-FINAL_low-res.pdf

Thwaites, K., Porta, S., Romice, O., & Greaves, M. (Eds.). (2007). *Urban sustainability through environmental design*. London, UK: Spon Press.

Towers, G. (2000). *Shelter is not enough: Transforming multi-storey housing*. Bristol, UK: The Policy Press.

Authors

Mariagrazia Agrimi
Department of Innovation in Agri-food, Biological and Forest Systems (DIBAF)
Tuscia University
Via S. Camillo de Lellis
01100, Viterbo
Italy
agrimi@unitus.it

S. Harris Ali
Faculty of Environmental Studies
York University
4700 Keele Street
Toronto, Ontario
Canada
hali@yorku.ca

Thémis Apostolidis
Laboratory of Social Psychology
University of Provence
29, Avenue Robert Schuman
13621 Aix-en-Provence Cedex1
France
themistoklis.apostolidis@univ-provence.frgrazia

Xosé Luis Barreiro Rivas
Faculty of Political and Social Sciences
Santiago de Compostela University
Avda. Dr. Ángel Echeverri, s/n. Campus sur
15782 Santiago de Compostela
Spain
xoseluis.barreiro.rivas@usc.es

Alena Bleicher
Department of Urban and Environmental Sociology
Helmholtz Centre for Environmental Research - UFZ
Permoserstr. 15
04318 Leipzig
Germany
alena.bleicher@ufz.de

Giuseppe Carrus
Department of Cultural and Educational Studies
University of Roma Tre
Via del Castro Pretorio, 20
00185, Rome
Italy
carrus@uniroma3.it

Renate Cervinka
Institute of Environmental Health
Medical University of Vienna
Kinderspitalgasse 15
1095 Vienna
Austria
Renate.cervinka@meduniwien.ac.at

Rittirong Chutapruttikorn
Faculty of Fine and Applied Arts
Bangkok University
Rangsit campus
Pathum Thani
Bangkok
Thailand
Rittirong.c@bu.ac.th

Francesca Cini
Department of Cultural and Educational Studies
University of Roma Tre
Via del Castro Pretorio, 20,
00185, Rome
Italy
francesca.cini@uniroma1.it

Zulmira A. C. Cruz Bomfim
Departamento de Psicologia
Universidade Federal do Ceará
Rua Adolfo Moreira de Carvalho 140
Edson Queiroz, Fortaleza-Ce
60811-750
Brazil
zulaurea@uol.com.br

Veriana de Fátima Rodrigues Colaço
Departamento de Psicologia
Universidade Federal Do Ceará
Rua Henriqueta Galeno, 960, Apto. 201
Bairro Dionísio Torres, Fortaleza, Ce
Cep 60135-420
Brazil
verianac@uol.com.br

Christophe Demarque
Laboratory of Social Psychology
University of Provence
29, Avenue Robert Schuman
13621 Aix-en-Provence Cedex1
France
demarque_christophe@yahoo.fr

Celina Dittmar
Faculty of Architecture
Federal University of Rio Grande do Sul
Porto Alegre
Av. Des. André da Rocha 111 apto. 31
Centro Histórico
CEP. 90050-161
Porto Alegre – RS
Brazil
celina.dittmar@gmail.com

Adina Dumitru
Faculty of Sociology and Psychology.
Blvd. V. Pârvan 4
West University of Timisoara
Timişoara 300223
Romania
adina.dumitru@gmail.com

Amalia Fernandez-Bilbao
Principal Consultant
URS/Scott Wilson
6 Greencoat Place
London, SW1P 1PL
United Kingdom
amalia.fernandez-bilbao@scottwilson.com

Francesco Ferrini
Department of Plant, Soil and Environmental Science
University of Florence
Viale delle Idee, 30
50019, Sesto Fiorentino (Florence)
Italy
francesco.ferrini@unifi.it

Andréa Filgueiras
Departamento de Psicologia
Universidade Federal do Ceará
Rua Fonseca Lobo, 333/apto. 101
Brazil
andreacfc@ufc.br

Helenira Fonseca de Alencar
Instituto de Psicologia da UFBA
Universidade Federal da Bahia
Rua Professora Guiomar Florense, 74, apto. 1402, Brotas, Salvador-Ba
CEP:40.279-750
Brazil
helenirafonseca@yahoo.com.br

Carmela García González
IES Val Miñor de Nigrán
Paradellas, 15. 36350-Nigrán (Pontevedra)
Spain
carmela.garcia@edu.xunta.es

Ricardo García-Mira
Faculty of Educational Sciences
University of Corunna

Campus de Elviña, s/n 15071 – A Coruña
Spain
ricardo.garcia.mira@udc.es

Matthias Gross
Department of Urban and Environmental Sociology
Helmholtz Centre for Environmental Research - UFZ
Permoserstr. 15
04318 Leipzig
Germany
matthias.gross@ufz.de

Abdul Samad Hadi
Institute for Environment and Development (LESTARI)
Universiti Kebangsaan Malaysia (UKM)
43600 UKM, Bangi, Selangor
Malaysia
asamad@ukm.my

Jörn Harfst
Leibniz Institute of Ecological Urban and Regional Development
Weberplatz 1
01217 Dresden
Germany
J.Harfst@ioer.de

Jan Hildebrand
Research Group Environmental Psychology
Saarland University
Campus A5.4, Room 332
66123 Saarbrücken
Germany
jan.zoellner@fg-upsy.com

Shaharudin Idrus
Institute for Environment and Development (LESTARI)
Universiti Kebangsaan Malaysia (UKM)
43600 UKM, Bangi, Selangor
Malaysia
dinn6358@gmail.com

Robert-Vincent Joule
Laboratory of Social Psychology
University of Provence
29, Avenue Robert Schuman
13621 Aix-en-Provence Cedex1
France
robert-vincent.joule@univ-provence.fr

Sigrun Kabisch
Department of Urban and Environmental Sociology
Helmholtz Centre for Environmental Research - UFZ
Permoserstr. 15
04318 Leipzig
Germany
sigrun.kabisch@ufz.de

Akikazu Kato
Graduate School of Engineering
Mie University
1577 Kurimamachiya-cho
Tsu City, Mie Prefecture
Japan 514-8507
kato@arch.mie-u.ac.jp

Peter Kellett
Global Urban Research Unit
School of Architecture Planning and Landscape
University of Newcastle upon Tyne
United Kingdom
p.w.kellett@newcastle.ac.uk

Fahed A. Khasawneh
Graduate School of Engineering
Mie University
1577 Kurimamachiya-cho
Tsu City, Mie Prefecture

Japan 514-8507
archfahedkhasawneh@yahoo.com

Judit Kovács
Department of Social and Work Psychology
Institute of Psychology
Faculty of Arts and Humanities
University of Debrecen
4032 Debrecen, Egyetem tér 1
Hungary
kovacs.judit@arts.unideb.hu

Kerstin Krellenberg
Department of Urban and Environmental Sociology
Helmholtz Centre for Environmental Research - UFZ
Permoserstr. 15
04318 Leipzig
Germany
kerstin.krellenberg@ufz.de

Christian Kuhlicke
Department of Urban and Environmental Sociology
Helmholtz Centre for Environmental Research - UFZ
Permoserstr. 15
04318 Leipzig
Germany
christian.kuhlicke@ufz.de

Anna Kunath
Department of Urban and Environmental Sociology
Helmholtz Centre for Environmental Research – UFZ
Permoserstr. 15
04318 Leipzig
Germany
anna.kunath@ufz.de

Raffaele Lafortezza
Department of Agricultural, Environmental and Territorial Sciences (DISAAT)
University of Bari
Via Amendola 165/A
70126 Bari
Italy
r.lafortezza@agr.uniba.it

Roderick J. Lawrence
Human Ecology Group
Institute of Environmental Sciences
Faculty of Social and Economic Science
University of Geneva
7 route de Drize
CH-1227 Carouge (GE)
Switzerland
roderick.lawrence@unige.ch

Letícia Leite Bessa
Faculdade de Psicologia / Centro de Ciências Humanas
Universidade de Fortaleza
Av. Santos Dumont, 6911, ap. 302, bl. Delta, Papicu. Fortaleza-Ce
Brasil
leticialeitepsi@yahoo.com.br

Mats Lieberg
Department of Landscape Architecture
Swedish University of Agricultural Sciences
P.O. Box 58
SE-23053 Alnarp
Sweden
mats.lieberg@slu.se

Heiko Lieske
Faculty of Architecture
Technical
University Dresden
Großmannstraße 1
01187 Dresden

Germany
heiko.lieske@tu-dresden.de

Gerd Lintz
Leibniz Institute of Ecological Urban and Regional Development
Weberplatz 1
01217 Dresden
Germany
G.Lintz@ioer.de

Luciana Lobo Miranda
Departamento de Psicologia
Universidade Federal do Ceará
Prof. Dias da Rocha 490- 503- Meireles - 60.170-310. Fortaleza-Ce
Brazil
lobo.lu@uol.com.br

Stefano Mastandrea
Department of Cultural and Educational Studies
University of Roma Tre
Via del Castro Pretorio, 20,
00185, Rome
Italy
mastandr@uniroma3.it

Dóra Medvés
Department of Social and Work Psychology
Institute of Psychology
Faculty of Arts and Humanities
University of Debrecen
4032 Debrecen, Egyetem tér 1
Hungary
medves.dora@arts.unideb.hu

Michael Mehaffy
Department of Architecture
University of Strathclyde
Urban Design Studies Unit
131 Rottenrow
Glasgow, G4 0NG
United Kingdom
michael.mehaffy@gmail.com

Elisabeth Michel-Guillou
Université de Bretagne Occidentale
Centre de Recherches en Psychologie, Cognition et Communication
(CRPCC – EA 1285)
20 rue Duquesne – CS 93837
29238 Brest cedex 3
France
Elisabeth.Michel-Guillou@univ-brest.fr

Ahmad Fariz Mohamed
Institute for Environment and Development (LESTARI)
Universiti Kebangsaan Malaysia (UKM)
43600 UKM, Bangi, Selangor
Malaysia
fariz@ukm.my

Ryuzo Ohno
Department of Built Environment
Tokyo Institute of Technology
4259 Nagatsuta-cho, Midori-ku,
Yokohama, Kanagawa 226-8503
Japan
rohno@n.cc.titech.ac.jp

Idilva Maria Pires Germano
Universidade Federal do Ceará
Rua Monsenhor Catão 948/302.Meireles
Fortaleza-Ceará
Brazil
CEP: 60.175-000
idilvapg@ufc.br

Sergio Porta
Department of Architecture
University of Strathclyde
Urban Design Studies Unit
131 Rottenrow
Glasgow, G4 0NG

United Kingdom
Sergio.porta@strath.ac.uk

Ruslan Rainis
Geography Section
School of Humanities
Universiti Sains Malaysia (USM)
11800 Penang
Malaysia
rruslan@usm.my

Irina Rau
Research Group Environmental Psychology
Otto-von-Guericke-Universität Magdeburg
Turmschanzenstr. 15
PSF 4120
39016 Magdeburg
Germany
irina.rau@fg-upsy.com

Kathrin Röderer
Institute of Environmental Health
Medical University of Vienna
Kinderspitalgasse 15
1095 Vienna
Austria
kathrin.roederer@meduniwien.ac.at

Ombretta Romice
Department of Architecture
University of Strathclyde
Urban Design Studies Unit
131 Rottenrow
Glasgow, G4 0NG
United Kingdom
ombretta.r.romice@strath.ac.uk

Fabio Salbitano
Department of Economy, Engineering, Agricultural and Forest Sciences and Technologies
University of Florence
Via S. Bonaventura, 13
50145, Florence
Italy
fabio.salbitano@unifi.it

Giovanni Sanesi
Department of Agricultural, Environmental and Territorial Sciences (DISAAT)
University of Bari
Via Amendola 165/A
70126 Bari
Italy
sanesi@agr.uniba.it

Ticiana Santiago
Faculdade de Educação
Universidade Federal do Ceará
Rua Waldery Uchoa, N. 1, Benfica
60.020-110. Fortaleza-Ce
Brasil
ticianasantiagodes@yahoo.com.br

Petra Schweizer-Ries
Research Group Environmental Psychology
Saarland University
Campus A5.4, Room 332
66123 Saarbrücken
Germany
petra.schweizer-ries@fg-upsy.com

Massimiliano Scopelliti
LUMSA University
Piazza delle Vaschette 101
00193 Rome
Italy
m.scopelliti@lumsa.it

Paolo Semenzato
TESAF Department
University of Padua

Viale dell'Università 16
35020, Legnaro (Padua)
Italy
paolo.semenzato@unipd.it

Abdul Hadi Harman Shah
Institute for Environment and Development (LESTARI)
Universiti Kebangsaan Malaysia (UKM)
43600 UKM, Bangi, Selangor
Malaysia
aharmans@gmail.com

Annett Steinführer
Institute of Rural Studies
Johann Heinrich von Thünen Institute (vTI)
Federal Research Institute for Rural Areas, Forestry and Fisheries
Bundesallee 50
38116 Braunschweig
Germany
annett.steinfuehrer@vti.bund.de

Rachma M. Syam
Department of Built Environment
Tokyo Institute of Technology
4259 Nagatsuta-cho, Midori-ku
Yokohama, Kanagawa 226-8503
Japan
rachma3382@yahoo.com

Gen Taniguchi
Graduate School of Engineering
Nagoya University
Furo-cho, Chikusa-ku, Nagoya
Japan 464-8601
tanigen@cc.nagoya-u.ac.jp

Antônio Tarcísio Reis
Faculty of Architecture - PROPUR
Federal University of Rio Grande do Sul
Porto Alegre
Praça Carlos Simão Arnt 21 apto. 202
Bela Vista
CEP. 90450-110
Porto Alegre – RS
Brazil
tarcisio@orion.ufrgs.br

Kevin Thwaites
Department of Landscape
University of Sheffield
Floor 13, The Arts Tower
United Kingdom
k.thwaites@sheffield.ac.uk

Peter Wirth
Leibniz Institute of Ecological Urban and Regional Development
Weberplatz 1
01217 Dresden
Germany
P.Wirth@ioer.de

Index